The Mass-Extinction Debates:

HOW SCIENCE WORKS IN A CRISIS

THE MASS-EXTINCTION DEBATES:

HOW SCIENCE WORKS IN A CRISIS

Edited by William Glen

Stanford University Press
Stanford, California 1994

Stanford University Press, Stanford, California

Printed in the United States of America

CIP data appear at the end of the book

Stanford University Press publications are distributed exclusively by Stanford University Press within the United States, Canada, and Mexico; they are distributed exclusively by Cambridge University Press throughout the rest of the world.

Preface

This book is the first to examine the arguments and behavior of the scientists who have for so long been locked in conflict over the competing hypotheses of mass-extinction cause. The debates—triggered in 1980 by the advent of the meteorite-impact hypothesis put forth by the Alvarez-Berkeley group—have grown into a broad interdisciplinary upheaval, one of the most dramatic in the recent history of science. The controversy raging around this issue has provided an exceptional opportunity to observe, firsthand, the workings of science that in quieter times are hidden from view.

In 1991, in the midst of the conflict, scientists embroiled in the debate came together for the first time with historians, philosophers, and sociologists who were trying to learn how science works in such a time of crisis. This diverse volume, which was seeded at that meeting—The Mass-Extinction Debates, July 12, 1991, Biannual Meeting of the International Society for the History, Philosophy, and Social Studies of Biology, Northwestern University—circles the conflict from a variety of vantage points in a novel mode of presentation, thereby to examine many aspects of the conflict. The contributors to this volume—essayists, interviewees, and panel discussants alike—were selected so as to represent the very wide range of thought underlying the conflict, and to provide balance in discussions of the conflicted issues. Most of the contributors also made presentations at the meeting, and some who were invited to the meeting but were unable to attend submitted written or oral contributions afterward. Several essays have been revised in order to encompass findings and arguments that have unfolded since the meeting, and my own initial chapter, which provides a historical overview of the ideas and evidence of science being debated, was expanded almost up to the date of publication.

The mass-extinction debates began slowly in 1979, but then quickened and broadened. By 1983 they had become a promising case for study. This volume is my preliminary report of that case, which has developed over more than a decade. The debates have expanded to entrain a variety of formerly isolated disciplines, forced unprecedented interdisciplinary collaborations, rejuvenated long-quiescent fields, given rise to significant bodies of new knowledge, and fostered the appearance of provocative theories within and across several disciplines, including both corollaries and alternatives to the impact hypothesis.

After completing *The Road to Jaramillo*, in 1982, I was left with the realization that many of the doubts and inadequacies that remained at writing's end were made inevitable by a lack of data that could only have been obtained in real time. That volume, a detailed, internalist history of the research programs in geophysics that proved continental drift and triggered the plate-tectonics revolution, was begun in the mid-1970's, several years after closure of that half-century-old conflict. I thus sought an opportunity to capture, as fully as possible, an ongoing theoretical dispute *as it unfolded*. I grasped that opportunity during the mass-extinction debates in 1983, but only after two false starts in doing histories of other current conflicts. One aborted attempt was on the expanding-Earth issue, which although very important theoretically, saw too few scientists informed or participating; the other study, also discontinued, examined the accretionary-terrane research program, which was mainly a conceptual adjunct to plate-tectonics theory. Although certain inadequacies of those two prior studies disposed me to put them aside and move on, they did provide considerable information on the behavior of scientists in conflict, information that served to corroborate much of what has been documented in my study of the extinction debates.

The extinction conflict has grown frenetically, from a single hypothesis into a concatenation of theories straddling a host of scientific disciplines, and has opened whole windows—where mere peepholes would have seemed all one could expect—onto the workings of science.

The promptness and good humor of the contributors to this volume simplified my task. Many of them are scientists who have participated in the debates, and over several years they have suffered repeated interviews and other inconveniences in helping me with this study, which is now in its tenth year. My work would not have been possible without the varied indulgences—which often included lengthy tutorials—of almost three hundred other natural scientists, journal editors, science writers, and historians, sociologists, and philosophers of science. Nor would this study have been possible without the support of the U.S. Geological Survey at Menlo Park, facilitated by Director's Representative George Gryc, and the unstinting aid and encouragement of its scientific and library staffs—especially the late Virginia Langenheim. Many earth scientists at both the Reston and Denver U.S.G.S. centers were also very helpful. Spencer Weart, Director of the Center for History of Physics of the American Institute of Physics, New York City, and Elihu Gerson, Director of the Tremont Research Institute, San Francisco, have kindly provided various forms of assistance over many years. Rewarding conversations with Léo Laporte have often served me well in this and other projects. The contributions to this volume by William Carver, Editor Emeritus at Stanford University Press, extend far beyond his superlative editing, which was vital in shaping the manuscript for publication. His counsel and grace, since the days of *The Road to Jaramillo*, have always smoothed the road. Aida and LeRoy Larsen deftly and kindly rescued me from the chasm of indexing. Heretics Club members are cautioned to remain calm upon discovering their ideas in my voice. My research has been funded by the National Science Foundation under grants SES-8420419, SES-8619545, and DIR-8821587.

William Glen
Menlo Park, California

Contents

The Contributors

John C. Briggs is Professor Emeritus of Marine Science, University of South Florida, and Adjunct Professor of Zoology at the University of Georgia. His published works include *Marine Zoogeography* (McGraw-Hill, 1974), *Biogeography and Plate Tectonics* (Elsevier, 1987), and a paleontologic overview of the mass-extinction debates: "Global Extinctions, Recoveries, and Evolutionary Consequences" (*Evolutionary Monographs*, University of Chicago, 1990).

Elisabeth S. Clemens is Assistant Professor of Sociology at the University of Arizona. Working at the intersection of organization theory and cultural analysis, she is primarily interested in the development of political institutions. Within that framework she now addresses theoretical questions about interdisciplinary politics in science. Since the mid-1980's, she has commented on the role of popular culture, disciplinary practices, and institutional constraints in the impact debates.

William A. Clemens has served as Chairman of the Department of Paleontology and Director of the Museum of Paleontology at the University of California, Berkeley. His studies focus on the systematics and evolution of Mesozoic and Early Tertiary vertebrate faunas and on extinction patterns at the Cretaceous/Tertiary boundary. He is a Fellow of the American Association for the Advancement of Science, the California Academy of Sciences, and both the Linnaean and the Zoological societies of London. Currently Professor of Integrative Biology at Berkeley, he serves as President of the Society of Vertebrate Paleontology and Vice President of the California Academy of Sciences.

S. V. M. Clube, a Fellow of the Royal Astronomical Society, has been Senior Research Fellow in Astrophysics at the University of Oxford, U.K., since 1984; his prior service of over 20 years was at Royal Observatories in Britain and South Africa. Interests in large differentiated comets led him to question whether the geological and historical records were explicable in terms of such bodies interacting with the Earth. His studies (some with W. M. Napier) have given rise to a variety of new insights in the fields of cometary and meteor science, and have strongly conditioned the course of the mass-extinction debates.

William Glen, a geologist turned historian, author of *Continental Drift and Plate Tectonics* (Charles C. Merrill Publishers, 1975) and *The Road to Jaramillo: Critical Years of the Revolution in Earth Science* (Stanford University Press, 1982), is writing a book based on his decade-long historical study of the mass-extinction debates. He serves as Editor-at-Large for Stanford University Press and Visiting Scientist/Historian, U.S. Geological Survey, Menlo Park, California.

Stephen Jay Gould is Alexander Agassiz Professor of Zoology and Professor of Geology at Harvard University, where he also teaches the history of science. Captivated by his cornucopian essays and his half-dozen award-winning books, which marry natural history to humanism, Gould's global audience—although likely unaware—will not be surprised to learn that his scientific efforts have occasioned 34 honorary doctoral degrees, three dozen academic medals and awards (a maiden MacArthur Foundation Prize among them), signal positions of professional leadership, and a score of memberships and fellowships that include the Royal Society of Edinburgh, the Linnaean Society of London, and the U.S. National Academy of Sciences.

Kenneth J. Hsü, Professor of Geology at the Swiss Federal Institute of Technology, has made some of the most gripping and controversial geological discoveries of our time. By drilling the floor of the Mediterranean Sea he demonstrated that the Sea had dried up entirely some 5 million years ago, and through geochemical studies, he provided crucial support for the postulate of a global catastrophe at the end of the Cretaceous Period. Author of some 300 research articles and more than a dozen books, he is the recipient of the William Hyde Wollaston Medal of the Geological Society of London, a member of Academia Sinica, and a Foreign Associate of the U.S. National Academy of Sciences.

David L. Hull is the Dressler Professor in Humanities at Northwestern University. He is currently President of the International Society for the History, Philosophy, and Social Studies of Science and Past President of both the Society of Systematic Zoology and the Philosophy of Science Association. He is a Fellow of the American Academy of Arts and Sciences and the American Association for the Advancement of Science. The author of four books, including the monumental *Science as a Process* (University of Chicago Press, 1988), he also edits a series of books on the Conceptual Foundations of Science for the University of Chicago Press.

Digby J. McLaren has served as President of the Royal Society of Canada, Director of the Canadian Geological Survey, President of the Geological Society of America, and President of the Paleontological Society. He is a Fellow of the Royal Society of London and a Foreign Associate of the U.S. National Academy of Sciences. In his presidential address to the Paleontological Society in 1970, he presented the first paleontologically based theory of mass extinction by meteorite impact and, during the past decade, has published widely to encourage the search for impact evidence at other extinction horizons.

David M. Raup is the Avery Distinguished Service Professor at the University of Chicago, with appointments in Geophysical Sciences, Evolutionary Biology, and the Conceptual Foundations of Science. His research lies primarily in the quantitative analysis of the fossil record, with emphasis on broad patterns of evolution. He has published several books, including two popular treatments of the extinction problem: *The Nemesis Affair* (Norton, 1986), which won an outstanding popular book award from the American Institute of Physics, and *Extinction: Bad Genes or Bad Luck?* (Norton, 1991). Raup has served as President of the Paleontological Society and is a member of the U.S. National Academy of Sciences.

J. John Sepkoski, Jr., Professor of Paleontology at the University of Chicago, received the Schuchert Award of the Paleontological Society for his research on large-scale patterns in the history of life. In addition to gathering census-style data on marine animals, he publishes on the evolution of biodiversity and the nature of mass extinctions. In 1984, he co-published with David M. Raup the profoundly provocative hypothesis that mass-extinction events follow a 26-million-year periodicity.

Herbert R. Shaw is a Senior Research Geologist at the U.S. Geological Survey, Menlo Park, California. He has served as Visiting Professor at the University of California, Berkeley, is a Fellow of the American Geophysical Union and the Geological Society of America, and holds the third Annual Ernst Cloos Award of Johns Hopkins University. Published across a broad theoretical front in geology and geophysics, he has pioneered in the application of nonlinear dynamics to the earth sciences. His latest book, *Craters, Cosmos, and Chronicles: A New Theory of Earth*, is forthcoming from Stanford University Press.

Leigh M. Van Valen is Professor of Ecology and Evolution, and of the Conceptual Foundations of Science, at the University of Chicago; he also edits two scientific journals. Among his 300 or so papers and monographs in science and philosophy lie his discoveries of the earliest known ungulates and primates, the origin of whales and several other groups of mammals, and what are now called fuzzy sets. He has solved conceptual problems such as the nature of homology, and has opened new paths in the study of extinction in paleontology and ecology.

The Mass-Extinction Debates:

HOW SCIENCE WORKS IN A CRISIS

Introduction

William Glen

The history of life on Earth, as recorded in the rocks, is punctuated by almost a dozen puzzling mass extinctions, great shifts in biotic composition that constitute the benchmarks of the geologic time scale. These great breaks in the continuity of the fossil record have invited speculation for over a century; explanations of all sorts have been suggested, but none has commanded consensus. Now, from the past decade of discovery and theoretical tumult, two new alternative hypotheses have emerged, each attempting to explain how most of life on Earth, including the dinosaurs, perished 65 million years ago in the best-known and most thoroughly studied of the mass extinctions. The "impactor" hypothesis bombards the Earth with a swarm of deadly meteorites, while the competing "volcanists," responding to the impactor hypothesis, call on explosive volcanic eruptions that trivialize any seen by humanity. In the interstices are other theories—and a persisting argument that none of the great extinctions was in fact particularly sudden at all.

This book, derived in part from a symposium convened in 1991 (see the Preface), examines the arguments and behavior of the scientists who are locked in conflict over these hypotheses. The science they pursue is alive with discovery, and their debates vibrate with conflict right from my opening chapter on the history of the arguments about competing extinction hypotheses and a host of other conflicted ideas that the debates have spawned. That sense of energy and grinding friction is sustained through successive chapters that address the organic character of scientific communities; how scientists responded to new theories, methods, data, and instruments; the origin, selection, and use of standards in the appraisal of a wide range of information; the variety of polemic styles; the careers and paradigms at stake; and how scientists, through it all, have gone about their business. The influence of theory choice and disciplinary specialty in shaping thinking and behavior emerges repeatedly in anecdotal accounts, historical examinations of scientific, philosophical, and sociological issues, interviews with leading scientific adversaries, and a concluding, overarching discussion by a balanced panel of embattled scientists and scholar-observers.

The decade-old debates have engaged a singularly broad range of scientific disciplines and revived long-stagnant areas of research, brought formerly isolated specialties into collaboration, redirected careers, and engendered remarkable new field studies, experiments, and painstaking reexaminations of old evidence. The conflict over the swarm of larger, new ideas spun off from the impact hypothesis itself—including periodicity of extinctions—has been reflected in more than 2,500 papers and books in just the last decade.

The impact hypothesis—as explosive for science as an impact would have been for Earth—has all but overshadowed the host of earlier causal postulates, save that of cataclysmic volcanism, and has called the canons of extinction studies and related disciplines into question. These debates over proliferating, interrelated theories have changed the literature of mass extinctions from a small mélange of randomly appearing articles—published before the impact hypothesis—to a growing and focused assemblage of materials from diversely comprised research programs that span paleobiology, biologic evolution, chemistry, geology, geophysics, and even astronomy and astrophysics. The study of mass extinctions has been catapulted to paleontology's frontier, and has become a catalyst for collaborations of interdisciplinary breadth as broad as any yet seen in the sciences. Conferences and symposia treating mass extinctions and their causes continue to attract growing audiences of scientists and scholars, and the media's preoccupation with the conflict has rejuvenated a dinosaur industry that now extends from four-hour-long TV documentaries to a tide of increasingly fanciful dinosaurian consumer products and entertainment. This upheaval presents scholars of both science and the media with an extraordinary case study.

Because only a few have been able to keep abreast of the burgeoning, frenetically paced literature, I open the volume with a historical review of the scientific content of the debates in "What the Impact/Volcanism/Mass-Extinction Debates Are About," which also serves to preface my larger essay, "How Science Works in the Mass-Extinction Debates." The latter is a précis of part of a forthcoming book that reports on a ten-year, longitudinal, in situ, historical study of the debates, which will focus on such questions as: To what extent do beliefs and predispositions of mind shape perception? Why are choices of theory, method, or standard that seem logical and inescapable to one intellectual cohort rejected by another? How do similarly trained and practiced scientists acquire contradictory views of the same reality? What figures in the tenacity and recalcitrance of the minority in the face of evidence that so firmly compels the majority? What is the balance between faith and fully informed reason (the two are often not easily differentiated) in "scientific" opinions? What roles do scientific doyens, journals, institutions, the media, and—by extension—the public play in shaping science? What factors of personal biography impinge on choices of enterprise, theory, and philosophy? And are such choices inevitably concatenated? At what cognitive level do philosophic tenets police observation and interpretation? Simply put, how and to what extent do the host of "subjectivities"—those that have already been identified and others that still beg definition—interact with "objectivities" to condition the workings of science? What is the value of history of science undertaken within the cofferdam that seals out sociologic, psychologic, and philosophic contingencies?

Focusing on the interplay of popular and expert scientific knowledge in "The Impact Hypothesis and Popular Science: Conditions and Consequences of Interdisciplinary Debate," Elisabeth Clemens illuminates the vital role of the media in shaping the character of the debates. She shows that rather than developing within the boundaries of single disciplines, the impact debates have involved collisions both of disciplines and of multiple research agendas, and that the researcher/participant in the debates is a member of numerous audiences: scientific specialties, the broader disciplines of science, and others, extending even to the public audience. Clemens employs citation analysis to address the social basis of interdisciplinary science and the reception of the impact hypothesis, and turns finally to the impact of the debates at the disciplinary, interdisciplinary, and popular levels. She cites Hilgartner (1990) to emphasize that "rather than displaying the sharp contrast between science and popularizations that have proven so ideologically useful for scientists themselves, the impact debate has been shaped by, and partially played out in, arenas intermediate to the poles of popular and expert, of exoteric and esoteric knowledge about science."

Three paleontologists, all at the center of the debates, then open views into the factory of science through personal experiences. Digby McLaren's "Impacts and Extinctions: Science or Dogma?" goes far in explaining the predictably negative reception suffered by his own impact hypothesis of extinction in 1970, through comparison with six empirically based general theories from decades past that were similarly rejected because, like McLaren's, their proposed causes were at variance with the prevailing canons. John Sepkoski then seats us at the dinner table to tell how his research on biodiversity providentially yielded data on patterns of extinction in, "What I Did with My Research Career." We learn how his mentors at Harvard—Gould, Kummel, E. O. Wilson, and others—shaped the paleontologic weltanschauung that conditioned his collaboration with David Raup. Theirs was a landmark joint effort that birthed the statistical hypothesis of periodicity of mass extinctions. Raup's "The Extinction Debates: A View from the Trenches" sketches a backdrop of paleontology's philosophic fundaments—especially those emplaced by Lyell and Darwin—against which he shows how he and his community were imbued with a mindset against impact, a gestalt so profound that rejection of the idea of impact-as-extinction-cause was *intuitive*.

But two less-intuitive, antipodal mindsets are next revealed in a flight to the heavens: Victor Clube laments canonical science's persistent disdain for exogenous influences on the Earth through a mini-review of historic notions of comets and their representation in human history, and shows how modernity is yet to escape its astrological past. He sifts both the historical archives and the current literature for the seemingly solid and empirically grounded reports and draws important conclusions about the cometary flux in "Hazards from Space: Comets in History and Science." His review of mankind's response, through the millennia, to how the visitations of cometary debris to the Earth have been variously received and cast in lore, legend, religion, and natural philosophies builds a bridge from little-known cometary science to the humanities; he makes clear that one must cross that bridge to assess properly the multi-hued reception of the impact hypothesis. Clube was the first to suggest the idea—now embraced by many—that multiple impacts bracket the Cretaceous-Tertiary mass-extinction boundary.

Clube's comets, along with asteroids, near-Earth objects, and the questions of periodic impacting and periodic extinctions are then swept up in Herbert Shaw's "The Liturgy of Science: Chaos, Number, and the Meaning of Evolution." His rich tutorial of solar/planetary-solar/galactic-Earth-coupled resonances shows—in opening a window on his own theory, in press as a book—how impacts have conditioned Earth's magnetic dynamo, mountain building, plate tectonics, volcanism, and biotic history. With historical emphasis, Shaw draws threads of philosophy, art, poetry, and information systems into a nonlinear-dynamical web of chaos theory that spans conventional periodicities, nonlinear periodicities, aperiodicities, and still other frequencies of earthly and cosmological processes and phenomena. The clockwork of the Cosmos, he holds, is imprinted in the solid Earth body through the ceaseless, patterned impacting of meteorites. Shaw's chaos-theory view of nature and culture impugns the complacencies borne of conventional analytic schemes.

Leigh Van Valen then guides us through the philosophic labyrinth of "Concepts and the Nature of Selection by Extinction" and asks, "Is Generalization Possible?" The complex journey includes definitional excursuses into several areas that seem to defy circumscription, and shows us once again how much room there is for maneuver in these debates. Van Valen forces examination of the high-stress metacryst, the hard kernel of mass extinction, against the larger, lesser-stress fabric of extinction in general, and through it all we read the gradualist. He shows the complexity of extinction through its varied definitions and characteristics, which helps to explain, as one paleontologist has noted, "why many paleontologists against the impact hypothesis believe as they do."

The starkly counterpoised visions of the natural world of Kenneth Hsü and John Briggs reveal just how philosophic mindset or gestalt conditions opinion, and help us to understand why disagreement in science is so often polarized. In "Uniformitarianism vs. Catastrophism," Hsü dissects the origins of substantive uniformitarianism and shows by examples how that philosophy has been a negative influence in earth science's history, in the current extinction debates as well as elsewhere. Hsü demands a central place for great catastrophes in the panorama of life. Briggs argues conversely, by an appeal to the extinction literature, in "Mass Extinctions: Fact or Fallacy?" that "there is no evidence that global mass extinctions (defined as short-term, catastrophic events) ever took place." He asks if, in the light of our present knowledge, what we are commonly calling mass extinctions did in fact occur very suddenly, or have global effects, or drastically reduce the world's species diversity. Briggs aims to deny the basis from which neocatastrophism seems to be rising.

The gradualist-catastrophist dialectic is renewed in two interviews that address questions in common on a number of scientific and philosophic issues. The telling responses of William Clemens, a leader of the paleontologic school that denies impact as the chief cause of the extinctions at the Cretaceous-Tertiary boundary, are juxtaposed with the views of the paleontologist Stephen Gould, who was among the few of his tribe to embrace the impact hypothesis at its advent. Clemens has long argued for the gradual, rather than geologically instantaneous, demise of the late Cretaceous dinosaurs, and was thus disposed against the idea of instantaneous extinction by any means. Early on in the debates, as a faculty colleague of the Alvarez group in the same lair at Berkeley, Clemens led the opposition to impact theory. From the outset, in contrast, Gould encouraged the Alvarez group for reasons that he himself has laconically noted: "Catastrophic extinction simply matched my idiosyncratic preference for rapidity, born of the debate over punctuated equilibrium" (quoted from Gould 1992; and see Chap. 2, this volume).

The contrapuntal tone of the two interviews is echoed in the repartee of the topically broad panel discussion that closed the mass-extinction symposium. It is this "Epilogue" to the volume that speaks to such questions as: Are mass extinctions part of a process continuous with other extinctions? Why was the impact hypothesis virtually ignored prior to its resurrection by the Alvarez group? To what extent do ascendant new ideas dignify less-justified related ideas? What is the role of modeling in the debates? How should editors of science journals handle conflict in science? Should philosophy and the social studies of science serve science—and if so, how? And what effect have the impact/mass-extinction debates had on neocatastrophism and neo-Darwinism? The panel included sociologist Elisabeth Clemens, philosopher David Hull, paleontologists John Briggs and Leigh Van Valen (who both oppose the impact-as-extinction-cause hypothesis), and paleontologists Digby McLaren, David Raup, and John Sepkoski (who favor the hypothesis); I convened the session and moderated the panel discussion.

1

What the Impact/Volcanism/ Mass-Extinction Debates Are About

William Glen

This essay offers a historical review of the scientific ideas and evidence at the center of the mass-extinction debates. It tells what the arguments are about, laying out the essentials before subsequent chapters take up the philosophy of those ideas, the logic and mode of their argumentation, and the behavior of the scientists involved. The participants and references I cite here are those that appear to have most strongly conditioned the course of the debates. Their number is necessarily constrained by the limits of this chapter; they are in fact but a select fraction from among the many contributing scientists and more than 2,500 references that subtend an interdisciplinary arc of breadth perhaps unprecedented in scientific conflict.

The current upheaval in science over the causes of mass extinctions was triggered in 1980 when the Berkeley team of geologist Walter Alvarez, his father Luis Alvarez, a Nobelist in physics, and nuclear chemists Frank Asaro and Helen Michel advanced the first testable causal hypothesis. Their idea was triggered by the apparently unearthly concentration of the element iridium within a distinctive, pencil-thick clay layer lying between limestone beds hundreds of feet thick in a gorge outside the northern Italian town of Gubbio. The clay bed was formed at the time of the dinosaur extinctions, 65 million years ago. The platinum-group element iridium, relatively abundant in certain kinds of meteorites, is virtually absent in the Earth's crust—measured in parts per trillion. It is a siderophile ("iron-loving") element that was dragged down by iron as the iron and the other heavy elements sank to the Earth's core early in the planet's history. The anomalous iridium-enriched clay layer, called the K/T, forms a boundary separating the older Cretaceous Period (geologic symbol K), with its reigning dinosaurs and conifer trees, from the Tertiary Period (T), which is dominated by mammals and flowering plants of modern aspect.

The astonished Alvarez team drew a dramatic conclusion from the anomalous iridium in the boundary bed at Gubbio. They proposed, in an article in *Science* in 1980, that the iridium had settled out of a global dust cloud kicked up by the impact of a 10-kilometer, iridium-rich meteorite. The same dust cloud, they suggested, had blocked the sun, chilled the planet, and killed the dinosaurs, along with 75 percent of all animal species in the seas and almost all the land animals that weighed more than 50 pounds. The initial and subsequent impact of the impact hypothesis on science is hard to overstate. The past 13 years have seen the publication of more than 2,500 papers and books touching on various aspects of the controversy, and the tide of publications continues to rise. Career goals have been refocused, dormant areas of research rejuvenated, tacit assumptions hastily reexamined, and scientists in widely separated fields swept into an unprecedented array of collaborative efforts. Ingenious experiments, field studies, and the development of new instruments now undergird an effort to resolve the many questions raised in these far-flung debates. The impact hypothesis has also prompted a number of fascinating related ideas; most notable among them is that mass extinctions are periodic, afflicting the Earth at regular intervals of 26 million years or so.

Today, out of a host of earlier attempts to explain mass extinctions, only one alternative to the impact hypothesis remains under serious consideration. The competing volcanism hypothesis suggests that the iridium anomaly and other evidence claimed for impact have resulted instead from massive volcanic eruptions. Paleontologists, in particular, have found volcanism useful in explaining certain puzzling features of the fossil record, such as the pattern of lesser extinction steps, or pulses, that seem to lead up to and follow the great kill at the K/T boundary.

In proposing eruptions on a cataclysmic scale, the volcanists thus encounter opposition of the same sort that proponents of impact have long faced (French 1990): both the impact and the volcanism hypotheses raise the specter of catastrophism, an outcast philosophy dislodged during fierce debates in the nineteenth century by the uniformitarianists, who viewed the exceedingly gradual workings of the present natural world as the key to the Earth's past. Before the advent of the impact hypothesis, a latter-day version of uniformitarianism had come to serve as the philosophical mainstay of the earth sciences; it was more accepting of uncommon events and forces than earlier versions had been, but was still not as well disposed to violent causes as to slow, gradual processes in the interpretation of Earth history (Gould 1965; Hallam 1989). Though the decade-long debates have dramatically elevated the role of impacting in shaping the planet and its inhabitants, catastrophes of extraterrestrial origin cannot quite be fit to any redefinition of uniformitarianism, whereas volcanism, however explosive and great its scale, can at least be viewed as a mere enlargement of a familiar earthly process. Articles by volcanists and paleontologists have sometimes closed with the telling sentiment that they find satisfaction in seeking an earthly cause before turning to the sky.

Evidence for an Impact

The Alvarez-group hypothesis was not the first suggestion of a catastrophic origin for mass extinctions. Earlier explanations had included supernovae, solar flares, and the Solar System's passage through the spiral arms of the Galaxy. Gradual causes such as sea-level changes, disturbed climates, seawater-chemistry fluctuations, and plate-tectonics effects had also had their day, but all now simply languish—some accorded residual status as secondary effects of impacting or volcanism. Digby McLaren had proposed, in 1970, that a meteorite caused a mass extinction 365 million years ago, long before the extinction at the K/T boundary. But McLaren's idea, which grew from painstaking analysis of the rock and fossil record, was not compelling. It was the iridium layer that made the Alvarez group's explanation like no other—it was based on an empirical finding from which testable predictions and a program of experiments could be written. After discovering the iridium anomaly at Gubbio and postulating a global dust cloud, the Alvarez group predicted that there should be similar concentrations of iridium elsewhere in the world at the K/T boundary. Within a few years, more than 50 such anomalies had been found, all over the globe; now over 100 are known (Alvarez & Asaro 1990; Smit & Hertogen 1980; Ganapathy 1980; Kyte et al. 1980; Orth et al. 1981; Kyte & Wasson 1986; Sharpton et al. 1992; etc.). The concentration of iridium in this global layer is typically between 10 and 20 parts per billion or roughly 1,000 times the average concentration in the rocks of the Earth's outer crust. This works out to about half a million tons of iridium spread over the planet—roughly the amount, according to the Alvarezes, that would be released by the explosion of a meteorite about 10 kilometers across.

In the decade since the advent of the Alvarez-group hypothesis, other strands of evidence have emerged to support it. Perhaps the most persuasive impact evidence, apart from the iridium itself, consists of signs interpreted as damage done by the meteorite to the Earth rock it struck. A 10-kilometer rock slamming into the Earth at a speed of, say, 72,000 kilometers per hour would deliver a tremendous shock. Bruce Bohor and his colleagues at the U.S. Geological Survey were the first to find grains of quartz in the K/T boundary layer that show the microscopic traces of such a shock: long, parallel planes, called shock lamellae, along which rows of atoms in the quartz crystal lattice have been dislocated (Bohor et al. 1984, 1987). Old hands in impact studies consider such multiple, intersecting sets of lamellae in quartz diagnostic of a meteorite strike. Until the Bohor group found them at K/T boundary sites—which amazed all—they were known only from the immediate vicinity of established meteorite craters and in rocks at nuclear-explosion sites.

Not only does an intense shock disrupt the crystal lattice of quartz, it can also change the crystal structure completely. In quartz, oxygen atoms form a tetrahedron around each atom of silicon, but at extreme pressures (greater than 8.5 gigapascals, or 85 kilobars) the atoms can be rearranged into a denser, octahedral form called stishovite. Stishovite is exceedingly rare in the Earth's crust; until recently it had been identified only at a few meteorite craters, and even there was found only in minute quantities. John F. McHone and his colleagues at Arizona State University thus bolstered the impact cause when they reported that they had isolated it from K/T boundary clay at Raton, New Mexico (McHone & Nieman 1989).

Constituting further evidence for impact are small spherical particles that most now believe—but far fewer had agreed early on—formed from the rapid cooling of molten-rock droplets splashed into the atmosphere by a meteorite impact. These spherules, which vary considerably in mineralogy, have been regarded by some as true impact products (tektites) that have been chemically altered after burial (Smit & Kyte 1984; Montanari et al. 1983, 1986; Pollastro & Pillmore 1987), but others have held that they were formed within the rock by chemical processes (Naslund et al. 1986; Izett 1987). Surprisingly,

Izett, who had earlier denied the tektite nature of spherules from certain North American K/T boundary sites, was converted by the spherules discovered later in the boundary layer in Haiti, which had an outer decomposition rind of smectite clay enclosing an inner core of unaltered tektite-like glass. The glass was reportedly (Izett et al. 1990; Sigurdsson et al. 1991; Kring & Boynton 1992) unlike that of volcanoes, a finding that greatly strengthened the case for impact. Argon dating of the portentous, unaltered, glass-cored spherules followed at the U.S. Geological Survey in Menlo Park in March 1991 (Dalrymple, oral comm.; Izett et al. 1991). The spherules were convincingly shown to have formed at the time of the K/T boundary, 65 million years ago.

The volcanists maintained (Lyons & Officer 1992), as they had for spherules discovered at other K/T boundary localities, that the Haitian glass-cored spherules could have been produced by a volcano. But when Joel Blum and Page Chamberlain of Dartmouth College (1992) did laser-extraction oxygen-isotope and major-element analyses of individual glass spherules, they found that the glasses contained a mixture of carbonate and silicate rock with oxygen-isotope ratios that they held "effectively rule out a volcanic origin for the Haitian K/T glass spherules."

Two other kinds of evidence for impact, both of them compelling, were proffered late in the game by David Carlisle of Environment Canada and Dennis Braman of the Royal Tyrrell Museum. In August 1991 they reported nanometer-size diamonds stratigraphically confined to the K/T boundary clay of Alberta, Canada, in a ratio to iridium (1.22:1) that was close to the value found in type C2 chondritic meteorites. They noted that a definitive answer to the question whether the diamonds were present in the impactor or formed by the impact can "best be addressed by an examination of the isotope ratios." Carlisle (1992) then did just that: he demonstrated that the carbon-isotope ratio in these diamonds is close to that of interstellar dust and incompatible with a terrestrial origin. A conservative, senior paleontologist who had denied a K/T impact in interviews spanning a decade remarked that he viewed the diamonds as finally convincing him of an impact, but he remained unconvinced that the impact was the main cause of the extinction. That has been the posture of most paleontologists who believe in impact(s)—they still subscribe to only part of the impact hypothesis (this point is detailed in "How Science Works," this volume).

After several years of laborious work, Meixun Zhao and Jeffrey Bada of the University of California, San Diego, had isolated, in 1989, both alpha-amino-isobutyric acid and racemic isovaline, two amino acids that are exceedingly rare on Earth, but common, by contrast, in carbonaceous chondritic meteorites. Found within the boundary clay at Stevns Klint, Denmark, the acids were within the same range as that measured in the Murchison meteorite, and were thus attributed to an extraterrestrial source. But the fact that the two amino acids were known also from earthly sources detracted somewhat from the authors' surmise of an extraterrestrial origin. Carlisle and Braman (1993a, b) then followed up on the work of Zhao and Bada; they searched their diamond-bearing K/T boundary bed in Canada for amino acids, using Zhao and Bada's analytical methods, and discovered 51 amino acids with carbon-isotope ratios compatible with interstellar material. All of the acids are known from carbonaceous chondritic meteorites, and 18 of them have no other known terrestrial occurrence. Most of the unearthly 18 show an amino-acid to iridium ratio that is "roughly equal to that in meteorites." Carlisle and Braman concluded that the diamonds and amino acids had been carried to Earth by a fragile comet (one made of rock and ice), since an asteroid is too strong to simply shatter and would have generated enough heat on impact to have destroyed the evidence of both the diamonds *and* the amino acids. (Carlisle joined with A. E. "Ted" Litherland of the University of Toronto to formulate a new two-stage theory of K/T extinction involving a supernova in concert with impacting; it was presented in a formal talk at the U.S. Geological Survey in Menlo Park on March 18, 1992, and is now in preparation as a book by Stanford University Press.)

The Impact Site(s)

Right at the outset, in 1980, the challenge was thrown down to locate the 200-kilometer crater predicted by the impact hypothesis. Such a crater, of course, would be the smoking gun. Few, however, paid attention then to what has since come to be seen as the most important candidate crater. Those who searched for the crater for over a decade were never overly optimistic about finding it, because they knew that much of the ocean floor had been pulled down into the mantle during the past 65 million years by the movement of tectonic plates. They were also aware that if the meteorite had struck a continent, the crater might have been hidden by subsequent erosion or destructive movements of the crust, or filled in by sediments or lava flows. To those who knew how incomplete and obscure the record of impacts on Earth was—such as Gene Shoemaker, the founder of the Astrogeology Branch of the U.S. Geological Survey, who counseled patience—the fruitlessness of the early search was not surprising.

Although iridium and shocked quartz characterize the K/T-boundary sediment layer at a large number of localities worldwide, the layer varies in other ways. The composition at all sites has suggested an impact into granitic rocks or quartz-rich sediments, which are typically continental; but at some sites the layer's chemistry also suggested a basaltic, oceanic-crustal target! Such mixed evidence could have been produced by multiple impacts into different kinds of crust, or by a single great impact into a site of diverse composition, such as a coastal area where the two types of crust meet (Kyte & Smit 1986).

An early candidate impact site was the Manson structure in Iowa. It seemed to have formed at K/T time, but it was too small—only 35 kilometers across—to have produced the postulated K/T catastrophe by itself, and it was wholly on continental rock, leaving the question of basaltic impact products in the boundary layer unanswered (Hartung & Anderson 1988). Officer and Drake (1989) questioned the age of the Manson structure and did not regard it as an impact crater, but rather as a "piercement feature, thrust upward through the overlying conformable" sedimentary-rock beds. Glen Izett, however, noted that there were unpublished seismic profiles across part of the Manson structure that indicated it to be "a classic impact structure" (oral comm., 6-6-90). The redating of Manson by Izett et al. in 1993 gave an age of almost 74 million years, which prompted Dale Russel (oral comm., 10-8-93) to note that he has recognized a faunal change of subcontinental proportions in both the marine and terrestrial vertebrates at the time of the newly redated Manson impact.

In 1988 the crater search turned to the Caribbean Basin, owing in large part to the bold surmise of Bourgeois and others (1988) that at sites near the Brazos River, Texas, an iridium anomaly and the paleontologically identified K/T boundary directly overlie an unusual sequence of beds that they attributed to deposition by a tsunami (or giant wave) that was generated by a bolide impact into the sea.

By 1990 a rash of evidence began to emerge, from around the Caribbean, that favored a site of diverse crustal composition, such as the edge of a continent. First, at a K/T boundary site on Haiti, A. R. Hildebrand and W. V. Boynton (1990) of the University of Arizona found shocked quartz and a half-meter-thick layer of what they considered to be altered tektites (the spherules first demonstrated to contain unaltered "dry, tektite-like" glass cores by Izett in 1990). They proposed that a giant meteorite had splashed down southwest of Haiti in the Colombian Basin region of the Caribbean, digging a crater 300 kilometers in diameter. Bohor and Russell Seitz, however, argued that the Haitian tektites could also have been produced by an impact just south of western Cuba, some 1,350 kilometers northwest of the Colombian Basin site. That part of Cuba is underlain by rocks of K/T age that look nothing like rocks elsewhere in Cuba. The rocks are in fact huge boulders, as much as 12 meters across, compressed in a layer that is as much as 350 meters thick near the south coast but thins progressively toward the north. Bohor and Seitz (1990) suggested that the boulders had been hurled onto Cuba by a giant impact, and that the coastline south of Havana may be the rim of the crater; Officer and others (1992), however, noted that several different field investigators believed the boulders were not impact debris, but instead had been formed in place. Impact layers thin away from their craters just as volcanic ash layers thin away from their vents. Thus if Bohor's belief that the impact layer thickens southward among North American sites were correct (some thought not), then both Hildebrand's and Bohor's sites seemed plausibly placed in areas to the south of the main mass of North America.

Late in 1990, another crater site was catapulted to prominence: at the tip of the Yucatan Peninsula in Mexico lies the almost-200-kilometer-wide Chicxulub crater (some now think it 300 kilometers wide). Both continent and ocean floor were seemingly excavated in its formation. It appeared to be an excellent candidate site: its size was about right, its age was possibly late Cretaceous, and its location appeared to answer the perplexing question about why the K/T boundary beds contained constituents seemingly derived from both continental and oceanic rocks. As early as 1981 Penfield and Camargo, working from geophysical evidence acquired in an oil search, had suggested that Chicxulub was an impact crater, but the site was more or less ignored until 1990, when active groups of impact debaters began to give it their attention (Hildebrand & Penfield 1990; Kring & Boynton 1992). A glassy rock—purportedly from the cooling of the melt produced by impact—taken from cores drilled at the Chicxulub site was isotopically dated in July 1992 by two different international teams (Swisher et al. 1992; Sharpton et al. 1992). The age they determined was precisely that of the K/T boundary—65 million years. In addition, the Sharpton team found an iridium anomaly of 13.5 parts per billion in the glassy rock. Chicxulub thus appears to contain the second known such anomaly definitely ascertained within an impact melt rock (Jansa and others, in 1989, had reported a siderophile anomaly from the younger Eocene Montagnais impact melt rock from Scotian Shelf, eastern Canada; and Lowe et al., also in 1989, had reported iridium, but stated that it was only *probably* present within microcrystals of spinel forming 3,400-million-year-old tektites).

The suggestion of Bourgeois and others (1988) that giant-wave (tsunami) effects were evident at the K/T boundary—a suggestion that met with much criticism—was reinforced in three articles by impactors: Swinburne and others (1991) and Smit and others (1992) concluded that a tektite-bearing, K/T boundary-section outcropping in northeastern Mexico at Mimbral contained a very distinctive sequence of beds that they interpreted as a tsunami deposit caused by the nearby Chicxulub impact. Simultaneously, Alvarez and others (1991) reinterpreted a deep-sea drilling core from Deep-Sea Drilling Project (DSDP) site 540 and suggested that the unusual and complex sequence was full of evidence for bottom-scouring by a giant wave; that it contained particles of tektite origin; and that the unit may correlate with the Haiti K/T-boundary section. But the volcanists Lyons and Officer, denying an impact-generated-tsunami cause (1992), suggested instead that "The Haiti, DSDP core 540, and Mimbral [K/T boundary] sections all represent redepositional sequences into deeper water by gravity flows and/or turbidity currents. . . ."

The Chicxulub site has repeatedly been referred to as the "smoking gun." It appears—in keeping with the surmise of Virgil Sharpton and Kevin Burke of the Lunar and Planetary Institute, Houston (1987)—to lie on a great circle that also touches a number of lesser craters, most of whose ages had been in doubt but approximated to a K/T boundary age. Surprisingly, I. A. Rezanov (1980), in his *Catastrophes in the Earth's History*, suggested that a K/T asteroid had fragmented to produce multiple craters extending from the twin Kara craters of western Siberia to the Ukraine to Libya along a perfect-arc line 27 degrees long, but Rezanov believed that the dinosaurs were already in decline and that the impacts merely "accelerated the process to some extent." The possibility that there was a linear trend to multiple impacts had been mused on repeatedly out of print all during the 1980's. Although most of the craters now being examined more closely were long-known, none was considered large enough to have produced the evidence claimed for the K/T event: Manson (35 km) in Iowa, Popigei (100 km) in eastern Siberia, the twin craters Kara

(65 km) and Ust-Kara (70–155 km), and perhaps even the Gusev (3.5 km) and Kamensk (25 km) twin craters near the northern Black Sea shore. The Avak impact structure, about 15–20 km across, at Point Barrow, Alaska—which seems to align with the other craters—has been assigned a best tightly constrained age of post-Albian to Pliocene time (between 95 and 3 million years ago; Kirschner et al. 1992); however, Arthur Grantz and Michael Mullin (oral comm., 6-9-91) think that "the pattern of intensity of deformation and the geometry of the crater suggest a late, late Cretaceous to early Tertiary age." At the start of the debates, in any case, no crater of K/T boundary age was known; there are now, with the determination of Chicxulub, "more than enough to have done the job," mused one group of impactors.

Stepwise Extinctions

Erle Kauffman (1988; University of Colorado), Gerta Keller (1989; Princeton University) and others have argued that the great terminal kill, marked by the big iridium spike, is preceded by a series of sharply defined smaller extinctions that occur over a few million years, forming a series of pulses, or a stairstep pattern. That idea was also supported by J. F. Mount and Stan Margolis (1986), then at the University of California, Davis, and others, who found that spikes of carbon and oxygen isotopes in the rock record, in some cases, coincide almost precisely with the occurrence of the stepwise extinctions. There are also sections at which the iridium anomaly at the K/T boundary is not sharply confined to the boundary layer, but instead seems to be "smeared," or to occur as "shoulders" both below and above the boundary layer, with lesser spikes of iridium concentration (Crockett et al. 1988; Asaro et al. 1988).

The idea of stepped extinctions at the K/T boundary was buttressed indirectly by the report of Orth and others (1987), who found two iridium-abundance peaks just below the 92 million-year-old Cenomanian-Turonian stage boundary; there, the lower peak coincides with the first in a series of five or six stepwise extinctions of ammonites and benthic organisms, and the upper peak matches the third extinction horizon. The boundary is not associated with tektites or shocked quartz, but the Orth group (1992, 1993) concluded that iridium and other elements at the Late Cenomanian horizon hinted at an impact source, while acknowledging that an endogenous cause was nonetheless possible.

Further indirect evidence for reassessing the K/T boundary had also been accumulating at the Eocene/Oligocene boundary interval. Alessandro Montanari (1990) lucidly summarized the data from several studies of fossils, chemical content, sedimentary magnetism, tektites, and age correlation for that interval of faunal change and multiple impacts. From those boundary beds, Isabella Premoli-Silva and others (1988) read two closely spaced impacts dated between 34 and 35 million years old that involved tektites; Gerta Keller and others (1987) interpreted a third impact at about 36 to 37 million years. But Joseph Hazel's (1988, 1989) radically different view of the same interval, based on 17 microspherule-bearing localities, showed not two or three, but nine or more microspherule layers spread over 1 million years. Although there were no "mass extinctions" during the times of purported multiple impacts, there were steplike changes in the oceanic biota.

Steplike happenings are also in evidence around the Late Devonian, Frasnian-Famennian extinction boundary that McLaren first attributed to impact in 1970. The boundary bed at Senzeilles, Belgium, contains both an iridium anomaly and some of the oldest tektite-like glass spherules known (Claeys et al. 1992). Claeys and his colleagues also reported that a short distance above the Frasnian/Famennian boundary in Hunan, China, there is an iridium anomaly associated with both spherules and a negative carbon-isotope anomaly; they therefore believe that a cluster of impact events is related to the Late Devonian extinctions, including the Frasnian/Famennian. The idea of an impact cluster in that interval seems also supported by Kun Wang and Helmut Geldsetzer (1993), who have reported microtektites, geochemical anomalies (including iridium), and Ken Hsü's (1987) "Strangelove-ocean"-like carbon-isotope anomalies from South China and Western Australia, at a horizon they date at "about 1.5 million years after the Frasnian/Famennian crisis."

The search by Wang and others (1992) at the Ordovician/Silurian boundary in China did not turn up evidence of steps, but at three localities they did find elevated values for iridium, other siderophile elements, and chalcophile elements, and noted that similar anomalies have been recorded at about the same stratigraphic level in Quebec, in the Yukon, Arctic, and Northwest Territories of Canada, and in Scotland. They attributed the anomalies possibly to a low-platinum-group comet, but more likely to a low sedimentation rate, reducing conditions, and a rapid transgression of the sea.

Evidence for extinction steps clustered within the timeframe of hundreds of thousands to a few million years continues to accumulate from the K/T and other extinction horizons. That trend is to the volcanists' delight, since they hold that such steps accord with their volcanism model. In terms of the K/T single-impact hypothesis, however, the stepped-extinction evidence must be seen as anomalous. How the impactors responded to that anomaly is detailed in the chapter on "How Science Works" (this volume).

Objection to the stair-step interpretation comes from Jere Lipps of U.C. Berkeley (oral comm., 1989; Signor & Lipps 1982), who believes "that apparent stepwise extinctions are an artifact of the Signor-Lipps effect, in which a progressive decline in sampling quality and differences in preservation approaching an extinction boundary can smear out even an instantaneous extinction event and make it appear to have been gradual or to have occurred in steps." Many paleontologists, including Karl Flessa (oral comm., 1991) and Philip Signor (oral comm., 1992), share Lipps's opinion. There are also impactors, eminent geochemists among them, who remain strongly opposed to any but the single-impact hypothesis. Nonetheless, the seemingly persuasive pattern of evidence for stepwise extinctions, coupled to the idea of multiple impact horizons spread over a timeframe of 1 to 3 million years, has moved many impactors to all but abandon their idea of a single great terminal-Cretaceous impact. Instead, most now embrace a revised hypothesis of multiple, serial impacts, due to a K/T comet swarm. The idea of such a swarm of comets was pioneered by Clube and Napier (1984), who proposed that it would produce "a series of spikes" over a few million years; Piet Hut and others (1985) similarly found a way to bombard the Earth with a dozen or so greater and lesser comets to accord with an interval of stepped extinctions.

The Extinction Mechanism(s)

The idea of a comet shower with a series of varied impacts spread over a million years or more seemed useful in explaining the apparent smaller stepped extinctions, but it was questionable whether their cumulative effects could have produced the great terminal K/T kill. Thus, some of those favoring multiple impacts early on called for a comet swarm that included an iridium-rich giant precisely at the K/T boundary—and the Yucatan Peninsula's great Chicxulub crater may yet fulfill that prediction.

A new twist on the multiple-impacts idea was proposed by Peter Schultz (Brown University) and Don Gault (Murpheys Center of Planetology), who argued (1990) that if a 10-kilometer meteorite were to strike the Earth's surface at 72,000 kilometers per hour and at a low angle (less than 10 degrees from the horizontal), it would break up into a swarm of fragments ranging in size from a tenth of a kilometer to a kilometer in diameter. The fragments would ricochet downrange; thus one such object might have excavated multiple craters. They have simulated such an event in the laboratory with a pellet gun and filmed the results: it appears that an oblique impact would eject enough debris into orbit to give the Earth a ring like one of Saturn's. Over time (it is not clear how much time) the debris would fall back to Earth, but meanwhile it would kill life on the planet by blocking out a significant amount of sunlight. The gradual descent of the impact debris long after the initial cataclysm might explain why some of the chemical evidence for a K/T impact is spread out in the stratigraphic record.

Schultz and Gault's work notwithstanding, the climatic effects of a K/T impact, after diverse modeling studies by several different groups, can still best be described as uncertain. The Alvarezes' original idea of a global stratospheric dust cloud that would block out the Sun for months or years and choke off photosynthesis is now widely regarded as too simple, and there is growing agreement on what to construct in its place. Suggestions have run the gamut of contemporary environmental concerns, from nuclear winter (a scenario that was actually inspired by the Alvarez hypothesis), to acid rain, to meteorite-metal poisoning of plants, to the greenhouse effect, and more. It is possible, and even likely, as suggested by increasingly sophisticated studies, that all of these effects and others come into play at one time or another after one or more impacts of the magnitude hypothesized for that at the K/T boundary (Roddy et al., 1987; Covey et al. 1990; Zahnle 1990; Davenport et al. 1990; Hsü & McKenzie 1990; Vickery & Melosh 1990; etc.).

The acid-rain idea has been advanced since 1981 by Ron Prinn and his colleagues at the Massachusetts Institute of Technology. In a series of detailed papers, they have argued that the plume of ejecta kicked up by a giant impact would shock-heat the atmosphere, and that the heat would cause nitrogen and oxygen to combine with water vapor to form nitric acid (Prinn & Fegley 1987). Raining out of the atmosphere in concentrated form, the acid would kill ocean-dwelling organisms and perhaps others as well. Percolating into the ground, it might also leach away fossilized remains; that, said Gregory Retallack and others (1987) at the University of Oregon, might explain the

3-meter interval of barren rock that, at certain sites, separates the youngest (and therefore uppermost) pre-K/T dinosaur fossils from the K/T boundary. The putative 3-meter gap (which varied from 1 to 4.5 meters at different North American sites) has often been cited by opponents of the impact hypothesis as evidence that the dinosaurs died out before the postulated impact. The belief long held by paleontologists that the dinosaurs' demise was gradual has also been contravened by a painstaking study of the distribution of their remains by Peter Sheehan and others (1991).

According to H. J. Melosh and his colleagues at the University of Arizona, the plume of vaporized rock ejected by the impactor—consisting of materials deriving roughly equally from the impactor and from the Earth's crust—would do much more than cause acid rain. The Arizona workers' calculations suggest that most of the plume would leave the Earth's atmosphere, and in the process it would take some of the atmosphere with it—in other words, the Earth would be stripped of part of its gaseous blanket. Much of the gas would be permanently lost, but the plume material itself would recondense into solid particles, and those particles would follow ballistic paths around the planet, like microscopic ICBMs. Ballistic trajectories, say Melosh and his coworkers, do a better job than stratospheric winds (the original Alvarez view, but now invoked by those who favor a volcanic-extinction cause) in explaining how iridium and shocked quartz from a single impact could be distributed at many localities around the Earth. What is more, as the particles reentered the atmosphere they would heat it even more effectively than had the original plume. At the Earth's surface, the thermal radiation from the decelerating particles would be on the order of 10 kilowatts per square meter for as much as several hours after the impact. Such a power level, Melosh and his colleagues note, is "comparable to that obtained in a domestic oven set at 'broil.'" Roddy and others (1987) modeled impacts by 10-kilometer bolides and found that most of their geochemical signature could be vaporized to escape from Earth. Vickery and Melosh's (1990) modeling also suggested that very-high-energy impacts may "neither produce as much acid rain nor leave behind a siderophile [iridium] signature in the boundary layer" . . . as [would] those of lesser energy. "Thus the lack of iridium enhancement at a mass-extinction horizon does not necessarily rule out an impact as the cause of the extinction."

The impact-induced thermal radiation would ignite wildfires that might surge over most of the globe, even into areas that were initially shielded from the heat blast by thick clouds. Purported evidence for such fires has been found at several K/T sites by Wendy Wolbach, Edward Anders, and their colleagues at the University of Chicago (Wolbach et al. 1985, 1988, 1990). The evidence takes the form of a high global concentration of charcoal and soot in the iridium-bearing boundary layer. They believe that "much or all of the fuel was biomass" because of the presence of retene, which is indicative of wood fires, and the isotopic composition of the carbon, which is close to that of natural charcoal and carbon from biomass fires.

Hansen and others (1987) found carbon black at Stevns Klint in Denmark in the gray chalk below the K/T boundary, and interpreted the carbon as having been deposited in pulses, which suggested that it could not have originated in a global fire but was instead produced by volcanism. Wolbach and others (1990) countered that Hansen's "carbon black" was really kerogen (organic matter from plants), a fact they held to nullify the argument for volcanic carbon. Some have said that the carbon concentration at the K/T boundary is exceptional only if one accepts the Alvarez hypothesis as true—that is, only if one believes that the K/T boundary layer consists of impact fallout deposited in a few months or years rather than across millennia. The carbon itself is not primary evidence for an impact, but the data of Wolbach and others (1990) and Herring (1990), combined, show that, on the basis of the charcoal values in more than 200 samples covering more than 70 million years, the K/T soot layer coinciding with a global marker appears singular. An abundance profile across the K/T boundary at Woodside Creek, in New Zealand, shows that iridium, soot, and elemental carbon all rise at the boundary by two to three orders of magnitude, and then return slowly to normal Cretaceous levels. The soot appears in the lowest one-third of a centimeter of the boundary clay, which Wolbach and others (1988) believe to be a signal that fires were burning "before the primary fallout had settled."

Another good way to heat up a planet, of course, is through the greenhouse effect, which was suggested as a killing mechanism by impactors almost from the time the impact hypothesis was first proposed, and even before that by volcanists. Recently John O'Keefe and Thomas Ahrens (1989) of the California Institute of Technology have provided experimental support for the impact-greenhouse scenario. They shot steel balls from a cannon into various types of rock at 7,200 kilometers per hour and measured the amount of carbon dioxide released by the impacts. They calculate that if a 10-kilometer asteroid were to slam into limestone or some other carbon-rich formation, it would increase the atmospheric concentration of CO_2 by a factor of from two to five almost overnight. The resulting heat would have devastated life.

Some of the plants that suffered most during the K/T catastrophe were single-celled marine algae, whose numbers may have dropped by as much as 95 percent. Whether or not the decline was caused by a climatic change, the decline itself may have had an important effect on climate. Michael Rampino and Tyler Volk (1988) of New York University have noted that some of the microalgae that were most affected, notably the chalky coccolithophores, are prodigious emitters of the substance dimethyl sulfide. In the atmosphere, dimethyl sulfide is converted into sulfate particles, which serve as the nuclei around which water condenses to form clouds. According to Rampino and Volk, a severe drop in the population of microalgae would drastically reduce the cloud cover long after the initial catastrophe. With more sunlight streaming in, and less rain falling, the Earth would heat up, and many land plants would die.

To summarize, there is no shortage of ideas on how an impact or massive volcanic catastrophe might kill organisms on the Earth. Publications treating postulated short- and long-term environmental effects are now legion, and regrettably, only a few of the many important ones have been touched on here. Wolbach and others (1990) summarized the most frequently cited literature on stresses produced by a K/T impact, but of course many of those stresses could be equally well attributed to massive volcanism. It appears that the students of the K/T extinctions, like the nuclear powers, have armed themselves with enough weaponry to extinguish all life on the planet many times over.

The Pattern of Extinctions

The selectivity of the extinctions at the end of the Cretaceous Period has been widely argued (Van Valen & Sloan 1977; Thierstein 1981; Clemens 1982; Emiliani 1982; Emiliani et al. 1981; Hsü 1982; Thierstein 1982; Surlyk & Johansen 1984; Sheehan & Hansen 1986; Hallam 1987, 1988; Stanley 1987; Kauffman 1988; Briggs 1990; Archibald & Bryant 1990; Johnson et al. 1990; Ward 1990; Wolfe 1990, 1991; Sheehan et al. 1991; etc.). Certain groups of land and ocean creatures are held to have expired gradually or stepwise in their approach to the K/T boundary; others seem to have made it through the crisis only to decline shortly afterward (Keller 1989); but many groups, such as the limey-shelled, microscopic, marine floaters (foraminifera and coccolithophores), as well as land plants in western North America and eastern Asia, appear to have suffered a catastrophic killing right at the boundary. McLaren (1989) lucidly analyzed such "knifeblade-sharp" killing boundaries in the rock record, emphasizing the total biomass that was exterminated rather than the number of taxa (species, genera, etc.) thus made extinct, since just a few survivors will keep a group off the extinction list, even if almost all of its members have been killed.

The extinction pattern in the oceans shows that tropical groups and those that fed on plant plankton were most vulnerable. Jennifer Kitchell of the University of Wisconsin at Madison and others (1986) suggested that the oceanic plankton, such as the diatoms and dinoflagellates that make up much of the Earth's biomass, survived in greatest numbers because of their resting-spore stage of reproduction.

On land the large reptiles (those over about 20 kilograms) suffered much more than their smaller lake- and stream-dwelling relatives. Perhaps large creatures found it more difficult to find shelter (Russel 1979). One might guess that land plants that reproduce by seeds, spores, pollen, or rhizomes—reproductive bodies that could lie under soil and thus be shielded from catastrophe—would have had a better chance of survival. And that is just what is found: after a short-term crisis right at the boundary, the flowering plants recovered rapidly (Tschudy & Tschudy 1986; Johnson et al. 1989; Wolfe 1990; etc.).

David Jablonski (1986) argued that mass extinctions differ from the normal or background extinction patterns that hold for most of geologic time. He thinks that the attributes that made for survivorship in normal times did not suffice during mass extinctions, and that only very broad geographical distribution of a group seemed to help it survive.

P. M. Sheehan and T. A. Hansen (1986) viewed the selectivity of extinctions at the K/T boundary as largely a function of a group's position within its food chain. They noted that the animals that became extinct, such as the dinosaurs, benthic filter feeders, and animals that lived in the water column, were in food chains that were tied directly to the living plant-food supply. In contrast, the less severely stressed animal groups that better survived the extinction fed not on living plants, but on dead plant material; these included bottom-dwelling marine scavengers and detritus (waste and debris) feeders, members of stream communities, and small insectivorous mammals. A catastrophe would have produced a long-lasting stockpile of organic waste for those who lived on it. On the continents, the doomed herbivorous dinosaurs probably ate living plants, but the small, live-bearing mammals that survived so well likely ate insects—for insects are hardy life indeed, and in the logical extreme may be why we are here today, evolved from our insect-eating ancestors.

Two of the life groups among the many that expired at the end of the Cretaceous Period are the nautilus-like ammonites, which dominated the seas of the Mesozoic Era, left a magnificent fossil record, and played signal roles in the rise of biostratigraphy and the growth of the geologic time scale, and the dinosaurs, which ruled the continents for almost 200 million years and have fascinated both science and the public. For more than a century the horizon at which those two groups disappeared from the rocks served as a marker for the upper boundary, or end, of the Cretaceous period. Both the ammonites (Ward 1983) and the dinosaurs (Sloan & Van Valen 1965; Clemens 1982; Sloan et al. 1986; etc.) were long thought to have expired gradually over hundreds of thousands to millions of years in their approach to the boundary.

Declining diversity patterns for them and other groups are still cited in evidence against sudden, catastrophic extinctions at the K/T boundary (Teichert 1986). But there are indications that with more careful and detailed collecting, the case for gradual decline in certain groups is becoming more difficult to argue. Ward and Wiedmann examined the Spanish coast at Zumaya in 1983; they had earlier concluded that the inoceramid clams there had undergone a gradual extinction. It thus came as no surprise when they found that the ammonites, too, had apparently begun their gradual decline toward extinction about 4 to 5 million years before the end of the Cretaceous—the last ammonite fossil at Zumaya having been found about 40 feet below the K/T boundary. But Ward and others (1986) continued searching for ammonites beyond Zumaya and found, contrary to their earlier surmise, that ammonites at other localities were fossilized in the rocks right up to the iridium-bearing K/T boundary. That find spoke clearly for a more sudden, rather than a gradual, extinction of the ammonites. Peter Sheehan and others (1991) did a painstaking, 15,000-man-hour study of family-level patterns of dinosaur ecological diversity in the Hell Creek Formation of Montana and North Dakota and showed that there was "no evidence (probability $P < 0.05$) of a gradual decline of dinosaurs at the end of the Cretaceous." More recent work tends to support a sudden extinction for many groups—especially those with certain lifestyles in selectively stressed environments—as we will shortly learn.

Jack Wolfe (1991) tendered an intriguing hypothesis, supported by unprecedented details, about the K/T event. His study appears to have been the first in these debates to involve experiments with living forms to aid in deciphering fossils. Water-lily and lotus-leaf remains in the boundary beds at Teapot Dome, Wyoming, show "freezing folds" that Wolfe believes resulted from impact-generated, sun-blocking debris in the atmosphere. He tested that surmise by duplicating the fossil folds in living leaves by freezing them. He further holds that the meteorite struck in early June, because the rocks contain seeds from the water lilies but none from the lotuses, a pattern that accords with the seasonal cycles of the plants. Stratigraphically above that horizon, another debris layer suggests that a smaller impact occurred, only months after the first, following the onset of a warm period (Wolfe 1990) that followed the freezing. Both Douglas Nichols (1992), Wolfe's Geological Survey colleague, and Leo Hickey and Lucinda McSweeny of Yale University (1992) have criticized certain aspects of Wolfe's hypothesis, but Wolfe has responded in print (1992) and remains steadfast. His novel idea is bound to promote searches for further evidence.

Periodic Happenings

The K/T mass extinction was not the only one in Earth's history, nor even the worst. That distinction goes either to the event that marked the transition from the Permian Period to the Triassic, some 250 million years ago, when as many as 96 percent of all species in the ocean vanished, or perhaps, as Philip Signor of the University of California, Davis, has described it (1992), to the extinction in the early Cambrian Period (end of the Botomian Stage, about 525 million years ago) that killed off 80 percent of the genera. There is evidence in the fossil record for at least nine other mass extinctions. One of the most interesting hypotheses to spring from the impact debates is the idea that these mass dyings have recurred with a regular period of 26 million years.

Periodic mass extinctions were first postulated in a virtually ignored paper of 1977 by Alfred G. Fischer of Princeton University and his student Michael A. Arthur. Although based on data culled from the fossil record of the past 250 million years, their postulate of a 32-million-year periodicity was not supported by a rigorous statistical analysis. In contrast to earlier ideas of periodicities relating to a number of geological phenomena and to processes that were attributed to astronomic causes, Fischer and Arthur guessed that still-little-understood cycles of convection within the Earth drove the extinction process.

The hypothesis of periodicity advanced by David Raup and John Sepkoski of the University of Chicago in 1984 arrived in the heat of the impact debates, was based on a pioneering statistical analysis of the record of 3,500 families of marine organisms (marine families were used, rather than land families, because they are much more numerous), arrested the attention of science, and triggered new research across a broad front, particularly in astronomy and astrophysics. Raup and Sepkoski then extended their analysis down to the more detailed taxonomic level of genera, analyzing 11,000 of them (1986, 1988). Their data set is being continuously updated and has now grown to more than 34,000 genera for which they have tabulated a diversity of stratigraphic and taxonomic information; and according to Sepkoski (1990), the 26-million-year cycle, evident as peaks in the extinction rate, appears even more strongly than it did in the original family data. Perhaps equally remarkable, a new extinction peak, recognized in 1990 in the Middle Jurassic Period, fell at a cycle time predicted by the hypothesis. But with very old rocks, such as those holding the fossils that allowed Philip Signor (1992) to recognize a great mass extinction at the end of the Botomian Stage of the Cambrian Period, uncertainties in dating preclude a statistical analysis that would test for periodicity.

Raup and Sepkoski's 1984 results stimulated a rush to find the cause of periodic extinctions. Within a short time three separate mechanisms had been proposed, all of which involved ways of gravitationally knocking comets out of their orbits beyond the outer planets and sending showers of them toward the Earth. In the first hypothesis, formulated independently by two different groups, the Sun has a faint, dwarf companion—dubbed Nemesis, or the death star, by the Berkeley Group—whose eccentric orbit carries it near the Solar System every 26 million years (Davis, Hut & Muller 1984; Whitmire &

Jackson 1984). In the second hypothesis, a tenth planet that lies beyond Pluto on a highly eccentric orbit passes through a hypothetical trans-Neptunian belt of comets every 26 million years (Whitmire & Matese 1985). In the third hypothesis, the Solar System's vertical oscillations through the plane of the Galaxy periodically bring it near dense clouds of interstellar gas that perturb the comets (Rampino & Stothers 1984a). Whether any of the three of these hypotheses can provide a meteorite shower of a duration in keeping with the extinction data—J. Sepkoski (*fide* Perlmutter et al. 1990) measured the duration of the extinction peaks at about 3 million years—is uncertain. It seems that both the galactic oscillation and Planet X models would produce comet showers of greater duration, but the effects of a Nemesis-induced comet shower would persist only 1 to 2 million years, according to Perlmutter.

One might surmise that the Rampino & Stothers theory (1984a; see above) calls for a mass extinction a million years ago that clearly did not happen, but they hold that interstellar gas clouds exist above and below the Galactic plane as well as in it, thus Stothers (1985) could predict only a "roughly 50-percent chance of detecting the Galactic period in the terrestrial impact record." Because of intrinsic scatter he noted that "we would actually not be able to explain a perfectly periodic pattern" (written comm., 12-4-90). The main problem with the other two theories is a lack of observational evidence: neither Nemesis nor Planet X has been found, although a group led by Richard Muller of the University of California at Berkeley continues to search for the death star (oral comm., 10-9-83).

Quite apart from the issue of a causative mechanism, the question of periodic impacting itself is still open, at least to date. In 1984 Rampino and Stothers (1984a) and independently Alvarez and Muller (1984) suggested that periodic impacting of the Earth could be read from crater ages that seemed to be clustered at the times of mass extinctions. But their samples consisted of too few craters, and the results left many workers unconvinced.

Raup and Sepkoski's methods have been criticized on the grounds that the fossil record is too incomplete, and the dating of extinctions too uncertain, to justify their claims of periodicity. But they offer an interesting rebuttal: such uncertainties, they argue, would merely introduce random noise into the data, and if a systematic 26-million-year signal emerges above such a noisy background, then there is all the more reason to believe the signal is real. This argument, however, has not won over all the skeptics. In an episode that perhaps epitomizes the polarization of views on periodicity, two members of the statistics department at the same university, after reviewing the Raup and Sepkoski data and analyses, found themselves opposed in the extreme: one said the finding of a periodicity was correct; the other said it was "junk."

Raup's continued interest in periodicity, and papers such as that of Negi and Tiwari (1983), which matched long-term periodicities of geomagnetic reversals and Galactic motions of the Solar System, led him to analyze the polarity reversals of the Earth's magnetic field, for which he found a 30-million-year signal (1985b). Raup shortly thereafter abandoned that conclusion when Timothy Lutz (1985, 1986), of the University of Pennsylvania, suggested that Raup's periodicity was merely an accident of record length. That, however, was followed by the surmise of Stothers (1986) that Raup was originally correct, and that Lutz's analysis was flawed. Lutz, Antoni Hoffman of the University of Warsaw, and others (1985, 1986), and Jennifer Kitchell and Dan Pena (1984), of the University of Wisconsin, and others have seriously disputed the statistical methodology by which periodicity has been identified in Earth's history. By now the flood of claims and counterclaims from a variety of experts sends the onlooker reeling (Raup & Sepkoski 1988; Stigler & Wagner 1988). Raup's informative tabulation (1991) shows that of more than a dozen differently authored papers evaluating the hypothesis of periodicity of extinctions, half found it acceptable, half did not.

Prompted by Raup's paper on periodic magnetic reversals, Richard Muller teamed with Don Morris (1986) to postulate that reversals of the Earth's magnetic field were caused by impact-induced glaciation. They argued that alteration of the rate of rotation of the crust and mantle by the formation of polar ice would induce a velocity shear across the Earth's liquid core, thus disturbing the geomagnetic dynamo that produces the magnetic field. A refined version of their hypothesis (1989) has been criticized by David Loper and Kevin McCartney (1990), who do not believe that impacts can cause reversals, and surmise that no viable model yet links impacts with reversals, or with the episodes of massive volcanic outpouring that in some cases seem to occur at the times of mass extinction; however, Rampino (1987), Alt and others (1988), Baksi (1990), and others continue to argue that there is a connection between impacts and flood-basalt volcanism (a subject—rejuvenated in 1992 by a new interpretation of certain "glacial" deposits as impact products—to which we shall return later on). Still other causes for magnetic reversal have been advanced since the advent of the impact hypothesis in 1979, but they seem not to have evoked much response.

The issue of periodicity was moved to a new arena in 1987 by Herbert Shaw (U.S. Geological Survey, Menlo Park). In a pioneering paper in *Eos,* he used recently evolved nonlinear dynamics methods, which differ from conventional statistical analysis, to search for patterns in several Earth processes that are repeatable but not necessarily symmetrical or equally spaced. With respect to the mass-extinction debates, he thinks it impossible to demonstrate valid periodicity patterns, because the data sets are too limited for both conventional and nonlinear analyses. He thus put questions of periodicity in a new light, but seemingly further still from resolution. It seems that resolution would be more likely if more statisticians became versed in nonlinear-dynamics methods, but unfortunately, as nonlinear guru Ralph Abraham of the University of California, Santa Cruz, remarked, "The bridge between the two approaches to data analysis is just beginning to be built." Bridge construction began to accelerate in 1988 and 1990 when day-long sessions at national meetings of the American Geophysical Union were devoted to nonlinear-dynamics applications to problems in the geosciences. Shaw continued to refocus his nonlinear-dynamics lens to encompass eventually the Cosmos, and to formulate a singularly broad, new theory of Earth history in book form (forthcoming, 1994). The theory deciphers the chronicle of Earth impacts from the cratering record of the Moon and planets and ties it to the dynamics of the Earth's mantle, global tectonics, volcanism, magnetic field, fossil record, and a host of other phenomena. The theory's arresting, predictive promise will be tested as data accrue from the host of research programs spanned by its vast conceptual reach.

The Volcanism Hypothesis: Impact's Alternative, Adjunct, or Partner?

Sulfur dioxide and other gases spewed into the stratosphere by volcanic eruptions would have cooled the planet for a year or two, extinguishing some species, and would then have gradually fallen out as acid rain, which would kill off others. Over many thousands of years, the carbon dioxide emitted by the volcanoes would have warmed the planet through the greenhouse effect. The end result would have been a series of selective extinctions spread over a considerable length of time. According to many who favor the volcanism hypothesis, such a pattern is similar to what the fossil record shows for latest Cretaceous time and immediately afterward.

The test of volcanism as an alternative hypothesis, of course, is its ability to account for the K/T boundary evidence that has been used to support the impact hypothesis. Start with the iridium. Iridium is rare in the Earth's crust, because it and the other siderophile elements were dragged into the core and the mantle by iron early in Earth history; thus a volcano that coughed up rock from deep in the mantle might also bring up iridium. Volcanoes that form along boundaries where tectonic plates collide, such as those that rim the Pacific, have shallow sources. But "hot-spot" volcanoes, which form in the center of plates—the Hawaiian volcanoes are good examples—are believed by some geophysicists to be fed by plumes of hot material rising from deep in the mantle, perhaps even from the boundary with the core. In fact, iridium was detected at hot-spot volcanoes as early as 1983, from measurements done at Kilauea in Hawaii by William H. Zoller of the University of Washington. The amount of iridium in the erupting rock was at least two orders of magnitude less than what is observed at the K/T boundary, but the concentration in the volcanic gases, and sublimates deposited by the gases, is significant. Zoller (oral comm., 4-27-90) and his colleagues analyzed the Kilauea emissions for the relative abundances of other rare elements, such as selenium and arsenic; they concluded that the elemental signature of the volcanic emissions is closer than that of meteorites to the one that is typically observed at the K/T boundary. However, Frank Kyte of UCLA (oral comm., 4-27-90) and C. P. Strong and others (1987) seem to reflect an alternative view that the volcanic-emissions data published to date do not necessarily support the volcanism cause. Finnegan, Miller, and Zoller (1990) now surmise that "The implications of iridium from Hawaiian volcanoes for the K/T iridium anomaly are not conclusive, but seem to weaken the volcanic explanation and strengthen the impact position."

Of the other volcanoes where iridium has been detected, one is particularly noteworthy. It lies on Réunion Island in the southwest Indian Ocean (Toutain & Meyer 1989). The hot spot that now feeds Réunion is believed to be the same one that created a vast formation of flood basalts called the Deccan Traps in western India. As early as 1972, P. R. Vogt suggested a connection between the Deccan, which is roughly a million cubic kilometers in volume, and the K/T extinction (Vogt, however, abandoned his own hypothesis and embraced the impact hypothesis at its advent in 1980; oral comm., March 1993). The Deccan, one of the largest known continental deposits of volcanic rock, was erupted in just a few great pulses over about a million years. The greatest pulse seems to have encompassed the time of the K/T mass extinction (Courtillot 1990; Baksi & Farrar 1991). Not surprisingly, the Deccan Traps have become a touchstone for the volcanists. Paleontologists have also looked to the Deccan for solutions since 1981, when the paleontologist Dewey McLean first proposed that such great volcanism had disturbed the carbon cycle and turned the Earth into a lethal greenhouse for many forms of life.

Iridium has also been found in the ash fraction of volcanic eruptions. Koeberl (1989) reported that the iridium in volcanic dust bands in Antarctic ice cores has surprising values of 4 to 8 parts per billion, which is comparable to that of some K/T boundary sections; and volcanic ash erupted by continental-type volcanoes in Kamchatka contained iridium at levels of 1 to 4 parts per billion (Felitsyn & Vaganov 1988).

Because the original volcanism hypothesis had only to explain the iridium at the K/T boundary, it was vital to the volcanists, early on, that it be demonstrated that volcanoes could draw up iridium from the Earth's interior; but after the discovery of shocked quartz at the K/T boundary by Bohor and others in 1984, the volcanists were faced with another difficult challenge. The conventional view is that the pressures needed to produce the characteristic features of naturally shocked quartz—especially multiple sets of intersecting shock lamellae—are generated only by meteorite impact, not by volcanoes. However, volcanists (not volcanologists) have argued that such pressures might be created in the magma chamber of an exploding volcano (Rice 1987). They note also that little is known about the ways in which a wide range of shock features are produced in quartz, especially at the high temperatures found in volcanic settings. Thus Neville Carter and Alan Huffman (Carter et al. 1990; Huffman et al. 1990) of Texas A&M University and other groups have tried to test this idea in laboratory experiments, but definitive, argument-constraining results still seem wanting.

S. L. de Silva and others (1990) concluded that "the maximum pressures associated with explosive volcanism are . . . orders of magnitude lower than those required to shock quartz. That seemed to be supported by Gratz and others (1992), who fired aluminum discs at granite heated to 600° C to simulate the stress (0.9–1.3 gigapascals, or about 3,000 atmospheres) of a volcanic explosion. They found no shock metamorphism in the sample and concluded that volcanism is not capable of producing shocked quartz.

Charles Officer of Dartmouth University, one of the authors of the explosive volcanism hypothesis (Officer and Drake 1983, 1985, 1989, etc.), and Neville Carter of Texas A&M University reported (1991) that they had found shocked quartz with "multiple intersecting sets of planar elements . . . [within] . . . several major, apparently *enigmatic structures* . . . of internal origin." That report, aimed at refuting the impact-diagnostic value of shocked quartz, was less effective than Officer and Carter hoped, because it could not be clearly demonstrated that high-grade shocked quartz was *unequivocally of volcanic origin*. But even if the impactors were to show that volcanic explosions can make shocked quartz, the response of the impact proponents can already be predicted. They will say that the Deccan basalts are not of the kind associated with explosive eruptions; the Deccan formation is thought to be the product of more gentle, Kilauea-type flows that were erupted in a series of pulses over a period of perhaps a million years. They will also argue that only an impact, not volcanism, could propel iridium and shocked quartz into ballistic trajectories around the globe to settle within the same thin sedimentary boundary layer (Alvarez 1986). The volcanists, instead, invoke explosive volcanism and giant dust storms to do the job of transport (Officer & Drake 1989).

The impact proponents will further point out that the iridium level in the Kilauea or Réunion emissions, when extrapolated to the size of the Deccan Traps, falls short by a factor of ten in the amount of iridium needed to produce the global layer at the K/T boundary. Some volcanists would probably agree that the Deccan eruptions were not by themselves enough to produce the K/T iridium layer or the mass extinctions. Instead, these investigators generally invoke a more loosely defined episode of global volcanism as the cause of the K/T crisis. There is some evidence, for example, that the Walvis Ridge, a submarine mountain range in the southeast Atlantic, was erected by hot-spot eruptions around the time of the K/T boundary (Officer et al. 1987).

Another type of support for the idea of global volcanism comes from mathematical models and simple physical models. Several volcanists have put forward models that attempt to explain not only mass extinctions but also other geological phenomena, such as the reversal of the Earth's magnetic field and the wandering of its poles, as effects of plumes rising off the core (Loper & McCartney 1986, 1988; Courtillot & Besse 1987). The key element of the models is a thermal boundary layer, a kind of blanket on the surface of the core, whose presence has been deduced from seismic data and whose thickness increases and decreases episodically. At a certain thickness in each cycle the boundary layer becomes unstable and lets loose a covey of hot plumes. Rising to the surface, the plumes trigger bursts of polar motion, volcanic and tectonic activity, and extinctions. Meanwhile, plumes of cold iron sink from the core-mantle boundary into the core, disrupting the currents that generate the geomagnetic field and causing the field to flip back and forth. According to David Loper of Florida State University, this sort of thing may happen every 30 million years or so.

A computer simulation, one of the first to model the mantle as a spherical body in three dimensions, suggests that hot-spot plumes do indeed rise to the surface from the core-mantle boundary (Bercovici et al. 1989); but some modelers caution that such a model—using sophisticated graphics and a Cray super computer—shows nothing fundamentally new about the physics of heat transfer and mantle convection.

Prospects

During the past few years new lines of evidence have convinced the majority of earth scientists, and many paleontologists, that a meteorite impact did indeed occur at the K/T boundary. But the debate about the killing mechanism(s) and modes shows no sign of ending soon. A satisfying resolution, of course, would be for everyone to agree that impacts cause extinctions by triggering massive volcanic eruptions that affect the host of variables that were invoked in the earlier hypotheses, such as sea-level and climate changes, or, as F. L. Sutherland (1988) of the Australian Museum has uniquely postulated, that simultaneously, and coincidentally, both impacting and volcanism caused the K/T mass extinctions—and he clearly stipulates that the volcanism he envisions was not triggered by impacts (this case, interesting from a social-studies viewpoint, is taken up in Chap. 2, this volume).

Ever since Thomas Ahrens and John O'Keefe (1983) of the California Institute of Technology estimated that a 10-kilometer bolide, as postulated by the Alvarez group, would produce an earthquake of about magnitude 13 on the Richter scale, many have thought that such a quake must have rung the Earth like a bell, relaxed an enormous amount of crustal strain, and set off every sleeping volcano on the globe. That idea was finally published in 1987 by Rampino and elaborated by Alt and others in 1988. Nonetheless, Jason Morgan of Princeton, G. Schubert of UCLA (oral comm., 1989), Mark Richards of U.C. Berkeley (oral comm., 1990; Richards & Duncan 1989), and others long published on mantle dynamics seem sour on the idea of impact-induced plumes. But those who invoked impacts as the cause of the great flood-basalt sequences were not dissuaded by the critics nor were the impactors' sympathizers; instead, they opened corollary lines of inquiry through reexamination of certain structures and sediment types that had long been regarded as paradoxical in certain respects.

Verne Oberbeck (NASA Ames Research Center) had been concerned with the role of impact-generated ballistic erosion and sedimentation, and with the geomorphic aspects of impacting, since he undertook his lunar studies in the early 1970's (Oberbeck 1975; Oberbeck et al. 1974; etc.). He was thus well prepared to entertain the problems—thrust to the fore by the impact debates—that led him and his colleagues to some remarkable and provocative surmises.

In March 1992, Oberbeck and J. R. Marshall reported on their several-year-old study of rock types that had long posed some problems of interpretation: tillites (sedimentary rock of glacial origin) and diamictites (sedimentary rock most often attributed to glacial action). Their findings suggested that the textures observed in tillites are also found in impact ejecta deposits formed on land surfaces or in the ocean. They also noted a number of other factors that justified an open-minded reexamination of the purported glacial origin of certain parts of the diamictite and tillite record. Among such factors was the difficulty of forming "glacial" tillites in the Precambrian, a time when hot climates prevailed; the coincidence of periods of "glacial" tillite production with periods of major biologic extinctions in circumstances that offered no apparent reason to attribute such extinctions to glaciation; the fact that at least one tillite deposit has already been redefined as impact ejecta after a crater was found at its center (Hambrey & Harland 1981); their discovery that "there is remarkable geographic and temporal association of tillites with flood basalts" and their analysis—based on their own research, reported in a second lengthy abstract in the same volume (Oberbeck & Aggarwal 1992)—that there is "agreement of tillite deposit thickness and thickness distribution with predicted [impact] ejecta thicknesses." The study that led to the last of these surmises was a natural outgrowth of Oberbeck's many earlier years of work on the effects of cratering and ejecta deposit distributions on the early evolution of life on the Earth and Moon. Oberbeck and Marshall believe that the sequence of tillite formation, flood-basalt eruption, crustal rifting, and breakup of Gondwanaland [the single protocontinent that gave rise to the present continents, beginning about 225 million years ago] are consistent with the hypothesis of the initiation of continental breakup from major impact events.

James Sears and David Alt had surmised (Alt et al. 1988) that sedimentary basins are so commonly circular because of impacts which, through pressure relief, generated partial melting that then flooded impact craters and capped them with flood-basalt sequences. Alt (oral comm., 3-26-93), aware that the ejecta blanket of the Ries crater in Germany lies on basement rock, noted that some had interpreted its striations to be glacial in origin, but that others took them to be the scars of low-flying, impact-ejecta bullets. Sears and Alt then turned to their own backyard of western Montana and northern Idaho to examine the roughly circular Belt sedimentary basin, and mentioned briefly, at the end of their paper (1992), that a nonsorted, terrigenous sedimentary layer with larger particles in a muddy matrix (a diamictite) found at the base of the Belt sequence showed shock-metamorphic features. They interpreted the diamictite as an ejecta blanket from the impact that had formed the Belt basin, and urged that tillites and diamictites at other localities and horizons be searched for impact-shock damage.

In October 1992, M. R. Rampino also questioned the wisdom of attributing a glacial origin to certain tillites and related deposits of several different ages in the Proterozoic and Paleozoic, because the climates thought to have prevailed at those times were too warm. He noted that many studies have concluded that recognized tillite sequences were actually the results of glaciomarine debris-flow deposits, but he thinks that solution problematic in terms of known depositional processes, rates, and scale. Instead, like Oberbeck and his colleagues, and Sears and Alt in part, Rampino believes that the purported glacial sequences "have the essential characteristics" of ballistic debris ejected by large impacts, and that they contain evidence of "probable shock-induced deformation"; are often associated with volcanic products; coincide with extinctions and other evidence of impacts; and typically lie at the base of rift-related basins "near known or possible impact structures."

These suggestions about the possible impact, rather than glacial, origin of certain tillites and diamictites, if verified, would likely require a reassessment of some aspects of the Earth's climatic record and other factors of Earth history; it seems unlikely, however, that the question will be decided overnight. At this early date, no scientific papers have yet appeared rebutting the proposals about impact-derived diamictites and tillites, but there appears to be strong opposition to such ideas from the community of scholars most experienced in the study of glacial processes and products and ancient climates; at a press conference immediately following Oberbeck's presentation of his ideas (at a scientific meeting in 1992), they were vocal in their opposition. Although it is too early for anyone to have assessed the response of glaciology and closely related subdisciplines to these ideas—ideas that necessarily impugn certain of their canonical tenets—the generalizations offered in Chapter 2 of this volume, which speak to the influence of subdiscipline on the reception of new ideas, would have predicted the response we have seen thus far.

But on an important subsidiary issue, there does appear to be some prospect of resolution: one of the key claims made by volcanists is that iridium and shocked quartz are not in fact concentrated in the K/T boundary layer, as impact proponents maintain, but rather are spread over a couple of meters above and below the boundary (Asaro et al. 1988; Crockett et al. 1988; etc.). If that were true, the iridium would have to have been deposited over hundreds of thousands of years, which would accord with the volcanic alternative. Impact proponents, however, challenge that claim, arguing that little is known of how iridium is transported in nature, how long it remains in sea and lake waters before it is removed by natural processes, or how far it can move around in the sediment or rock after deposition. Such a lack of understanding provides fertile ground for argument. Izett and others (1990) found that the clay rind of the glass spherules of the K/T boundary layer in Haiti was severely depleted in rare earth and other elements—and such depletion indicates unexpectedly high elemental mobility. Colodner and others (1992) studied the mobility of some platinum-group elements, including iridium, in marine sediments. They inferred that it is unlikely that the processes they observed could produce elemental spikes of K/T boundary magnitude, but noted that the elemental redistributions they observed might account for the lesser enrichments found at other extinction horizons. Their concluding call for diverse further research on elemental mobility after deposition merely emphasized how little of what might limit argumentation is understood.

After several years of disagreement about the occurrence and stratigraphic distribution of both iridium and shocked quartz in crucial sections around Gubbio, Italy—the original site of the Alvarez-group find—Officer's Dartmouth group and the Berkeley group agreed, in 1990, to participate in sampling the sections and splitting the samples for blind testing at various laboratories. The whole effort is being overseen—with the enthusiastic approval of both camps—by Bob Ginsburg of the University of Miami. By fall 1992 the work was almost completed and resolution was expected by winter. Ginsburg (oral comm., September 1992) *has* noted that although not enough shocked quartz had been collected to determine its stratigraphic distribution, "the iridium results looked promising." Whatever the outcome, it is unlikely that it will suppress the extinction-cause aspect of these extremely diverse and complicated debates, or address the seeming coincidence of one or more impacts and the eruption of the Deccan flood basalts. The debates have been extraordinarily fruitful—so fruitful that there seems little reason to hope that they will be brought to an end soon.

The issues raised by evidence for the seeming coincidence of a bolide impact, massive volcanic eruptions, and one of the greatest extinctions of life in Earth history continue to hold the interest of both the public and science. That interest, already broad, has been expanded to encompass the broader questions whether *other* mass extinctions might be associated with episodes of impacting and/or massive volcanism, and whether such episodes might be periodic. The debates have also revitalized the centuries-old question of how to balance catastrophism against gradualism, have forced reexamination of the nature of the fossil record, and have redefined the role of mass extinctions in evolution. They teach us again that the success of a hypothesis, or its service to science, lies not simply in its perceived "truth," or power to displace, subsume, or reduce a predecessor idea, but perhaps more in its ability to stimulate the research that will illuminate the bald suppositions and areas of vagueness that lie hidden in unchallenged canons, and to promote—from self-evoked evidence—the formulation of other hypotheses. By this standard the iridium-birthed impact hypothesis, whatever its eventual fate, has already served science exceedingly well, and so too have the hypotheses it has spawned, both corollary and opposing.

Coda

Caroline S. Shoemaker, Eugene M. Shoemaker of the U.S. Geological Survey in Flagstaff, Arizona, and the astronomer David H. Levy of Tucson, using the 0.46-meter telescope on Mt. Palomar, California, made an astounding discovery on March 24, 1993. They found about twenty glowing, icy rock objects arrayed in a linear train about 200,000 kilometers long. The objects are fragments of a great single comet—now named Comet Shoemaker-Levy—that was pulled apart by gravity in its passage within 100,000 kilometers of Jupiter. The comet fragments, ranging from about 10 to 2.5 kilometers across and moving at about 60 kilometers per second, are unable to escape Jupiter's pull and will probably impact the planet in July 1994. Donald K. Yeomans of the Jet Propulsion Laboratory in Pasadena, California (oral comm., 9-9-93), notes it should take about 5.5 days for the entire train of fragments to collide, and the mid-point date of the collision sequence should be on July 21. Shoemaker noted in an interview (Ronald Cowen, Science News, 6-26-93) that the impacts on Jupiter's nightside will not be seen from Earth, but should become visible a few hours later as the impact sites rotate into view. Such a scenario may be similar to the catastrophe postulated for the K/T boundary extinction, and a variety of effects have been estimated, including the creation of further Great Red Spots on the Jovian surface. The Galileo spacecraft is out in front of Jupiter and is equipped to observe the predicted cataclysm. Although each impact flash on the Jovian surface is expected to be so small (on a planetary scale) and fast that it is likely to appear in only a single Galileo camera pixel (Michael H. Carr, U.S. Geological Survey, Menlo Park, Calif., oral comm., 9-14-93), the eagerly awaited impacts, unique in science, may still tell of things unimagined about the Earth's history.

2

How Science Works in the Mass-Extinction Debates

William Glen

The written record was often accurate; it was rarely true.

—*Bernard Ostry, 1977, "The Illusion of Understanding: Making the Ambiguous Intelligible,"* Oral History Review, *p. 7*

Study of the mass-extinction debates, from early on, began to suggest ideas about how scientists behave in doing science, and what seems to impede or promote intellectual change. In what follows I will discuss some of the inferences it seems to me one might reasonably make. These inferences, based in large measure on direct observation of unfolding events, are congruent with unpublished surmises I made over a decade ago in studies of the plate-tectonics upheaval (Glen 1975, 1981, 1982), the debates over expanding-Earth theory, and the accretionary-terrane research program.

The plate-tectonics study (Glen 1982), not begun until 1974, by which time plate theory was firmly emplaced as geology's paradigm, revealed frequent inconsistencies between respondents' accounts of their own intellectual postures prior to the paradigm change, and what others disclosed to me about the stances those respondents had taken. Those interview accounts, offered in the post-revolutionary milieu of the mid and late 1970's, were marred by apocryphal remarks and other disparities that frequently posed intractable problems. Furthermore, few former anti-drifters—who had of course constituted the vast majority of the pre-revolutionary geological community—were forthcoming in discussing their personal intellectual histories.

Most presented themselves as having been more open-minded and equivocating toward the debated issues prior to the revolution than in fact they had been.

Although my historical study of the drift controversy was undertaken in the years immediately after its resolution, and although most of the main players were available for interview and much of the primary historical material was still accessible, a number of trenchant inconsistencies precluded the presentation of certain larger inferences at that time. In contrast, documenting the history of the current mass-extinction debates, *as it has unfolded,* has afforded me historiographic opportunities that were absent in doing plate history.

What I offer here is a preliminary report of a continuing longitudinal examination of the mass-extinction debates, which by now already encompass a nexus of conflicted theories and a host of ancillary ideas developed over more than a decade. And the growth and ramification of the debates continue unabated.

This study of the mass-extinction debates includes in situ observation of research, interviews (including multiply corroborated accounts of important episodes by scientific participants within hours or days of happening), attendance at important meetings, large and small, and collection of other primary historical data. The resulting archive holds a bibliography of more than 2,500 publications, hundred of preprints and referee reports, notes, letters, proposals, a question-set survey obtained from 700 members representing nine scientific societies that span the debates by discipline, and records of more than 500 hours of interviews (mostly tape-recorded with running tables of contents) with almost 300 scientists.

The nuances of behavior, subtleties of meaning and intent, tacit knowledge, and other fine-grained details observable at first hand in doing this current history were unobtainable in the study of the drift controversy, even though it was undertaken just a few years after that controversy's closure. Only after I had done both kinds of history did their profound differences in data quality become clear. Here I present some early conclusions that are likely to excite interest in a more detailed presentation of supporting data, and this essay is in fact merely a partial précis of a large book now in preparation, one that will detail the evidence underpinning the conclusions drawn here, and draw additional ones.

The Hypothesis of the Alvarez-Berkeley Group: Presentation and Elaboration

It never crossed my mind that . . . the brachiopod groups I worked with expired suddenly by modern K/T boundary standards. I thought that the brachs went out suddenly, but that "suddenly" . . . in 1967 meant a few million years, which was considered geologically sudden.

—*Martin J. S. Rudwick, interview of 11-5-90, tape 1, side 1*

All hypotheses of mass extinction prior to that of the Alvarez group were untestable. These ideas, including some of impact cause, extend back to Pierre de Maupertuis' suggestion of 1742 that "comets have occasionally struck the Earth, causing extinction by altering the atmosphere and oceans." A few later authors of the eighteenth and nineteenth centuries, seemingly ignorant of de Maupertuis, reformulated his suggestion with little further elaboration. Among them was the French astronomer Laplace, who wrote in 1797 that "a meteorite of great size striking the Earth would produce a cataclysm that would wipe out entire species and destroy . . . all the monuments of human history." This century, too, has seen a rash of impact-extinction theories, the most notable being those of M. W. De Laubenfels (1956), who calculated the physical effects of an impact and invoked the modern Tunguska bolide (or exploding meteorite) event in Siberia as an analogue to explain the dinosaur extinction. René Gallant's (1964) ambitious book, *Bombarded Earth: An Essay on the Geological and Biological Effects of Huge Meteorite Impacts*, is a catastrophist's tour de force focusing mainly on how a meteorite strike would alter Earth's rotation and physical environment, but digressing to a précis of the paleontologic evidence for geologic instantaneity of extinction at several different horizons. Gallant was aware of what the more recent reexaminations of the nature of mass extinctions indicate today, that the survivors of global catastrophes ". . . would be not the 'fittest,' but the 'lucky.'"

The first paleontologically sophisticated hypothesis of impact cause for a mass extinction (370 million years ago) was advanced by Digby McLaren in his presidential address to the Paleontological Society in 1970. Because of untestability, a strong anti-catastrophic community gestalt, and a general lack of familiarity with bolide impacts, all of the impact hypotheses prior to that of the Alvarez-Berkeley group in 1980 were more or less ignored. Geology students at a university that was likely typical were told to read the McLaren presidential address—at the very close of which the diplomatic author seemingly slipped in the impact hypothesis—but not to pay attention to that "silly meteorite idea." Endogenous cause was the heart of the zeitgeist: impacting lay beyond the pale.

The diverse cluster of hypotheses that in decades past sought to explain mass extinctions, especially at the K/T boundary, varied greatly in their acceptance by the paleontological community. Prior to the advent of the Alvarez-group hypothesis there had in fact been little attempt to address extraterrestrial hypotheses, such as a supernova or impact, in a systematic way, whereas hypotheses of terrestrial causes, such as climate or sea-level change or, in particular, plate movements, were consistently accorded that scrutiny. During the 1970's several influential essays centered plate tectonics as the cause of major extinctions, and an extensive literature grew up articulating plate movements with several aspects of the fossil record (Valentine & Moores 1970; Valentine & Eldridge 1972; etc.). By the early 1970's, plate theory was emplaced as geology's overarching conceptual scheme and, not surprisingly, became a strong factor in shaping resistance to the impact hypothesis. Plate theory seemed also to have played a role in deflecting attention, in 1972, from Peter Vogt's pioneering Deccan-volcanism theory of faunal poisoning at the K/T boundary.

The cluster of frequently invoked, uniformity-friendly postulates for extinction that had long been discussed, evaluated, and placed on the shelf of respectable candidates may be said to constitute collectively a significant part of the paleontologic paradigm. And they are what the current hypotheses of impact and volcanism challenge.

The mass-extinction hypothesis of the Alvarez-Berkeley group, advanced in 1980 (see the preceding contribution, this volume), was based on the discovery of an unprecedented, anomalous concentration of the platinum-group element iridium in a distinctive, pencil-thick clay layer, 65 million years old. That layer, discovered near the northern Italian town of Gubbio, precisely marked the second greatest extinction of life in Earth history; it included the dinosaurs, many other terrestrial animal groups, and approximately 75 percent of all oceanic species, which at that time had comprised the bulk of the Earth's animal biomass. Since iridium is virtually absent in the rocks of the Earth's crust, its anomalous concentration at the K/T boundary was attributed to an extraterrestrial source. In 1979, the Alvarez group first invoked a supernova cause to explain the iridium, but after negative evidence for the supernova was found in the same year, the hypothesis was revised in 1980 and a lethal, iridium-rich, single meteorite replaced the supernova. The single-event impact hypothesis was further altered a few years later as evidence accrued for multiple, lesser extinction horizons, spread over 1 to 2 million years near the K/T boundary; that led to the formulation of a hypothesis of multiple impacts spread over enough time to accommodate the multiple extinctions.

The single-impact meteorite hypothesis of 1980 was based in part on two unspoken assumptions, contained two major retrodictive (past-predictive) components, and because of the highly specific form of evidence from which the hypothesis grew—siderophile-element anomalies in purported meteoritic ratios—made predictions from which a course of experiments could be formulated wherewith to test the impact component of the hypothesis. Thus:

1. The first assumption of the hypothesis—which was only weakly suppositive—was that there was a mass extinction at the end of the Cretaceous Period of magnitude and global extent sufficient to cause it to stand out from the "normal" or "background" pattern of extinction. The idea of such a mass extinction had long since been accepted by the great majority of paleontologists, but because a very small group did not accept it—or any other mass extinction (see Briggs, this volume)—the idea of a mass extinction at the K/T boundary was considered to contain at least a minor assumptive element.

2. The second and much stronger assumption of the hypothesis was that the mass extinction was instantaneous (poorly defined "geologic instantaneity" ranges from days to hundreds or even thousands of years). The vast majority of paleontologists still do not accept the idea of instantaneity of extinction, but instead hold strongly—or really assume, since definitive evidence for temporal constraint on extinction is wanting—that mass extinctions require hundreds of thousands or millions of years.

3. The first retrodictive component called for a 10-kilometer-wide meteorite to collide with the Earth at the instant before the K/T mass extinction.

4. The second retrodiction held that lethal biospheric effects of the meteorite impact were the major cause of the mass extinction.

5. The central prediction, around which research programs were organized almost instantly, so as to test the hypothesis, was that siderophile anomalies (e.g., high levels of iridium)—and, implicitly, companion forms of impact evidence distributed through impact-generated ballistic trajectories—would be found globally within, or stratigraphically proximal to, the K/T boundary layer.

The last of these predictions was extremely bold, for several reasons, but especially because no highly constrained, widely accepted, geologic reference frame, whether theoretical or empirical, for the production, distribution, and preservation of bolide-impact products existed that could serve to guide such a prediction. The evidence advanced by the *impactors* (subscribers to the impact hypothesis) as "diagnostic of impact," such as anomalously concentrated siderophiles, shocked quartz, and tektites, was immediately opened to a host of questions about origin and distribution. Additionally, only a few in the geological community, aside from those in specialties such as planetary and lunar geology and impacting studies, had had meteoritic impact impressed on them as an important agent in shaping anything other than the very early history of Earth. Conventional wisdom in introductory historical-geology courses—in any of the dozens of textbooks of the last century that I have searched—did not even discuss impacting. If alluded to at all, impacting was typically relegated to the opening chapter on the initial formation of the Earth or the Moon. In such versions, meteorites never got to play a significant role in shaping the fossil record, almost all of which is contained in rocks formed only during the last 600 million years or so of the planet's 4.6-billion-year history.

As predicted by the impact hypothesis, iridium anomalies were found within the K/T boundary layer at more than 100 localities globally, in both marine and terrestrial rocks, during the 1980's. Further, independent lines of evidence also arose—some strong, others extremely tenuous. Within the first five years after the impact theory's advent the two forms of evidence that most strongly influenced assessment of the impact component of the impact hypothesis, other than siderophile element anomalies, were quartz grains bearing multiple, intersecting sets of shock lamellae (now known from more than 30 localities), and spherules strongly suggestive of glass tektites that are formed from impact-melted rock droplets (known from more than 60 localities).

In the last three years, at a few localities around the Caribbean basin, geologists have found spherules with a clay rind enclosing unaltered glass cores of tektite-like composition. Most now attribute them to impact for two reasons: their compositions have been closely related to the target rocks at the leading purported candidate impact site on the Yucatan Peninsula, and the oxygen-isotope ratios were demonstrated to preclude the possibility that they are volcanic (Izett et al. 1990; Blum & Chamberlain 1992). That latter demonstration was particularly damaging to the cause of the volcanists (who embrace massive volcanism as the K/T extinction cause), since they had argued throughout the debates that the spherules in the boundary interval were of volcanic, not impact, origin.

In 1991 Carlisle and Braman published an important new line of evidence for impact: in a K/T boundary bed in Alberta, Canada, they found microscopic diamonds (3–5 nanometers) in a ratio to the iridium (1.22:1) that is close to that found in type C2 chondritic meteorites (elemental ratios previously had indicated a meteorite, but not of a specific type); in 1992, they reported a second impact-supporting discovery to an audience at the U.S. Geological Survey in Menlo Park: the bed that held the diamonds also contained 51 amino acids, of which 18 are peculiar to carbonaceous chondritic meteorites.

The location of the killer K/T impact crater, postulated to be 200 kilometers wide, was never made a central issue by impact's critics because it was understood by all that many craters on Earth have been destroyed or obscured by earthly processes such as erosion, sedimentation, volcanism, and subduction into the mantle by plate tectonics. But from time to time during the early years of the debates the anti-impactors raised the question of crater site. Where was the smoking gun? By the summer of 1992, the evidence of Carlisle and Braman, and of Blum and Chamberlain, in concert with other lines of evidence, brought many new adherents to the impact camp. However, far from all of those who have been convinced of one or more impacts at the K/T boundary believe that impacting is necessarily the single or primary extinction cause.

The roughly circular Chicxulub structure at the northern tip of the Yucatan Peninsula, in southern Mexico, is certainly almost 200 kilometers wide (and may be as wide as 300). It had appeared, by about late 1990, to be a promising candidate impact site because of its size, and although its age had by then only been approximated, its crustal makeup was in accord with the perplexing mineralogy of the K/T boundary layer that seemed to contain both continental and oceanic impact debris. The year 1992 saw a number of papers that seemed to confirm the Chicxulub crater, on the northern coast of Yucatan, as the killer. It was isotopically dated at 65 million years—thus precisely of K/T boundary age—and iridium was found in the impact melt rock in situ at the crater. Chicxulub's great, Alvarez-theory-satisfying, 200-kilometer diameter (perhaps even 300) encompassed continental rocks that matched the precisely dated, unaltered glass-tektite cores from the K/T layer around the Caribbean basin; the Chicxulub impact had also excavated iron-magnesium-rich crust that could have supplied the oceanic-rock evidence in the boundary layer. That flurry of reports from different Caribbean sites during 1991 and 1992, treating the purported products of impact by composition, age, and structural context, went far toward converting many—even some of the leading impact opponents—to belief in an impact. *But few of the paleontologists who converted to impact expressed their belief that impact was the main cause of the K/T boundary extinction.*

The Reception of the Impact Hypothesis

For what a man had rather were true he more readily believes.

—*Francis Bacon, 1620, Aphorism xlix,*
Novum Organum

Because the geological community was largely unfamiliar with the science of meteoritic impacting, a skeptical response to the impact hypothesis could have been predicted. A massive impact's instantaneous global effects fly in the face of uniformitarianism, geology's philosophical mainstay (Gould 1965, 1989; Marvin 1990), and the community regarded the iridium anomaly as an unfamiliar, "black-box," product. Still another factor chilled the greeting of the impact hypothesis: it had been derived mainly by disciplinary aliens. Westrum (1978) spoke to that point in his study of anomalies in science and surmised that "The treatment of reports about alleged anomalous events is . . . shown to be related to the scientific community's concerns about protecting its internal processes from external interference"—and such clearly obtained in these debates.

Prompted by the Alvarez-group hypothesis, Bruce Bohor, a geologist with the U.S. Geological Survey, submitted a proposal in 1981 for a Gilbert Fellowship to search for shocked quartz at the K/T boundary. He was rejected by a diversely constituted committee of referees within the Survey—among them was even an expert in shocked quartz. Bohor was turned down a second time on resubmittal the following year. At that time, almost any geologist regarded a needle-in-the-haystack search as more promising than the totally unprecedented and largely exotic enterprise that Bohor proposed. Shortly after the second rejection, and without the sought support, he nonetheless made the landmark discovery of shocked quartz in the K/T boundary bed (Bohor et al. 1984). Such was the tenor of that time, but Bohor's seemingly miraculous find of shocked quartz, evidence that was familiar to and well understood by the geologic community, went far in attracting the attention of many who had initially viewed the "black-box" iridium of "alien" nuclear chemists as somehow dismissible.

Scientists often lumped all of the components of the impact hypothesis into a single, inseparable unity, accepting or rejecting the hypothesis in its entirety without differentiating the impact-as-impact component from the impact-as-mass-extinction-cause element. That failure to discriminate was exposed often in interviews and questionnaire responses, and sometimes in publications. Before being asked more specific questions, fewer than half of the interviewees responded in such a way as to separate their reception of the two components.

A Priori and Post Priori Understandings of the Issues

That the opinions of scientists are typically well informed was not borne out in this study. More often in evidence was a consistent, widespread pattern of unambiguous response to both general and specific questions by scientists who were poorly informed (which in this context means in ignorance of the few dozen most frequently cited papers treating the central issues in the debates). An ignorance of older, more fundamental information was also commonly evident, but more often what was seen was unfamiliarity with the daunting new literature; publications on the many topics of the debates mounted at a pace that virtually precluded one's keeping abreast.

Only a few interviewees abstained from answering for admitted lack of information, and some were cavalier in disclosing the weak basis for their opinions. Upon being questioned, one National Academician unhesitatingly chose one of the mass-extinction hypotheses. When asked why he was so disposed, he remarked, "Nothing really, but I know [one of the authors of one of the extinction hypotheses], and he convinced me of it." The interviewee and his hypothesis-author friend had done their doctoral studies under the same dissertation adviser. Harry Overstreet (1949) concluded that the "ability to hold a suspended judgment" was the most useful criterion in judging *The Mature Mind*—an uncommon virtue both in and out of science.

During studies of continental drift and drift history in the 1970's (Glen 1975, 1982), it became apparent to me that scientists commonly claimed, after the fact, to have been better informed than they actually had been about debated issues that had eventuated in ascendant, important ideas. For example, most interviewees claimed to have kept up with the continental-drift debates during the half-century before the proof appeared, in the form of confirmation of seafloor-spreading, which thus triggered the plate-tectonics revolution in 1966 (Hallam 1973; Glen 1982). Actually, few earth scientists were well informed about the evidence for drift prior to 1966. Drift typically was not taught, or was barely touched on, in the undergraduate curriculum, and was most often regarded with disdain, ridicule, or neglect in American graduate schools. That was certainly the case I witnessed at Berkeley, Harvard, and Stanford in the late 1950's and early 1960's. Bailey Willis's widely cited, drift-impugning piece of 1944 in the *American Journal of Science*, titled "Ein Märchen" ("A Fairy Tale"), makes clear the low esteem in which drift was generally held (see McLaren, this volume, p. 127).

The same phenomenon is apparent in the mass-extinction debates: scientists commonly claim to have been better versed early on than they actually were in issues that were later resolved to provide a new datum, method, or theory; such has been borne out through hundreds of interviews over the past decade. For subject matter that figures to become historically or scientifically important, this seemingly natural inclination to appear better informed than one was seems to be heightened. In general, during 1983 and 1984, prior to the almost simultaneous advents of the hypotheses of periodic extinctions and periodic impacts—which greatly enlarged the audience that followed the extinction debates—few of the earth scientists I questioned at random were informed to any significant extent about the Alvarez group's impact hypothesis or about McLean's volcanism hypothesis. Latter-day interviews, when systematically arrayed against the earlier interviews, reveal contradictory retrospectives; the resculpting of memory, typically in small increments over the long term, to accord with the larger demands of mental life is a phenomenon with important implications for historians (Spence 1982; Rubin 1986). (How members of a research team reshaped their memories while producing a *collective* account of an important discovery that differed in important ways from the earlier, interview-recorded, *individual* accounts of the same team members will be detailed in my forthcoming book.)

Such disparities of memory and other worrisome aspects of data collection that mount with the passage of time had concerned me greatly in doing drift history, and I wondered about the significance of data gathered by other investigators *after* a revolution had realized its goal, in such valuable studies as those of Nitecki and others (1978). But Nitecki and his collaborators apparently shared my concerns, and outlined them in 1980 (Lemke et al.), concluding that "Ideally, the processes of acceptance of a new scientific theory should be studied by direct observation, while they are happening."

The Effect, on the Reception of the Impact Hypothesis, of Prior Published Commitment to an Extinction Theory

Scientists who, at the advent of the impact hypothesis, were the published authors or published supporters of alternatives to impact were, in all but one of the cases examined, disposed against impact (the exception was Peter Vogt, the first author of a volcanism-extinction hypothesis). They included authors or advocates of various extinction hypotheses offering sea-level changes, climatic changes, plate-tectonic movements, volcanically induced greenhouse effects, or other agents as causes. By contrast, the paleontologist Stephen Jay Gould, perhaps earth science's most widely read spokesman, was intellectually predisposed to welcome the impact hypothesis straightaway, in part because his unorthodox theory of punctuated equilibrium (Eldredge & Gould 1972), which attempted to explain the "lack of expected [gradual evolutionary] patterns during normal times" (Gould 1984; Gould, this volume), articulated well with impact theory. Gould thus provided welcome and vital encouragement through sustained communication with the Alvarez group early on, when only the iridium evidence was at hand and paleontologic backlash against the theory was strong. Dale A. Russel was one of the very rare vertebrate paleontologists to embrace impact theory at its outset; he too was predisposed to the theory, having earlier published in favor of a supernova mass-extinction cause (Russel & Tucker 1971) and having argued for the sudden, rather than the widely accepted gradual, demise of the dinosaurs (Russel 1975).

The Effect of Discipline on Reception

The effect of discipline on theory assessment was strong for most disciplines; and some disciplines overwhelmingly predisposed their members to one particular hypothesis. Still, a few disciplines did so only barely, or did so indiscernibly.

Although most paleontologists denied both the impact and the impact-as-extinction-cause components of the Alvarez-group hypothesis during its first few years, the discovery of several new lines of evidence for impact, which accrued over the subsequent half-decade, convinced many of an impact. Most of those converts, however, did not, and still do not, subscribe to impact as the major cause of the mass extinction. There were a very few paleontologists, including a few doyens, who were initially disposed favorably toward some connection between impact and mass extinction, and who remain so. That small minority increased somewhat as further evidence for impact grew.

Some paleontologists, best represented among the vertebrate specialists, did not accept the idea that there was a mass extinction at the end of the Cretaceous, or indeed any mass extinctions at any time. Instead, those who doubted the phenomenon of mass extinction typically held that the incompleteness of the rock and fossil record and/or the unusual chance combination of a number of variables had been misinterpreted to give the impression of great taxonomic change in a brief span of time, change that was mistakenly perceived as a mass extinction (Briggs 1990, 1991, and this volume). My own thesis adviser in paleontology, J. Wyatt Durham of U.C. Berkeley, whose presidential address to the Paleontological Society in 1963 explicated "The Incompleteness of the Fossil Record," does not believe in the existence of mass extinctions (oral comm., April 1993). As late as 1982, Newell noted that "The concept of mass extinction varies with different investigators. We are faced with a very loose definition and we may even sometimes follow false leads." Such questions about mass extinction and other fundaments of paleontology and stratigraphy, which had languished during times of "normal science," were being catapulted to scrutiny, willy-nilly, by the impact hypothesis. The heat of the debates thus seemed able to ignite the least volatile of topics.

Cosmo- and geochemistry, planetary geology, impacting studies, and closely related fields inclined their members in much greater numbers than did other disciplines toward sympathy for the whole of impact theory. This finding was not surprising: in general, scientists from fields that had dealt with the dynamics of meteorite impacts or with their effects and products were more receptive to the impact hypothesis. Resistance to the hypothesis seemed inverse to familiarity with impacting studies.

Most volcanologists (specialists in the processes and products of volcanism) did not favor the explosive-volcanism hypothesis (which was formulated by Charles Officer and Charles Drake, neither of whom is a volcanologist), but neither did any exceptional proportion of volcanologists subscribe to the alternative impact hypothesis, in its entirety or even in part. The volcanologists opposed to the explosive-volcanism hypothesis cited especially the inability of volcanoes to produce pressures adequate to the generation of impact-induced shocked quartz with multiple sets of intersecting shock lamellae. All the volcanologists interviewed regarded the postulate of volcanically generated shocked quartz as untenable in the terms of their working volcanologic paradigm. That poor reception of the volcanism hypothesis by the volcanologists clearly demonstrates the fact that disciplinary specialty is a significant conditioner or predictor of theory reception, and of the assessment of evidence. Such a role for disciplinary specialty emerged repeatedly in this study for other disciplines as well.

Paleontology was one of the fields in which the influence of *sub*-discipline on the reception of novel ideas and hypothesis choice was clear. Many paleontologists viewed the nature, and especially the time frame, of the K/T mass extinction in terms of the fossil groups in which they specialized. Different kinds of organisms suffered different fates in the extinction, as read from the fossil record; thus specialists in the groups that were most affected (e.g., oceanic, planktonic calcareous microfossils) generally regarded the environmental effects of the K/T-boundary extinction as more severe, the temporal frame as briefer, and the cause(s) of the mass extinction as more powerful and effective than did their colleagues who specialized in groups that were only little affected (e.g. oceanic, deepwater bottom-dwellers). Specialists in the most severely affected fossil groups were the most likely to embrace the impact hypothesis in its entirety. Those scientists who were very narrowly focused in their studies, i.e. those without broad knowledge of a variety of taxonomic groups and/or without diverse field experience in biostratigraphy (especially field experience at sharply defined extinction horizons) were less likely to embrace any part of the impact hypothesis. There were, however, a significant number of paleontologists for whom the influence of fossil group or subject specialty (disciplinary gestalt) was not recognizable in the choosing process.

A greater proportion of vertebrate paleontologists, as contrasted with specialists of any other subdiscipline within or beyond paleontology, rejected the impact hypothesis (including the impact component as well as the impact-as-extinction-cause component). Traditionally, vertebrate paleontologists are more biologically oriented than are most other paleontologic specialists, and the continental deposits in which the bulk of their fossils are found are rare as compared to the marine deposits that comprise most of the sedimentary record. Vertebrate fossils, too, are very rare, as compared to invertebrates. A single invertebrate phylum, the Mollusca, with its clams, scallops, snails, squids, chitons, etc., accounts for most of all the fossils ever collected, and microfossils such as foraminifera, radiolaria, diatoms, coccoliths, and spores and pollen are legion in certain rock beds. The nature of the vertebrate fossil record was therefore likely to imbue its students with a gestalt somewhat different from that of other paleontologic specialties. That idea was addressed by several paleontologists, only one of whom was a vertebrate paleontologist and all of whom were sympathetic to impact as extinction cause but generally requested no attribution. They remarked that the mindset and disciplinary bent of the vertebrate paleontology (VP) group constituted a subculture different from paleontology as a whole. More than one nonvertebrate paleontologist in fact used the term "VP ghetto."

Similar things have been said, however, about other groups of subspecialists in paleontology, notably the micropaleontologists. Micropaleontology is viewed by Lipps (1981) as discrete and "alienated" from paleontology as a whole. It is a subdiscipline separated by its unique service and connection to geology. Rather than being mainly or largely concerned with organisms as once-living things and focusing on their biology, micropaleontologists have served mainly in reconstructing ancient environments, locating stratigraphic position, and dating rocks. Such service derives largely from the use to which the many planktonic microfossil forms lend themselves. They are very widely distributed in life, and they rain down on the seafloor to become entombed in the greatest range of marine environments. They thus often provide escape from the problems and constraints encountered by those who work with typically more localized bottom-dwelling and nearshore faunas. Widespread geographic distribution, vast numbers, and short species life contribute greatly to the usefulness of planktonic forms in the dating and correlation of rocks—and power in earth science has long accrued to those who could construct a temporal framework. In a coherent geologic study, all else hangs on the linchpin of time; petrology, structure, geomorphology, and the host of particulars accumulated in almost any field study are set aside until they can be ordered in time. During most of this century, micropaleontology's identity has derived largely from its position at the center of high-resolution biostratigraphy and correlation.

The early history of micropaleontology was centered in the study of foraminifera: amoeboid, mostly hard-shelled protozoans of enormous utility in dating and interpreting ancient environments of deposition. Foraminifera lent themselves, through recovery from drill cores, to studies in oil exploration and development. The vast majority of micropaleontologists have thus been employed in the oil industry and feel more closely allied to geology than to biology. Not surprisingly, very few micropaleontologic studies have treated ecologic, biologic, or evolutionary problems, as have so many studies in most other fossil groups. Micropaleontology has suffered an even lesser input from biology in general, and its neglect of fully dimensioned paleobiologic studies has contributed to its isolation from other branches of paleontology.

Micropaleontologists are thus imbued with a gestalt recognizably different from that of specialists in other fossil groups. Such a mind-set, one that was probably closer to that of geology in general than to paleontology per se, was a strong factor in the assessment of evidence about impacts or extinctions and the reception of new ideas. Moreover, the microfossil record is unique; it can typically be read more closely and in greater detail than those of other groups and thus often reflected great taxonomic changes that were interpreted as "knife-blade-sharp" boundaries. That was especially true for the calcareous, planktonic microfossils that suffered greatly at the K/T boundary as it has been read at a number of localities.

Micropaleontology's history engendered a supra-personal or institutional mindset, as has been addressed by both Fleck (1935) and Douglas (1986). That mindset systematically inclined micropaleontology's members toward the distinctive pattern of response that was found: micropaleontologists, more than other paleontologic subspecialists, were sympathetic to impact and more likely to assess newly emergent evidence as supporting impact. To varying extents, specialties in any discipline will possess mindsets (Fleck's "thought styles") that are unique in certain respects. Institutions, which would in Douglas's view include a research tradition or paleontologic specialty, do appear to respond in self-preservational, self-aggrandizing, and other organismlike ways, including resistance to change. The institution, after all, is brought into being and empowered by the subscription of its members to a fund of commonalities of almost any sort, ranging from magic to law to tradition to belief in any organized body of knowledge. The institution's members relinquish freedom, time, money, and other resources in return for the host of benefits that the institution provides: authentication of specialized identity, access to funding, channels to employment, honorific rewards, networks of communication, etc. The idea of an institutional supra-personal intelligence or mind has naturally suffered criticism, since analogies almost always beg license in their comparisons of generically different wholes on the basis of partial similarities in properties or functions. But such criticism may derive more from semantic inadequacies in drawing the analogy between organismic and institutional minds than from incomparability or lack of similarity in the behaviors that such "intelligences" or "minds" engender (see Gould interview, this volume).

For many groups in these debates, it was often possible to predict how the "institutional" mindset, or what I term the "collective gestalt," would shape the responses of its members to the issues in question; for example, I never encountered an impact-studies specialist who did not embrace the impact hypothesis of extinction in its entirety, whereas only very few vertebrate paleontologists did. The power of the institution—and the disciplinary specialty is indeed an institution—to influence many forms of individual behavior was clear. Fleck (1935, p. 10) thought that "Truth is not 'relative,' and certainly is not 'subjective' in the popular sense of the word. It is always, or almost always, completely determined *within a thought style.* One can never say that the same thought is true for A and false for B. For if A and B belong to the same *thought collective,* the thought will be either true for both or false for both. But if they belong to *different* thought collectives, it will *just not be the same thought*! It must either be unclear to, or be understood differently by, one of them."

Styles of Argumentation

> This interpretation aroused much resentment, for many scientists unconsciously deplore the resolution of mysteries they have grown up with and have therefore come to love.
>
> —*Peter B. Medawar, quoted by Luis W. Alvarez in a personal letter of 3/1/85 to Daniel E. Koshland of* Science

The styles of argument of the mooters have been shaped largely by the history of their disciplines and their evidentiary sources. However radical or catastrophic their hypothesis, from the outset the impactors have mainly invoked canonical standards and knowledge in advancing their several lines of evidence. The impactors claim that their empirical evidence lends itself to the formulation of programs of experiments by which to test a series of clear predictions that follow from parts of the impact hypothesis.

Impact theory predicted that impact products were distributed globally by ballistic transport; that prediction evoked hope or expectations for many that commonalities, such as similar stratigraphy and marker beds, concentrations of iridium, etc., would be found at K/T boundary sites worldwide. Impact's opponents focused on the great range of variation in the character of the boundary interval around the world, arguing that the profundity of the differences found from place to place betokened a cause that was neither globally uniform nor instantaneous in its effects. The impactors countered that since local differences in sedimentary processes, sediment sources, and post-depositional changes (diagenesis) naturally produce unique stratigraphic intervals at widely separated places, then such differences would likely alter or mask the effects of an impact at far-flung K/T boundary sites. The ratio of locally derived sediment to that from purported ballistic transport would also vary with distance from the impact point, and with other poorly understood variables of the impact process. The impactors believed that impact-generated global signatures would become increasingly readable through the turbid overlay of local processes of formation, diagenetic changes, and variations as a function of distance to impact target, etc., as more localities were studied and more methods and instruments, permitting ever finer discrimination, became available. Such a hope was unconstrained, partly because the global effects of a massive impact lay virtually outside the realm of both stratigraphic theory and experience.

The impactors appealed to classical stratigraphy and sedimentation to explain that different depositional environments and diagenetic histories would have produced widely varying records. They also stressed the differences to be expected, and focused on the unprecedented commonalities. In contrast, the anti-impactors trumpeted the differences between K/T boundary sites. They focused mainly on the variations in siderophiles, spherules, mineralogy of the boundary layer(s), and other purported lines of evidence for impact, and argued that such evidence was engendered by earthly causes. The anti-impactors also invoked volcanism and locally anomalous rates and conditions of sedimentation and other candidate processes by which to concentrate chemicals or minerals such as to have produced the range of variations encountered at the many K/T boundary sites. The first charge of the anti-impactors was to argue and demonstrate a lack of the ubiquitous, globally uniform effects implied by an impact; the second was to open alternative mechanisms to possibility, by which to produce the various lines of evidence claimed as uniquely diagnostic of impact.

The questions raised about the nature and geographic variation of the boundary interval—especially by the anti-impactors, since such questions best served their brief—impugned a wide range of suppositions that had lain long hidden in established principles and methods of stratigraphy and related disciplines, and thus prompted much research at unprecedented levels of refinement. The need to reassess what had long been accepted at a lesser level of specificity required the development of new methods, techniques, and instruments, the punctilious reexamination of old, classical field sections and laboratory specimens, and exploration for new sources of evidence and data.

The most notable example of new instrumentation was the design and building, by the Alvarez group, of a coincidence spectrometer that enabled the rapid determination of the siderophile content of large numbers of rock samples collected at very short stratigraphic intervals through K/T boundary sections from many localities. The automated instrument processed in ten minutes what the older methods required hours or days to process. The effort to design and build that machine had been prompted by one of the earliest criticisms of the impact hypothesis, one that had been heard in coffee klatches right at the advent of the new hypothesis: "How do we know that an anomalous, siderophile-rich bed such as that of the K/T boundary is so uncommon; how many vertical feet of sedimentary beds have actually been examined elementally at the level of parts per trillion to demonstrate iridium's rarity?" Only a great number of samples taken every few centimeters or so from extended stratigraphic sections representing a significantly long time interval at a number of distantly separated localities could serve to answer such a gnawing, fundamental criticism. The question abated as empirical data demonstrating iridium's rarity in the stratigraphic record accrued through the efforts of several research groups over the decade of the 1980's. The paper of Frank Kyte and John Wasson (1986) of U.C.L.A. was an early, particularly influential one that showed, through a deepsea-core stratigraphic interval interpreted to represent tens of millions of years, an iridium anomaly present *only* at the K/T boundary. However, since sedimentation rates in the deep sea are exceedingly slow, such a core was not considered a likely repository for the host of lesser signals in the depositional record that are more likely to be found in strata deposited rapidly; rapid sedimentation, most likely in continental and marine nearshore deposits, is analogous to the higher tape velocity of a high-fidelity magnetic recorder.

The volcanists' arguments against the impact hypothesis led to still other fruitful new lines of research, but the first charge of the volcanists was clear: how to produce a global iridium anomaly without impact. In 1983, airborne particles that had spewed from an orifice of the Kilauea volcano in Hawaii were shown to be enriched in iridium 17,000 times the value found in Hawaiian basalt (Zoller et al. 1983). A series of exchanges then ensued in the literature, in which the volcanists argued that although the iridium in the erupting Kilauean basalt rock was at least two orders of magnitude less than that of the K/T boundary layer, it was highly concentrated in gases and sublimates. A further study by Olmez, Finnegan, and Zoller (1986) stressed the point that no iridium had been detected in numerous studies of volcanic gases, with the exception of that documented in the Zoller et al. paper of 1983. Their conclusion thus appeared to undermine the volcanists' ability to argue their case from the evidence of an empiric volcanic-iridium source.

But uncertainties about origin, mechanism of transport, and chemical differentiation of magmas in volcanism gave the volcanists license to sustain their use of modern volcanic iridium as a datum in their brief. They were buttressed in 1989, when Toutain and Meyer found that sublimates from the Piton de la Fournaise volcano in the Indian Ocean were anomalously rich in iridium, and suggested that such hot-spot volcanoes could be related to the events at the K/T boundary. Modern volcanic iridium is perhaps still a touchstone of the volcanists. But not only did Piton de la Fournaise provide an iridium source, it also is situated over the hot spot in the Indian Ocean that is thought to have erupted the million-cubic-kilometer Deccan Traps volcanic sequence as the Indian tectonic plate passed over it. That great Deccan volcanic pile was seemingly spewed out in a few great pulses within about a million-year-interval that straddles the K/T boundary (Baksi 1990). Although the environmental effects of such a vast lava flow are not precisely calculable, the flow certainly seemed a plausible alternative to comet showers to explain stepped extinctions and the purportedly stratigraphically "smeared" record of iridium evidence proximal to the great iridium spike.

New information bearing on the boundary sequence proliferated at a daunting rate in the second half of the 1980's. By 1989, more than 2,000 professional publications addressing some aspect of the diversely comprised debates (Glen 1990) had appeared. From early on, the various citation indexes lagged greatly in tabulating, or missed completely, a great number of references, and this problem expectedly worsened as the diversity of fields and subtopics enlarged.

Although the K/T boundary debates have been under way for more than a decade, and the volume of literature almost precludes one's keeping abreast, no globally comprehensive, detailed, general review monograph has yet appeared. Many of the extant reviews focus on specific categories of boundary-bed contents, such as spherules, shocked quartz, iridium content, carbon soot, or certain fossil groups (an exception is Izett's monograph of 1990, which is topically diverse but restricted to the interior of western North America). The more general and more broadly inclusive works—there are more than a dozen such—are brief and partisan (an exception is Sutherland's, of 1988, an unusually lengthy and balanced one, which I discuss shortly under "Gestalts, Polemics, and Conversions"). But there are a few large, topically broad anthologies that include authors from the different extinction-theory camps; these allow the convenient comparison of diverse evidence and arguments in a single volume (Silver & Schultz 1982; Sharpton & Ward 1991).

Scientists often, and the media almost always, failed to differentiate between the separate ideas of the hypothetical impact and its postulated lethal environmental consequences. There is a further difficult question, reformulated in more specific ways immediately after the impact hypothesis appeared but seldom addressed on balance: What properly constitutes the stratigraphic record of the K/T boundary mass extinction? Is it the single horizon at which the greatest number of taxa disappear, at a knife-blade-sharp stratigraphic shift that is contiguous with the K/T boundary layer(s) of variable thickness (thickness was limited to a few centimeters until 1991) containing the bulk of the iridium signature, shocked quartz, and other purported evidence for violent change? (That important stratigraphic question is further complicated by vigorous debate over temporal uncertainty about the boundary layer or layers.) Or is the K/T boundary a stratigraphic interval extending through tens to hundreds of centimeters of sedimentary beds representing hundreds of thousands or even 1 to 2 million years? Such a larger boundary interval is supported by a stepped series of variable extinctions stratigraphically coincident with stable chemical-isotope spikes (Kauffman 1984, 1988; Keller 1989; Mount et al. 1986; etc.). Moreover, at some localities the iridium and other purported evidence for catastrophe have been reported as occurring through a stratigraphic interval *of a few meters* that is bisected by the great iridium spike and the horizon of greatest taxonomic change (Asaro et al. 1988; Crockett et al. 1988). The "shoulders" of the iridium spike, especially, have given rise to debates in which the volcanists offer long-term, pulsed volcanic infusion of iridium, while the impactors note that post-depositional transport of elements would produce such a "smeared" stratigraphic distribution. The mobility of various elements within the sedimentary column, especially those of the platinum group, which includes iridium, is not well enough understood to limit argumentation closely.

The idea of a stepped extinction pattern at the K/T boundary was indirectly bolstered by evidence for steps found at other horizons: Orth and others (1987) found two iridium spikes that matched two of a series of five or six stepped extinctions at a 92-million-year-old stage boundary; and from a number of diverse, partly contradictory studies summarized by Montanari (1990), stepped faunal changes can be read through a 1- to 3-million year interval around the Eocene-Oligocene boundary (34–37 million years ago), which is also punctuated by a number of separate tektite horizons, each indicating impacts (see "What the Debates Are About," this volume).

It was only after the debates were under way for several years that the stepped-extinction pattern, superjacent and subjacent to the K/T terminal boundary layer(s), came to prominence and became a compelling point for the anti-impactors, led mainly by the volcanists. Scientists seldom made this or other issues clear in their discussions of the impact hypothesis, and news writers and editors were typically remiss. But as the evidence for possible multiple impacts mounted, even the reporter Richard Kerr of *Science*, who had followed developments closely, drew attention (in 1988) to the diverse published evidence seemingly not compatible with the hypothesis of a single impact. There is an extreme range in opinion regarding the nature of mass extinctions, extending from those earlier mentioned who deny even their existence (see Briggs, this volume), to those who accept them in the oldest sense as representing faunal and floral change within the time frame of mountain-building episodes (orogenies extending over a few million to tens of millions of years), to those who attribute them to geologically instantaneous catastrophes (effected within days to years). Here, as in countless other areas of geological conflict, the lack of a precise temporal framework for process and product simply precluded dialectic closure.

The volcanists, of course, seized on the idea of stepped extinctions, since it seemed to accord with the episodic, repetitional behavior of volcanoes and detracted from the idea of a single great kill at the end of the Cretaceous Period. In originally formulating their hypothesis, the impactors had focused on the single-impact event marked by a major extinction horizon. Because the claim of a stepped extinction pattern weakened the impact hypothesis—which originally posited only a single, great holocaustic event—the volcanists sought sedulously to bolster their argument for stepped extinctions. After an extended and strenuous defense of the single-impact hypothesis, marked by acerbic confrontations, some impactors, including the Alvarez group, came to recognize the evidence for stepped extinctions as compelling. The lesser extinction horizons with their stable isotope signatures seemed to vitiate the single-impact idea, as did also the "iridium shoulders" super- and subjacent to the main iridium spike, which had been argued long before publication (Asaro et al. 1988; Crockett et al. 1988). The impactors were thus increasingly pressed to acknowledge and accommodate the growing anomalies. They accordingly altered their hypothesis to include multiple impacts, through the device of comet showers spread over 1 to 2 or more millions of years. That crucial revision—which is examined shortly vis-à-vis standards of appraisal—was not embraced by certain other impactors. Instead, that small, stalwart group, with several notable geochemists among them, denied the need to invoke comet showers and continued to attribute all the evidence at the K/T boundary to a single bolide impact, or to multiple, all but instantaneous impacts.

The impactors' evidentiary appeal, different in certain respects from that of the volcanists, has been mainly to prevailing criteria by which to judge the origin of their data. For example, the siderophiles (iridium, etc.) in the K/T boundary bed must have come from the sky, since no endogenous, earthly cause is known to concentrate them ubiquitously in sediments on the Earth's surface; global iridium distribution at the same horizon could only be accomplished through ballistic transport unique to a bolide impact; and shocked quartz with multiple sets of shock lamellae could only have been produced by meteorite impact or an atomic-bomb explosion. It has been generally held by volcano scientists (volcanologists)—most of whom are opposed to the volcanism hypothesis—as well as by impactors, that a volcanic explosion is a decompression event in which the pressure is not nearly great enough to produce high-grade, shocked quartz. Gene Shoemaker of the U.S. Geological Survey, an authority on the Solar System's record and dynamics of impacting, has likened a volcanic explosion to the uncorking of a bottle of champagne, during which the pressure never rises above that present immediately prior to the movement of the cork. Most earth scientists and all volcanologists interviewed agreed with that view, but there are a few, including Alan Rice of the University of Colorado (1987), who do not think the uncorking analogy applicable to certain types of volcanic episodes which, they hold, can produce the same effects as those claimed for an impact.

The volcanists—in contrast to the impactors, who, notwithstanding their radical hypothesis, have argued mainly from conventional standards—have focused instead on the vagaries and suppositions that lie hidden in and beyond the standards of appraisal and orthodox knowledge. Volcanists have cited, as examples, the lack of experimental data on the generation of shock features that would strictly disallow a volcanic source; the mantle as a possible source for the iridium anomalies, supported by siderophile-rich volcanic gases and sublimates of the sorts that have been discovered in modern volcanoes and by the high iridium levels found in ice cores and volcanic ashes; and the possibility (or even likelihood) that both qualitatively and quantitatively, geologically infrequent, cataclysmic, prehistoric volcanism of a magnitude having no analogues in our brief human history has occurred. The volcanists held, in the best tradition of uniformitarianism, that recorded history is an inadequate sample of geologic time from which to formulate certain volcanologic generalizations; by that stroke they simultaneously gainsaid orthodox volcanologic precepts damaging to their brief, impugned the application of volcanological strictures and constraints to the whole of the geologic record, and triggered the formation of new research efforts across a broad disciplinary front. The resulting decade-long, sometimes painfully acerbic, but diversely fruitful, dialectic could not have unfolded as quickly or as productively in the absence of the volcanist campaign—nor, of course, would the volcanists' reexamination of volcanologic precepts have been undertaken in the absence of the impact hypothesis.

Standards of appraisal, definitions, and suppositions that had been serving stratigraphy and paleontology well over much of their history were called into question by the higher levels of resolution demanded by new evidence and by experiments addressing the new hypotheses. New questions drove new research at a frenetic pace to produce new evidence, which then raised further new questions—and at times the upheaval seemed a positive feedback loop. And as the debates between impactor and volcanist enlarged, the languishing "quasi-paradigmatic" hypotheses, such as sea-level changes, climate changes, and plate movements, which had accumulated during the past century and had generally come to be regarded, singly or multiply, as ineffectively or only partially explanatory of mass extinctions, drifted further from center stage.

Modeling

> [Niels H. D.] Bohr never trusted a purely formal or mathematical argument. "No, no," he would say, "You are not thinking; you are just being logical."
>
> —*O. R. Frisch,* What Little I Remember, *1979*

Impactors and volcanists have both argued from physical and mathematical models, but in different ways and different contexts.

Prior to the advent of the Alvarez-group hypothesis, impacts had already been recognized throughout the Solar System (Shoemaker 1977; Shoemaker et al. 1979). The connection between impacts and their products—craters, shocked quartz, tektites, stishovite, etc.—had entered the empiric realm. Thus modeling for the impactors, irrespective of uncertainties about the mathematical codes, methods, and models employed, was simply a means of elaborating and refining the impact-effects aspect of that hypothesis. Perhaps the strongest direct tie, next to that of generating high-grade shocked quartz by an atomic-bomb explosion, was made in 1989 (Gostin et al.), when for the first time a cosmogenic siderophile element concentration was found within a widely dispersed impact-ejecta blanket surrounding the Acraman crater in South Australia. This was the first demonstration that an impact can produce a sedimentary iridium peak.

In contrast to models of impacting, the models of deep-mantle plumes—which the volcanists hoped would buttress their cause—were hypothetically more tenuous, and the cause-and-effect link between plumes and their postulated products was not as secure as that between impact and its effects.

The hypothesis of mantle plumes was first advanced by J. Tuzo Wilson in 1963 in his attempt to explain the origin of the Hawaiian Islands, and the idea was elaborated by W. J. Morgan in 1971. Although mantle plumes acquired the stature of a working hypothesis through the ascendancy of plate theory by the early 1970's, many still regard them as extremely tenuous constructs or deny them completely. Early on, mantle plumes grew to acceptance largely from the interpretation of their purported surface effects: age-graded, linear chains of mid-oceanic volcanoes called thread ridges or nemataths, with the youngest island of each chain, such as the island of Hawaii, situated above the active plume. Further evidence suggesting a deep-mantle source for volcanic plumes comes from latter-day studies of volcanic isotopes (Zindler & Hart 1992). Some continental flood basalts, however, have features that reflect, rather than deep plumes, a shallow source in the subcontinental mantle (Gallagher & Hawkesworth 1992). The impactors counter the plume-model argument for the volcanism hypothesis by noting that mid-oceanic, nonexplosive volcanism and its products—from which the concept of plumes mainly grew and became elaborated—offer little support for the massive, violent volcanic episodes of the sort postulated by latter-day volcanists, which entail a cause-and-effect regime seemingly different from what nemataths bespeak. Whether or not the nematath is indeed being built by the passage of a tectonic plate over a thermal plume that extends from near the base of the mantle all the way to the lithosphere, as has been modeled, is open to serious question on several counts detailed in an extensive literature.

Mathematical modeling of plumes was first undertaken in the mid-1970's (Yuen & Schubert 1976). Its resurgence in the mid-1980's, apparently prompted in large part by the mass-extinction debates, marked a favorable turning point for the volcanism cause, however poorly constrained the plume models were purported to be. The volcanists were pleased that a growing number of highly competent modelers were being published on a feature that served their cause, notwithstanding that such modeling (Loper & McCartney 1986, 1988; Courtillot & Besse 1987; etc.) invoked processes and violent effects that have not been tied empirically to known volcanic products. Plume modeling appears in fact to have grown into almost a separate hypothesis in its own right—which is precisely what at least one of its authors told me by 1988. The plume modelers—aside from the fact that many of them view their efforts as contributing to the volcanism-extinction hypothesis—are part of an extended research tradition that has attempted, from the time of Lord Kelvin in the late nineteenth century, to fathom the role of mantle convection in carrying heat from the interior to the surface of the Earth. They are acutely aware that during the heady days of the rise of plate theory, mantle convection too easily appropriated the role of major, driving force for plate motion. Latter-day assessments, by contrast, emphasize other important forcing functions: mantle-convection models have invoked linear-stability theory and boundary-layer theory, numeric calculations, laboratory simulations, and more, and earlier, simple two-dimensional box models have yielded to three-dimensional convection in both boxes and spherical shells.

The 1960's view of mantle convection—with its more or less steady, vertical, sheetlike upwellings and downwellings positioned under the ridges and trenches—has been replaced by more sophisticated patterns of ephemeral, shallow upwellings beneath ridges, and more stable and deeper downwelling under trenches, with cylindrical, rising plumes feeding hotspots (Schubert 1989). But despite almost a quarter century of research, much of mantle dynamics remains an enigma with a host of unanswered fundamental questions: it is not even yet known if the mantle is divided into two separate convective realms, an upper and a lower, by the 670-km-deep discontinuity that seismic studies have identified. The plume modelers, who have tendered intriguing and stimulating ideas, were limited by that host of uncertainties, which have not significantly abated since their outset.

As growing complexity restricted an understanding of mathematical modeling to very small groups within disciplines, the reservoir of potential referees shrank. Some nonmodelers became children of faith, and others, lacking full understanding, simply disavowed models as constructs unworthy of their trust. New methods such as mathematical modeling, especially those that required great effort to master, have generally had a cool reception. An example of the latter was evident in seeking referees for both Shaw's (1987) article, which pioneered the application of nonlinear-dynamics analysis to the earth sciences, and his (1994) book (forthcoming), which, prompted by the impact debates, presents a radical new theory that reinterprets the effects of impacting on the Earth through the lens of chaos theory. Very few in 1986 in the earth sciences were able to use nonlinear analysis; at that time the conceptual bridge between conventional mathematics and chaos theory had not yet been built. Shaw's career digression to master these exotic new methods, however heuristically and predictively promising, marked him as a curiosity among his peers—for support is seldom offered those who move beyond the discipline's deepest salients.

Although, for many, impact theory triggered the idea that a great meteorite strike might shake the Earth to produce crustal relaxation on an undreamed of scale and set off great earthquakes and volcanism, it was not until 1983 that the impact of a 10-kilometer bolide was calculated to produce an earthquake of about 13 on the Richter scale. That is an energy release almost one million times greater than the strongest earthquake ever recorded (Ahrens & O'Keefe 1983). Although many anticipated a rash of articles drawing connections between such an impact and its cataclysmic quake and other effects, little was published. Naturally, the volcanists did not pursue that topic—they religiously avoided it—since their hypothesis would then be subsumed, willy-nilly, by that of the impact hypothesis. But surprisingly, neither did the members of the Alvarez group, who have never published a word about impact-produced volcanism nor brought that idea up in formal meetings, though of course they have discussed it among themselves.

A small number, all of them impactors, eventually did call on impacts to produce flood-basalt volcanism, and they continue to do so (Hildebrand & Boynton 1987; Rampino 1987; Alt et al. 1988; etc.); but the majority of those who have addressed the problem hold that the energy involved in the production of such massive flood-basalt bodies as the Rajmahal, Deccan, or Siberian is orders of magnitude beyond that imparted by impacting. That later surmise is based at least in part on the idea that the great flood-basalt bodies originated from the activity of thermal plumes that are held to have originated over millions of years near the core-mantle boundary, in the thermally unstable layer called the D-double prime. Some impact modelers surmise that lithospheric excavation by even a major impact could not give rise to the sustained lava supply required to build the great flood-basalt sequences; but others disagree.

Volcanism theory in its original, post-Alvarez form (McLean 1981), which I term the "non-explosive volcanism hypothesis," did not entail great explosive volcanism but merely a carbon-cycle perturbation of long duration triggered by the Deccan Traps volcanism. And even in its modified form, which I term the "explosive volcanism hypothesis" (Officer & Drake 1983, 1985), and which included great explosive episodes, volcanism-extinction theory did not at first have to accommodate the additional problem of providing impact-level shock pressures, because the shocked-quartz evidence for impact was not found until almost a year later (Bohor et al. 1984).

Rice's (1987) defense of the volcanism hypothesis against the shocked-quartz evidence was born of his earlier experience with the quenching of metallic materials in foundries; that experience had prompted his postulate that the precipitous overturning of basalt and rhyolite layers in a magma chamber could produce explosion pressures high enough to generate shocked quartz. But here too—as with Loper and McCartney (1988), who proposed a mechanism similar to Rice's for the production of shocked-quartz-generating basaltic-magma explosions—a large analogical leap from the foundry to the magma chamber was required, and there was no demonstrable way to connect such a postulated mechanism with a known product, as could the impact modelers. Loper and McCartney (1988, p. 971) were well aware that "Even now [the endogenous model] is not as far along as the Alvarez model was in its infancy, for we know even less about the Earth's interior than we know about meteorites. The nature of activities in the lowermost parts of the mantle are known only through mathematical and experimental studies, and the surface effects are mostly speculative."

Beyond the doubt engendered by the poorly bound coefficients of the plume modelers lies the prior question of how much value is to be found in physical models whose components are only barely analogous to their natural-system counterparts. And when empirical knowledge of the natural system is wanting to start with, as it is with plumes, uncertainty about the model mounts. Some note that with a lessening sense of modeling's limitations (Posey, 1989) over the last three decades, modeling of all sorts has become more widely practiced. A recently published physical model attempted to show how large globs of less dense, more buoyant molten rock (diapirs), derived by heating at the base of the mantle, escape to rise upward every 20 to 30 million years to erupt explosively at the surface. The model was built within a glass vessel 15 centimeters tall: a thick layer of highly viscous corn syrup was placed above a thin layer of water that was dyed black in order to more easily observe its movement, and the two layers were separated by a silk membrane to enhance their discreteness. The water stood for the low-viscosity, low-density molten layer at the base of the mantle, and the corn syrup represented the denser mantle. Because of the great difference in physical properties between the two surrogate substances, even without heating, diapirs of water formed, broke off from the parent layer, and rose through the overlying corn syrup. A news writer in a very widely circulated science magazine seized on the published article and photographs to note in his caption that "the basic [mantle plume] theory has been vindicated by laboratory experiments."

David Loper and Kevin McCartney, the authors of the simple experimental model, both knew the virtues and limitations of such a model. Although critics have denounced them as simplistic, such models have often yielded vital clues to the workings of complex processes, and guides to further data-gathering and experimentation. Simple models have a long history of invaluable service in opening new channels of thinking. Such models, however, have never served to homologate theories. Science-news writers are a varied lot—the best seem indispensable, but others are sometimes the victims of their own enthusiasm or the stress of following a fast-paced science. Still, even the scientists themselves, when excited at the edge of discovery, sometimes allow their models to speak too loudly (more on this in the discussion of the "Media," below).

Theory Choice and Standards

The internal standards of the subject . . . may be but the method according to which a particular madness is being pursued.

—*Paul K. Feyerabend,* Philosophy of Science: A Subject with a Great Past, *1989*

Standards of appraisal in science have been variously defined: they may consist of principles, methods, techniques, or other rigorously "authenticated" conceptual components of the systematic body of knowledge constituting a discipline, and may include information of high status or critically placed data that in quiet times enjoys near immunity from doubt and forms part of the knowledge system.

In several areas of the mass-extinction debates it can be shown that impactors and volcanists invoked different standards of appraisal and weighted the same evidence differently. This disparity is particularly well demonstrated by the published arguments and interview responses of the mooters in the shocked-quartz controversy (Bohor et al. 1987; Carter & Officer 1989; Officer & Drake 1989; Grieve 1990). The volcanists claimed that shocked quartz with multiple sets of intersecting shock lamellae could be produced under certain conditions by volcanic explosion. That claim, never previously made, amounted to an outright rejection of the long-held, community-wide standard that natural shocked quartz is uniquely diagnostic of a meteoritic impact origin.

Standards are rooted in the disciplinary matrix, and are often difficult to treat as discrete entities. The conceptual changes that are accelerated during upheavals of the sort fomented by the impact debates include standards that change, seemingly in spurts, as whole disciplines evolve. In the impact debates, standards appeared only little more stable than other parts of the system from which they derive and which they condition, and the challenge to standards heightened as crisis mounted. The ephemerality of standards repeatedly observed in this study suggests that if standards are used to measure rates and kinds of conceptual change in science, circularity of reasoning is assured.

Irrespective of the extent to which differences in the choice of standards, rigor in their application, or the weighting of pertinent evidence may have affected hypothesis choice, *once embraced, the cho-*

sen hypothesis became overwhelmingly the strongest predictor of the future selection and application of standards in the assessment of evidence.

Standards of appraisal were the most frequent targets of attack in the debates. For example, the volcanists held that "there is no proof that only an impact can generate shocked quartz"; they noted that "stepped extinctions and iridium tails with lesser peaks super- and subjacent to the main K/T iridium spike cannot accord with a single-bolide impact" (this claim was supported in part by the stepped-extinction and stable-isotope spike pattern); and they claimed that "ballistic-trajectory transport is not the only means by which to distribute the iridium signature found at the K/T boundary" (the volcanists here offered high-altitude transport of volcanic clouds as the mechanism of distribution). It was as if the impactors were blind to even the strongest elements in the cognitive framework of the volcanists, and vice versa—Kuhn's term, "incommensurability of viewpoints," is not adequate to the descriptive task here. Standards of all sorts floated in and out of question and power.

The evolutionary history of the stepped-extinction idea (see "What the Debates Are About," this volume) shows that *stepped extinctions began life as an anomaly of the single-impact hypothesis and then eventually evolved into the primary force driving the formulation of a new hypothesis designed to replace the original hypothesis.*

I propose that a standard may have a history of gestation and application as was found in the stepped-extinction episode. What ultimately became a standard of appraisal started out as anomalous data that grew to constitute a serious problem for the single-impact hypothesis; that problem defied solution by the challenged hypothesis, and within the community that embraced the single-impact hypothesis there ensued a localized state of anomie (normlessness; Merton 1951; Ben-David 1971); after a few years, the insoluble problem gained cognitive esteem as a normative assumption; as data generally regarded as supporting accrued, the normative assumption acquired increasing stature; and what had initially been a not very threatening anomaly had grown into a standard of appraisal. That newly evolved standard then played the major role in forcing revision of the challenged single-impact hypothesis of geologically instantaneous mass extinction. The single-impact hypothesis was replaced by a hypothesis of multiple impacts spread over 1 to 3 million years; the multiple-impacts hypothesis seemed to explain the multiple extinction horizons; and stepped extinctions as a standard of appraisal were thus accommodated in the new theoretical framework.

But a generalization that freezes stages where in fact the boundaries between stages are not sharp is an artifice. Not only does such a generalization (or taxonomy) arbitrarily divide what is more reasonably seen as a dynamic continuum but it also requires a corollary: *such an evolution—from anomaly to hypothesis-impugning standard—varied in rate as a function of the rate of change in the discipline in which the evolution took place*. The moment during the evolutionary sequence at which the anomaly evolved into a standard powerful enough to force thinking, within the single-impact community, about the possibility of theory revision was strongly conditioned by discipline: the paleontologist/impactors were the first to entertain theory revision; that seems expected, since the stepped-extinction anomaly originated in their own discipline; in contrast, the nuclear chemist and geochemist/impactors were generally among the last to be moved to theory revision (several still eschew the evidence for stepped extinctions and continue to argue the single-impact or multiple-instantaneous-impacts hypothesis).

When the stepped-extinction idea was in its incipiency, most of the impactors denied it on various bases: inadequate sampling; the Signor-Lipps effect (Signor & Lipps 1982), which explains why extinctions may appear more gradual than they are; methodological errors in deriving the evidence; etc. But the concept of stepped extinctions (Kauffman 1984, 1988; Keller 1989) was buttressed as further evidence mounted: osmium-isotope ratios (Luck & Turekian 1983); stable-isotope spikes at lesser extinction horizons (Mount et al. 1986); and the problematic "shoulders" on the main iridium spike that bear lesser iridium spikes and shocked quartz (Asaro et al. 1988; Crockett et al. 1988). The idea also gained support from the less direct, but inescapable, evidence from sources other than the K/T horizon: Orth and others (1987) found two iridium peaks just below a 92-million-year-old extinction boundary, the lower of which coincided with the first of a series of stepwise extinctions, while the upper matched the third of the extinction steps; and Montanari's (1990) summary of diverse earlier-derived data across the Eocene/Oligocene boundary, an interval of significant change in sea life over a few million years, shows multiple impacts, as evidenced by tektites and microtektites. Although the latter cases did not involve a "mass extinction," they seemed to accord with models for the time frame of comet showers (or volcanic activity) and, with only little license, invited visions of sequential bombardment by still larger bolides that could produce a whole series of significant extinction steps.

From the outset, the impactors had held the extinction to be represented by the pencil-thin boundary layer, and thus to have been geologically instantaneous (a much thicker boundary bed was not found until a decade after the impact theory's advent). But in time the stepped-extinction idea attained such commanding stature that it could not be ignored by the impactors—and increasingly, many of them found a single impact untenable. Eventually, the evidence that had constituted an anomaly for single-impact theory—inexplicable data sets derived by different methods supporting stepped extinctions—evolved into a standard for reappraising the impact hypothesis and ultimately forcing its revision. The revised version of the hypothesis was formulated by a diverse group that included an astronomer, geologists (including Walter Alvarez, a formulator of the single-impact hypothesis), and paleontologists (Hut et al. 1985, 1987). It proffered a comet shower with multiple impacts extending over a few million years so as to accommodate the newly evolved standard of the stepped-extinction pattern.

Although two hypotheses of multiple impacts had been published earlier, they both arrived before most of the earthly supporting evidence for stepped extinctions had been found. One was a briefly presented theory of instantaneous multiples (Kyte et al. 1980), and the other, an extensive one posited by astrophysicists (Clube and Napier 1984), had the multiple impacts spread over enough time to accord with the then-prevalent notion of a drawn-out dinosaur extinction (Clemens 1982). A most compelling line of evidence for multiple impacts—the recently described multiple craters of K/T boundary age—if found earlier, would have contributed greatly to the revision of the single-impact hypothesis (see Endnote 2.1).

Wide variation in the use of standards and the assessment of evidence was found even among subscribers to the same hypothesis; the variance was expressed by differences in their choice of standards, the rigor with which they applied standards, and how they weighted evidence. The impactors provided an example: the growth of stepped extinctions to the stature of a standard helped force the revision of the single-impact hypothesis into one of multiple impacts, as described above, by a group of authors that included Walter Alvarez and several early supporters of the single-impact hypothesis. But some other impactors who had provided signal evidence early on in support of the single-impact hypothesis remained stalwart in the face of the stepped-extinction evidence and continued to deny the multiple-impacts hypothesis. The stratigraphic/elemental evidence that the abiding single impactors had discovered through their own research (Kyte & Wasson 1986, etc.), and the fact that they had authored the first instantaneous-multiple-impacts hypothesis in 1980 (Kyte et al.), appear to contribute to their rejection of the long-term multiple-impacts hypothesis. Examine that rejection in terms of the conclusion drawn earlier here that *scientists who, at the advent of the impact hypothesis, were the published authors or published supporters of alternatives to impact, in the cases examined, were all (with only one exception) disposed against impact*. As late as January 1993 both Frank Kyte and John Wasson (oral comm.) did not consider that compelling evidence for stepped extinctions had been demonstrated, and they still see no evidence for multiple impacts.

Collins (1974, 1985) and similarly Margolis (1987) emphasized that "Any particular . . . specialty grouping within a larger field will develop habits of mind specialized to that group. . . . People within the group sometimes act as if X was true, sometimes as if X was false, and seem not to notice the inconsistency. Phenomena inconsistent with such shared commitments come to be habituated." Collins (1985, pp. 86–88) shows that the selection and use of standards and arguments about standards is an "interesting *social* process."

Gestalts, Polemics, and Conversions

> I would say usually, theories act as straitjackets to channel observations toward their support, and to forestall data that might refute them. Such theories cannot be rejected from within, for we will not conceptualize the potentially refuting observations.
>
> —*Stephen Jay Gould, "Dinosaurs in the Haystack,"* Natural History, *1992*

Subscribers to any of the competing extinction hypotheses, almost always, seemingly culled, shaped, or ordered both old and new evidence at some level of cognition to accord with their chosen hypothesis. In their oral and published arguments, subscribers also employed polemics that sometimes included the use of obsolete data and/or the omission of important contrary evidence. One may simply pick any of the many review articles written over the last several years, by the protagonists of either camp, to find that no single such article approximated to balance, whether in space or in words, between the volcanism and impactor arguments. Some were almost devoid of references to the most damaging adversarial papers. For the most part, the partisan character of such review articles approximated to the polarity of the two papers juxtaposed in the October 1990 issue of *Scientific American*, one written by authors of the impact theory (Alvarez & Asaro), the other by a champion of the volcanism cause (Courtillot). They are fine opposing legal briefs, and that is the way science has always worked. Polemics sometimes included much beyond rhetoric and "solid" evidence; in some cases remarks made in the course of formal presentations at national meetings impugned the basic competency of theoretical adversaries.

It was thus with great surprise that I read Sutherland's review (1988) of the impact/volcanism debates, which he sent me in 1990. His paper seemed the exception to my surmise that it was extremely difficult to find a broad and balanced overview of the debates in any single article, or even in several articles, written by scientists. Sutherland's is a seemingly unique case in these debates, a clearly nonpartisan review of the important evidence and arguments for and against both the impactor and volcanism hypotheses. But his paper also supports my earlier surmise that theory choice strongly conditions the selection and application of standards and assessment of evidence. Sutherland concluded his paper by very briefly stating his own hypothesis, *that both volcanism and impacting caused the K/T extinctions, and that they did so coincidentally—he does not believe that impacts triggered the volcanism.* His very balanced presentation of the evidence in support of both the impact and volcanism hypotheses is thus a prerequisite to his own hypothesis, which validates and subsumes both volcanism and impacting. Sutherland's simultaneous and equal embrace of those competing hypotheses engendered his uniquely hybrid gestalt, one that allowed him to invoke standards, assess evidence, and present a lengthy, very balanced overview of the debates, *because that is the brief for his own hypothesis.* His equal affection for both children allowed him nothing less than equal consideration of both. He was alone in his own bipartite-theory choice, and thus removed from the generalizations earlier argued here regarding the effects of single-theory choice. Sutherland's hypothesis is in fact a third hypothesis, but one composed mainly of the other two hypotheses.

In a few cases, immediately following field observations at critical localities, scientists converted to an alternative hypothesis or reversed firmly held opinions about the nature of the K/T-boundary environmental and biological effects or other phenomena. These were active researchers who had published partisan papers on the cause of the K/T boundary layer and later switched to an alternative hypothesis while, or immediately after, examining well-exposed K/T boundary rock layers that they regarded as compellingly informative. A few forthright respondents were able to express in interviews precisely how they had assimilated new information through their own direct observations of the complex diversity of data arrayed in situ—in total stratigraphic context. They spoke of how such empiric, contextual evidence simply compelled a new view, the birth of which displaced the old—while still at the field site in two cases. They regarded the new view as far beyond what had been, or seemingly could have been, envisioned by them up to that point on the basis of more abstract data, however extensive.

Such rare, direct observational experiences were singularly influential in altering views on debated issues: they appeared to trigger the intellectual equivalents of perceptual/conceptual (gestalt) shifts. There were similar cases early in the drift controversy in which European or North American geologists, the products of communities with strong anti-drift sentiment, traveled to observe directly the myriad factors that constituted the geologic, in situ, reality of critical sections cited in support of drift by Wegener and others. Kenneth Caster of the University of Cincinnati was such a case. As a young man he had traveled abroad, had seen the Gondwana sequence—and returned a confirmed, lifelong drifter, pleading for decades with his fixist detractors to observe the evidence in the field that he believed would "compel" acceptance of Wegener's arguments. It is by now an old saw that, early on, the geologic communities astride the best field evidence contained most of the world's drifters. Archibald Geikie's dictum about learning geology through the soles of one's feet was confirmed long before the Caster case.

The bearing of any single research finding on theory choice is open to many interpretations. The cognitively and historically specific meanings that scientific communities later attach to such findings are seldom evident at the time the original work is done. Galileo's telescopic observations and Michelson and Morley's measure of the effect of ether on the speed of light surprisingly seemed not to have had a profound influence, and certainly not a decisive one, on their contemporaries; neither did J. Tuzo Wilson's epochal recognition of transform-fault motion move geology closer to the acceptance of continental drift theory. The recognition of those signal contributions came only after other compelling evidence allowed broad acceptance of the theories the prior contributions had addressed.

In the case of drift, conversionary proof came not from those like Wilson who were addressing the drift question, but instead from graduate students at Lamont Observatory who *inadvertently* found the magnetic data that served instantly to confirm seafloor-spreading theory in 1966. Seemingly magical, numerical ratios were matched between the widths of magnetized blocks of the seafloor crust, the recently assembled geomagnetic-polarity-reversal time scale, and magnetic-polarity intervals found in marine sediment cores. The simple, immediately comprehensible congruence of those ratios lent itself to presentation in diagrams so visually simple and directly representative that reality seemed barely abstracted or surrogated. Most in earth science were awestruck on first sight. What they saw in those diagrams constituted an interlocked framework of such conceptual cohesion and symmetry that even the authors of the hypothesis that had predicted it could not have anticipated the near-perfect, legible form in which the effects of multiple, independent processes had been written and revealed themselves (Glen 1982).

The compelling force or conversionary power of different forms of data was evidenced repeatedly as the mass-extinction debates progressed. A clear example lay in the discovery of shocked minerals at the K/T boundary extinction horizon. Their discovery (Bohor et al. 1984), four years after the advent of the impact hypothesis, was crucial in convincing many geologists to accept the impact component of the Alvarez hypothesis (the mass-extinction effects are another matter). The shocked-quartz data, unlike the iridium anomalies, were derived through a methodology that was fully familiar and understandable to geologists; were offered by a well-credentialed member of their own geological community; were cast in their professional language; and were published in the most widely read, most prestigious American science journal. Although the iridium anomaly fell like a hammer blow and was the most compelling evidence for impact theory for most who embraced the theory, for many others it was merely a bunch of elemental abundances and ratios offered up by unfamiliar "black-box" nuclear chemists.

For certain fields the finding of the shocked quartz was at least as influential as iridium in capturing attention and "forcing" an altered, favorable view of impact theory. The elements and phenomena of each discipline differed in the extent to which they lent themselves to observation, ordering, analysis, and comprehension—and differed in how readily they could be read by students in distant disciplines. *Such interdisciplinary differences suggested that models of science, when not confined to specific disciplines or closely related groups of disciplines, are likely to suffer in application.* The template drawn from a skiff will not fit the hull of a schooner. Bauer (1992) is among those who recognize that different sciences frequently do not resemble one another to a great extent.

Paleontology, for example, differs from nuclear chemistry in its expectation of rates of progress—and especially in its tolerance for the foreseeably insoluble, the heuristic power of algorithms, the predictive success of generalizations, the ostensibly clear and precise, levels of instrumentation and technical support, the relative importance of "historical contingency," and a host of other tools in the limitless reservoir of tacit knowledge. The myriad singular particulars of a historical subject like paleontology, which most often defy both the formulation of strong predictive generalizations and precise experimentation, engender a mindset and framework of expectations that is wholly unlike that of physical science, in which powerful generalizations, drawn from a limited number or even only few recurrent particulars, often guide successful experiments. Such differences foster wholly different gestalts.

Even among the members of a given community, there are wide variations in their resistance to a powerful new idea that seemingly threatens an "older tradition of normal science." A spectrum of credence in a given new hypothesis can be exhibited by one faction of a community at the same time that another remains significantly polarized. That open-minded scientists might waver in their beliefs as arguably telling evidence unfolded alternately from opposite theoretical camps has been repeatedly demonstrated in these debates; such was especially the case from 1986 to 1989, when the volcanic-iridium evidence mounted, and the plume modelers began to bolster the cause of volcanism. But those who vacillated as the stream of data alternated to favor first one hypothesis and then the other were in fact a small minority, and not to be found at all in certain specialties—most participants held tenaciously to the hypothesis they had chosen at the opening of the debates. For example, I found neither in planetary geology nor in impacting studies anyone who ever wavered from the impact-as-extinction-cause component of the hypothesis, nor in vertebrate paleontology anyone who converted to embrace it.

Only long-practiced stratigraphers and field geologists come to recognize how the basically historical, nonrepeatable character of the stratigraphic record eludes attempts at uniformity of classification. Ten stratigraphers asked to divide an extensive, lithologically and stratigraphically complex section into members, formations, and groups would likely submit ten different schemes of order. And such classificatory differences would mount at the finer level of discrimination required to identify beds in a subtly graded sequence, or in trying to recognize slight disconformities (breaks in the continuity of deposition). Geologic maps of the same area by different field geologists almost always show significant differences. Such differences derive in part from "play" in the system of field mapping (perhaps the most diversely complex interpretive exercise in geology) and stratigraphic science, which is understood in significant part at the level of tacit knowledge. Such uncertainties must surely have underpinned the geologic community's license to dismiss Wegener's claims of intercontinental stratigraphic continuities in the drift controversy. Stratigraphic units that are laterally constant in most—let alone the totality—of their properties over broad regional areas are unusual, and are very rare intercontinentally.

Because of the difficulties of recognizing lateral changes of every sort within rock units, stratigraphic science passed through a tortuous early history. Such lateral variations or facies may be manifested in fossil content, lithology, or even time value. Although the Baron Cuvier had already, at the turn into the nineteenth century, hinted at homotaxy (the idea of identical faunas being entombed in rocks of different ages and thus giving rise to different vertical successions of taxa at different localities), and although T. H. Huxley by 1861 had noted explicitly that the same faunas in different rock strata are not necessarily of the same age, it took almost a century for the community to fully assimilate the fact that the time value of a stratigraphic unit that *was* laterally constant in its lithology and fossils could nonetheless vary in age from place to place. As the land sinks or the sea rises, the sea may march across the land to deposit identical sediment—and in some cases, identical, long-lived fossil species—at different times at different places, and thus build a laterally continuous and homogeneous unit that varies continuously in its time of formation, in the direction of the sea's advance. Lacking short-lived, time-diagnostic fossil taxa, such a rock unit, if undatable by radiometry, would contain no intrinsic evidence to signal the fact that its age changes laterally from place to place.

The unreliability of most sedimentary beds as time markers made rare, geologically instantaneous, air-fall ash beds from explosive volcanism extremely valuable. They often serve as stratigraphic guides in correlation because of their ubiquity and unique mineralogic fingerprints over vast regional areas; but with rare exceptions they too are limited in their usefulness to only parts of single continents. Magnetostratigraphy, based on the temporally unique magnetic imprint of the Earth's magnetic field on rocks, is a powerful global correlation tool, but it too has its limitations: in almost all cases it must be guided by both fossils and radiometric dating. Even given the entire fund of techniques, methods, and principles of correlation extant, there was still, in the past decade, widespread uncertainty about correlating marine rocks of K/T boundary age with their continental contemporaries, even where both sections were richly fossiliferous, because the two sections were almost always mutually exclusive in time-diagnostic fossils. Such difficulties formed the basis for the claim of many, through the first half of the debates, that it was not possible to establish the synchroneity of K/T boundary extinctions on land with those in the sea.

It was not until 1986 that Saito and others recognized in Hokkaido a sudden change in the spore and pollen flora precisely at the base of the K/T boundary clay, as determined by planktonic foraminifera. That Japanese horizon, which contained both marine and terrestrial fossils, was held to correspond with the fern spike at the K/T boundary in the terrestrial rocks of the North American interior discovered by Robert Tschudy of the U.S. Geological Survey and others in 1984. Tschudy's fern spike occurred precisely at the level of the great iridium spike. At the time, few knew of, or seemed to pay attention to, that profoundly important Japanese stratigraphic section in Hokkaido, even though its significance was reemphasized by Anthony Hallam in 1988.

Against such a historical backdrop it can be seen that *stratigraphic science had never before claimed—as did the impactors—the presence of a globally ubiquitous stratigraphic marker horizon containing components of fingerprint value that came into being through processes entailing impact-driven, ballistic-trajectory transport.* The reception of the impact hypothesis was conditioned by the geologic mindset about "play" and uncertainties, the inexact nature of a mainly historical science, and a long record of unsuccessful search for precise, global, stratigraphic-correlation tools. How likely, then, to those called upon to assess it, was the existence of a globally correlative rock layer containing exotic materials that had formed through processes alien to the history of their science? Indeed, not even the planetary geologists, who had regularly classified and mapped impact-ejecta layers such as that of Mare Imbrium on the Moon, had ever demonstrated such a layer to be planet-wide (Wilhelms 1987, 1993).

Margolis's (1987) cognitive-dynamics view of "paradigm shifts" seems to accord in large part with the conversionary episodes documented in this study, and also with cases in drift and plate-tectonics history (Glen 1982). Such documentation suggests that Laudan and Laudan (1989) have likely been too harsh in their remark that ". . . neither Kuhn nor Feyerabend . . . has anything constructive to say about how theory agreement arises in science. Both are reduced to talking psycho-babble (for example, Kuhn's 'gestalt switches' and 'conversion experiences,' Feyerabend's 'bandwagon effects')." Phenomena of mind that are widely known and documented but cannot yet be more precisely explained in terms of extant systems of knowledge, or accommodated by philosophic models of science, ought not be dismissed as "psycho-babble," but rather examined more closely and in new ways. The "bandwagon effect," exacerbated by the rapid pace of the mass-extinction debates, was strongly in evidence in this study; it was also documented in vivo in studies of the accretionary-terrane research program (Le Grand & Glen, 1993) and will be further detailed in my forthcoming book on the mass-extinction debates.

Many aspects of this study, and of the ascendance of plate-tectonics theory, support Kuhn's view that a paradigm shift—*sensu latu*, the switch to the alternative hypothesis—carries with it, for the individual, a fundamental change in cognitive values. Indeed, the term "cognitive value" seems inadequate in conveying the sense of profound change apparent in the entire range of perception and response following a conversion experience.

Magisters, etc.

> There is nothing more difficult to take in hand, more perilous to conduct, or more uncertain in its success than to take the lead in the introduction of a new order of things, because the innovator has for enemies all those who have done well under the old condition, and lukewarm defenders in those who may do well under the new.
>
> —*Niccolò Machiavelli,* Il Principe *(1513)*

Earth science has repeatedly been encumbered by deferring to the "magisterial authority" (S. Toulmin's term) of the "restricted sciences," i.e. those sciences that emphasize the use of a small number of powerful laws in matters of great theoretical significance (Pantin 1968). Lord Kelvin invoked the second law of thermodynamics in order to attack the law of uniformity, and dated an earth so youthful as to preclude the development of life within the time frame of the geologists. Sir Harold Jeffreys "proved" by the laws of physics that granite ships could not sail through basaltic seas, and lectured that rocks could not possibly be reliable carriers of fossil magnetism by which to reconstruct the ancient magnetic fields of the earth. Other physicists had earlier claimed that Argand's interpretation of massive overthrust sheets in alpine structure was contravened by physical law. There were still other magisters from the restricted sciences who pontificated in matters seemingly beyond the ken and tools of earth science. Many geologists were thus imbued with healthy skepticism about conclusions "imposed" on them by those whose language they did not speak. Luis Alvarez, the Nobelist in physics and co-author of the iridium-backed impact hypothesis, appears indeed to have suffered a reception that was conditioned in part by those unhappy aspects of earth science's history.

It is clear that Alvarez encountered that widely prevalent attitude of resentment most strongly among those branches of earth science farthest removed from his own discipline. Many of the earth scientists from specialties not familiar with impacting, meteoritics, cosmo- and geochemistry, planetary geology, or related fields had little sympathy for an outlandish idea that came from black-box disciplinary aliens. It was not unlike the greeting that paleomagnetists often got from geologists in the 1950's and early 1960's when they were plotting polar-wander paths to support continental drift (Irving 1988) and developing the time scale of geomagnetic-polarity reversals (Glen 1982). Nor was it unlike, ten years earlier, the apprehension of many paleontologists, whose science had formulated the fossil-based, relative, geological time scale, that their scale would be found wanting by the early atom-counting geochronologists, whose "absolute" dates were calibrated in years by the use of alien methods and instruments. And indeed those radiometrists—realizing the worst fears of paleobotanists, whose fossil plants told much about ancient environments but little about dating—reversed the long-established relative sequence of certain fossil floras (Evernden et al. 1964; Glen 1981, 1982).

Magisterial authority was seen to play a role also at the level of the individual scientist. It is clear that the leadership of the various debates bearing on the K/T mass extinction was in the hands of only one or a very few magisters, to whom the community members deferred and to whom, all too frequently, they turned to form and reform their opinions on evolving issues. This seemingly common phenomenon was first described by E. Katz and P. F. Lazersfeld (1955), from observations of physicians and the general public. Every social world recognizes leaders to whom its members turn for information and advice; the phenomenon is general. Everett P. Rogers (1982) summarized the literature on this two-step flow of communication, from the world to the magister and then from the magister to members of the social world.

Reliance on the magisters also appeared not so much a matter of the ascendancy of dogma within a broadly and nearly uniformly informed community, but rather a matter that most scientists simply did not have the time, energy, or interest to keep abreast of issues that required the continual assimilation of rapidly burgeoning data, and were thus moved increasingly to a reliance on the magisters (and even on the media, as we will shortly learn). One of this study's most surprising finds was that many scientists, including eminent ones, were poorly informed but nonetheless quick to volunteer strong opinions on their choice of theories and other issues *on the basis of a single magister's position*. Such opinions, moreover, were proffered during *formal* interviews.

One might expect the position of the magister to be strengthened during such a rapidly evolving controversy. The growth of published literature in the natural sciences, which shows a doubling rate for many disciplines of about seven years over the last two decades (Garvey and Tomita 1972; Glen 1989), has drastically eroded scientists' ability to keep up with their own literature. And perusing articles far afield, however important they may be, has become even more difficult. But the rate at which new literature was generated in several topical areas of these debates far exceeds this average for any entire single discipline. This situation should only strengthen the position of the magisters, part of whose role *is* to stay abreast.

Paradigms Key Collaborations

> In the twelve and one-half years of the impact debates only one paleontology graduate student has ever sought me out to talk about anything—that was David Fastovsky [a confirmed impactor].
>
> —*Walter Alvarez, interview of 2-23-91*

A central force driving and channeling the collaboration observed in these debates was a mutuality of commitment either to impact theory or to volcanism theory; but with one or two exceptions members of one theoretical camp did not collaborate with those of the other. Collaborative efforts—aside from the circumstance of theoretical upheaval as documented here—are typically fueled by a complementarity of intellectual or technical resources or personality, and, generally, propinquity. Subscription to the same theory of extinction, by contrast, prompted members of unrelated or only distantly related fields to collaborate, and thus to redirect their research during an interval that may be viewed as an incipient crisis or upheaval. These collaborative groups, more diversely comprised than any alluded to by Kuhn or others, are, however, recognizable by the criteria of special conferences, the distribution of draft manuscripts, and networking among them in both communication and citations. Before collaboration they shared insignificantly in Kuhn's "disciplinary matrix" or community paradigm (the fund of commonalities: theories, methods, techniques, instruments, and a host of subcultural elements) held by scientists working in the same field. They were little likely to have the commonality of approach to the same facts, methodological rules, or cognitive aims and values that one might expect of members of a team laboring in concert. But they did come to share a belief in the need to work toward the resolution of problems of the broadest significance to several branches of science—for the profundity of the puzzles and hypotheses, and the height of the stakes, were clear to them.

These collaborative teams were bonded by a commonality of purpose in addressing long-standing, classical problems that impinged on the philosophic core of earth science. And the escalating publicity that the debates fostered provided an unprecedented audience for their publications. It was here, among scientists of vastly different traditions, where poor communication and little mutual understanding might be anticipated, that shared views of newly formed theories, as well as interpersonal and intergroup rapport, were so much in evidence. The intellectual convergence that fostered such commensurable views among scientists of vastly different disciplines reinforced my surmise from earlier studies that *prior subscription to a hypothesis is the chief conditioning factor in the cognition of facts, methods, and standards of appraisal.* The embrace of a hypothesis appears to condition cognition to an extent that precludes hope for the success of Gilbert's (1886) or Chamberlain's (1890) methods of multiple working hypotheses. They both counseled that entertaining more than one hypothesis at the same time contributed to objectivity and balance in investigation, but I have never been convinced that any scientist has been able to formulate hypotheses that he or she believed to be of equal weight, or, if that were possible, to remain nonpartisan and conduct research toward the elimination of all with equal dispassion. Johnson (1990) and Locke (1990) appear to share that view.

The diversely comprised volcanist teams seemed similar to those of the impactors in their circumstances of formation, division of labor, and a number of other operational characteristics. Collaborative teams, both of impactors and of volcanists, sometimes included members who said or implied that they were not clearly disposed to any theory, but who in every case, in interviews, inadvertently revealed distinct partisanship.

The effect of the reigning paradigm was demonstrated in another way in 1984 and 1985, when dozens of seismologists of varied research interests visiting the U.S. Geological Survey Seismic Data Library in Menlo Park were asked how they ordered and applied their data, toward what end, in the examination of the problematic subduction zone at trenches, where plate theory holds for subduction at a compressional juncture, but expanding-Earth theory denies compression and subduction and invokes a mainly gravity-driven mechanism. There was still, in the early 1980's, a certain amount of open questioning about the difficulties of a collisional scenario, because there were, unexpectedly, nearly undeformed sediments in some of the trenches. All of the seismologists questioned responded that they had never considered ordering and testing their data for the subduction zone—or any other seismically active structure—in terms of the expansion model or any other model save that of plate theory. Only a very few were even familiar with expansion theory's major premises. When the few who seemed somewhat familiar with expansion were asked if expansion might more satisfactorily address certain problems than does plate theory, largely derisive commentary about expansion's authors ensued. How typical, then, is the true believer in science? What is the power of the ruling paradigm, or of any paradigm to which one subscribes? Bohm and Peat (1987), while examining this problem broadly, concluded that "[the paradigm] exacts a price in that the mind is kept within certain fixed channels that deepen with time until an individual scientist is no longer aware of his or her limited position." Is it really possible to remain uncommitted, hold a suspended judgment, and live Overstreet's (1949) life of *The Mature Mind*?

The growth in the number of scientists, representing several very different disciplines, who have refocused their research aims in order to collaborate, and thus to examine the validity of the impact and volcanism hypotheses, constitutes an example that seemingly stretches the boundaries of what Kuhn (1970) and others define in discussions of community structure in science. Kuhn's model does not address the problem of the intersection of distant and unrelated disciplines that we find in this history. In contrast, the views of Strauss (1978), Gerson (1983, forthcoming), and Latour (1987), of science as constituting a changing network of alliances brought into being and driven by a complex of forces, seems well demonstrated in this study.

Gerson's application of Strauss's "social world perspective" takes an approach similar to Latour's in explaining intergroup conflict. That model seems partly applicable here, in the sense that different disciplines became concerned with the same problem, and debates arose because different groups held different perspectives and made different assumptions about the nature of the data, the methods used to produce the data, the model or interpretive framework used to decipher the data, and the standards used to evaluate the models, data, and so forth.

Among the three categories of "social world processes" that Gerson (1983) defined to aid in understanding how lines of work develop and change was "intersection processes." These processes regulate the ways in which subworlds and lines of work cross and recross to diffuse the elements of different research traditions. The dynamics of intersection are conditioned by skills, prestige, friendships, money, technical resources, propinquity, etc., but since the central axis on which scientific worlds turn is problem-solving, specialized capability is crucial. Such was the case in the mass-extinction debates, where intersections (collaborations) arose mainly through mutualities of need for techniques and specialized knowledge and skills. *But in almost every case the intersectors had already subscribed to the same hypothesis of mass extinction.*

The very diverse nature of the questions raised in the mass-extinction debates called a singularly broad range of disciplines into collaboration. Such teams, engaged in data-gathering, experimentation, the production of new instruments, or the formulation or revision of theory, might include members representing several different disciplines from among geology, geophysics, stratigraphy, paleontology, mineralogy, volcanology, impact studies, geochemistry, nuclear chemistry, astrophysics, astronomy, statistics, and mathematical modeling. Such collaborative groups, channeled by mutuality of need to intersect from different "worlds," coalesced seemingly overnight in many cases, driven by the rapid pace and extraordinary levels of competition in the impactor-volcanist rivalry.

The Media

> Astronomers should leave to astrologers the task of seeking the cause of earthly events in the stars.
>
> —The New York Times *editorial of April 2, 1985, "Miscasting the Dinosaur's Horoscope"*

Although the media varied significantly in the extent to which they adopted subjective views, none can be said to have been scrupulously nonpartisan; sometimes their views were astonishingly polarized, or even fatuous (see the quote from *The New York Times* above—which was written not by a science writer, but by an editorialist). The *National Geographic* issue of June 1989, in contrast, simply presented impact as the cause of the dinosaur extinction. At that date the debates over cause were still heated, and most dinosaur paleontologists argued for a gradual decline prior to the K/T boundary, as defined by the iridium horizon; many continue to do so in spite of the new evidence for sudden dinosaur extinction advanced by Sheehan and others (1991).

Many, if not most, science journalists and editors clearly relied, in the long term, on privileged sources of information that tended to promote a sustained imbalance in their presentation of the debated issues, but their mastery of the science they treated was generally satisfactory, and in a few cases remarkably good. Most science journalists do have significant training in science. Interviews with science reporters and editors expectedly revealed great variations in their understanding of the full range of science entrained in the debates, the amount of time they devoted to background and preparation of articles of nearly equal length on the same subject, and the deadline pressures under which they worked. Only a few of the science writers were well-informed across the broad range of topics that the debates encompassed. Such conditions made most journalists particularly vulnerable to the polemics of strongly opinionated scientists, and, as usual, the magisters were called upon to play signal roles in informing the media. Clemens (1986; and this volume) has shown well, mainly from a sociological viewpoint, how both professional and popular scientific publications played significant roles in shaping certain aspects of the debates.

Dorothy Nelkin (1984) and others have demonstrated that although historically the relationship between scientists and journalists has been quite negative, there are indications that it has improved considerably in recent times, that the profession of science journalism is growing, and that its level of competency is rising. But the journalist's need for brevity, the press of publication deadlines, and a workload that precluded the possibility of truly representative sampling of opinion has led to wide variation in completeness, accuracy, and balance in treating diverse, complex subjects in these debates. Because the explosive rate of publication across a research front that may be unprecedented in its disciplinary diversity was daunting enough even to the researchers, we may well pity the journalists! Beyond the imbalances that understandably may arise from not keeping up with the literature, some science writers and editors (including the staffs of certain journals of science) have over several years treated this conflict by writing consistently partisan articles and editorials. I inquired of the author of a totally partisan article published in a journal sponsored by a federal agency why he had not interviewed and quoted scientists from the opposing camp. His reply: "I simply wanted to present the debates from [the one viewpoint]." I later wrote him, inquiring when the journal planned to present the opposing view, but no answer and no balancing article appeared.

The producers of television "science documentaries" such as *Nova* and *National Geographic* have in recent years produced increasingly sophisticated and elaborate presentations on aspects of the debates, presentations that have captured large home audiences, become widely used in education, and been viewed, as well, by much of the scientific community. Of the two such "documentaries" that I have examined on the impact-volcanism controversy, one was flawed in its facts and somewhat partisan; the other was more factually correct, but astonishingly presented the controversy, at the time of broadcast about seven years ago, in 1987, as more or less closed.

Much in keeping with what I learned in earlier studies still in progress on the accretionary-terrane research program and the expanding-Earth theory, many of the scientists I interviewed in this study who were willing to offer assessments of the ideas in debate based their views on hearsay, i.e., on the reading of scattered abstracts, editorials in *Science, Nature,* and other general journals, science-news articles in magazines such as *Discover, Science 85,* and *Science News,* newspapers, and even television "documentaries." The difficulty in staying abreast of the professional literature in these debates has made scientists especially dependent on review articles, and vulnerable to poorly informed and biased commentary from many quarters.

The media must be considered not only as to their effect on community opinion in science, which strongly conditions the refereeing process in funding research, but also the extent to which science reporters act as conduits for the exchange of research information among scientists. Journalists are often privy to research results prior to their publication, and even among the most scrupulously careful and professionally ethical, power of that sort carries implicit dangers. Doing current, in situ history reveals that science reporters play significant roles in the workings of science and the "making" of science history: respondents with whom I had established good rapport over several interviews often sought my opinion on issues in question, perhaps more to learn what others in the community were saying out of print than to fathom my intellectual posture. Such exchanges, in more than one case, resulted in the formation of new collaborative teams. Although the aims and purposes of historians of current science differ from those of science journalists, both professions cater to the scientist's need for recognition and response, and often play significant roles in conditioning the course of research.

Summary of Conclusions

Within a decade, the upheaval ignited in 1980 by the impact hypothesis transformed the literature of mass extinctions from an unfocused, sporadic collection of papers that virtually ignored suggestions for extraterrestrial cause, and treated suggestions for endogenous cause only marginally better, to an integrated, cornucopian body of literature reflecting all sorts of possibilities. Research programs organized seemingly overnight spawned research teams whose members, often from formerly isolated disciplines, virtually redirected their careers in order to address the captivating, high-stakes issues. *Almost always, the members of a given collaborative team had previously embraced the same extinction hypothesis.*

The generally skeptical, even poor, reception that the impact hypothesis received initially might have been predicted, for any of several reasons: such an instantaneous catastrophe flew in the face of earth science's reigning philosophy of uniformitarianism; it was based on a form of evidence alien to the community charged with its appraisal; it invoked a causal mechanism that was unlikely in terms of canonical knowledge; and it was proffered mainly by specialists outside of earth science and paleobiology.

It became clear early on that *irrespective of which mass-extinction hypothesis a scientist chose, the chosen hypothesis became the strongest predictor of how the scientist would select and apply standards in assessing evidence bearing on all such hypotheses. Somewhat weaker correlations appeared between disciplinary specialty and choice of hypothesis, and between disciplinary specialty and how evidence was assessed.* Such correlations varied with level of subspecialty: the strongest correlations appeared in the most restricted subspecialties, the weakest in the broadest categories of science practice. And the gestalt or mindset seemingly engendered by subscription to a particular extinction hypothesis overrode the intellectual predispositions attributable to disciplinary specialty.

The vast majority of paleontologists rejected impact theory straightaway, and most who later came to believe in one or more impact events still deny that impact(s) are the main extinction cause. Many paleontologists viewed the duration, severity, and other aspects of the K/T mass extinction in terms of the fate of their own fossil groups at the extinction boundary; specialists in severely affected taxa generally supported the idea that an impact caused the mass extinction. The subspecialty communities within paleontology differed markedly in their opinions on the cause of mass extinction: vertebrate paleontologists, for example, were generally opposed to impact-as-extinction-cause; in contrast, micropaleontologists, especially those treating planktonic calcareous forms, were most often oppositely inclined.

Most cosmo- and geochemists, planetary geologists, impacting specialists, and those concerned with Earth-crossing comets and asteroids embraced impact theory in its entirety. Familiarity with impacting studies appeared to foster sympathy for the impact hypothesis; most volcanologists (volcanic specialists), however, did not accept the volcanism hypothesis—which was not conceived by volcanologists—but neither did any unusual proportion of volcanologists favor the alternative impact hypothesis.

In all cases examined save one, published authors of alternatives to the impact hypothesis of mass extinction, as well as published supporters of these alternatives, opposed the impact hypothesis at its advent. Irrespective of their discipline, scientists rarely failed to embrace one of the various mass-extinction hypotheses, however poorly informed they were.

The use of obsolete data and/or the omission of contrary evidence almost always punctuated published and oral arguments, and opposing views were never treated at equal length. The gestalts (mindsets) or cognitive frames of members of the opposing theoretical camps seemingly precluded mutually congruent viewpoints on any of the important debated issues, or even the assessment of evidence. So deaf did they appear to each other's arguments that Kuhn's (1970) view that such adversaries suffer an "incommensurability of viewpoint" seems understated.

Impactors and volcanists routinely invoked different standards of appraisal and weighted the same evidence differently. The application of standards and the weighting of evidence varied widely, even within the same theoretical camp. The impactors argued mainly from canonical standards and claimed that their hypothesis—backed by empirical evidence—facilitated clear predictions with implicit directions for testing. The prediction of impact-generated, global, ballistic transport of impact products raised expectations for many that impact evidence in addition to iridium would be found at K/T boundary sites, and these expectations furnished great impetus for the rapid formation of diversely comprised research teams.

Impact's opponents focused on the great range of variation in the character of the boundary interval around the world, emphasizing that the differences found from place to place betokened a cause that was neither global nor instantaneous. Demonstrating these differences became a central mission of the opponents of impact. Unlike the impactors, who in the main advanced orthodox evidence for their radical hypothesis, impact's active opponents—mainly the volcanists—sought to undermine a wide range of suppositions and vagaries that had lain long hidden in established principles and methods; the volcanists impugned orthodox standards, and thus prompted much research at unprecedented levels of refinement. The call for higher resolution required, for both camps, the development of new, or the acquisition of theretofore unacquired, methods, techniques, and instruments; such needs were often fulfilled through the formation of appropriately composed collaborative teams. These great new efforts spawned proposals to various funding agencies for research grants. When such proposals were refused, undercurrents of opinion about the partisan bent of this or that funding agency spread within opposed theoretical camps.

During the mid-1980's the postulate of stepped or multiple extinctions near the K/T boundary evolved, through a series of recognizable stages. The evidence for stepped extinctions was initially viewed as an anomaly within the broader terms of the single-impact hypothesis and, as such, was more or less dismissed early on by the impactors. But as supporting data accrued, it became an increasingly serious problem, and stepped extinctions approached the status of a normative assumption. Finally, with still further affirmative evidence at hand, stepped extinctions and all that they came to imply, in terms of impugning single-impact theory, evolved into a standard of appraisal that forced a reassessment of the nature of extinctions at the K/T boundary. At that point *the newly born standard of multiple extinction steps—which had begun life as an anomaly in terms of the single-impact hypothesis—evolved into the primary forcing function that drove the reformulation of the single-impact hypothesis into a hypothesis of multiple impacts spread over enough time to accommodate the extinction steps.*

In all cases, the leadership of the various factions engaged in the debates was clearly in the hands of only one or a very few senior leaders, who exercised all but magisterial authority. Such doyens were most frequently sought out, by both their own communities and the media, for their opinions on the debated issues; this was reflected in both the publications of science and the media of the public. The rapid pace of the mass-extinction debates, with which only few could keep abreast, seemed to add to a general reliance on the magister. Such two-step communication, from the world to the magister and then from the magister to the world, has been documented in other studies of conflicted ideas.

Definitive closure has not been reached on any of the many issues entrained in these debates, but the vast majority of earth scientists are now convinced of at least one impact at the Cretaceous/Tertiary boundary, and a substantial number are inclined to think in terms of multiple impacts, either instantaneous or spread over 1 to 3 million years; however, far from all who subscribe to impact(s)—especially among the paleontologists—view impact(s) as the chief cause of the mass extinction(s).

Research programs organized in the past decade to address the many issues that have arisen in the course of this upheaval continue to generate new publications at a surprising rate. Modeling of impacts and mantle plumes, and of their effects on a number of internal, crustal, and biospheric processes, grow increasingly sophisticated, with diverse research approaches showing promising convergence in their conclusions. The search for impact sites, which includes the re-examination of many long-enigmatic structures and the remapping and dating of great flood-basalt bodies, continues to be actively pursued. Mass-extinction horizons of all ages are being scrutinized with methods, techniques, and instruments that did not exist just a decade ago, and the broad character of the debates has forced unprecedented interpenetrations of long isolated subdisciplines. Meanwhile, the debates continue, sustaining—for historians and sociologists and philosophers—the opportunity to study the workings of science *on the ground,* during a time of conflict over several theories that span multiple disciplines and promise still more remarkable findings.

3

The Impact Hypothesis and Popular Science: Conditions and Consequences of Interdisciplinary Debate

Elisabeth S. Clemens

The interplay of popular and expert scientific knowledge characterizes the impact debate throughout the decade of its history. Yet in much commentary, this incessant cross-fertilization has been treated as a form of contamination or media sensationalism, rather than as constitutive of the debate itself. But without something beyond individual disciplinary specialties to unite researchers around a common set of questions, an analysis of the impact debate would inevitably echo the earliest formulations of the impact hypothesis itself: the random collision of two bodies in a void.

Just as proponents of the impact hypothesis have gone in search of the astronomical mechanisms driving such impacts, an adequate account of the impact debate must consider the shared beliefs, institutional settings, and social relations through which this interdisciplinary research effort has been constituted. Rather than displaying the sharp contrast between science and popularization that has elsewhere proven so ideologically useful for scientists themselves (Hilgartner 1990), the impact debate has been shaped by and partially played out in arenas intermediate to the poles of popular and expert, of exoteric and esoteric knowledge about science. As the philosopher of science Ludwig Fleck suggested:

> Let us begin the discussion . . . by considering *popular science*. This furnishes the major portion of every person's knowledge. Even the most specialized expert owes to it many concepts, many comparisons, and even his general viewpoint. . . . We shall presently have the opportunity of repeatedly finding items of popular knowledge from other fields within the depths of these sci-

ences. Such items have set the standard for the content of expert knowledge and have determined its development for decades. (1979: 112)

Most obviously, the debate is grounded in the widely accepted view that it addresses an important and comprehensible question: What caused the extinctions 65 million years ago at the K/T boundary, or, more popularly, what suddenly killed the dinosaurs? Though these questions are compelling within the context of popular knowledge about science, they have also been judged mistaken by many of the disciplinary specialists in the history of life: "For quite a while now, orthodox paleobotany has maintained the extinctions were spread out over tens of thousands of years or more" (Bakker 1986; see also Benton 1990: 387–89). So if popular science helped to constitute a wide audience for the debate, this foundation also helps to explain many of the specific characteristics of the debate's evolution: the broad scope of interest outside paleontology and the initial aversion to the hypothesis by disciplinary experts in the history of life, the consequent disarticulation of both research and theorizing across various fields, and, not least, the resistance of proponents of the impact theory to any reformulation of the central question of what killed those dinosaurs. Finally, by examining the institutional positions of different disciplines and their influence on the generation of popular knowledge about science, we can begin to speculate about the impact of the impact debate itself. If the distribution of popular knowledge about science helps to explain the pattern of participation in this debate and the emergence of a set of focal questions, the different institutional positions of the disciplines, in turn, have influenced their ability to reshape the "knowledge about science" shared by the general scientific community as well as by broader popular audiences.

Who Cares about Dinosaurs? The Social Basis of Interdisciplinary Science

The first puzzle is not who has accepted the impact hypothesis as "fact"—i.e. a well-supported statement about empirical reality—but why so many people found this an interesting question in the first place. After all, the polite disinterest of even family and friends is a more typical reaction to a description of one's research. But among the things that "most people know" is an established way of asking the question of the K/T transition, one which owed as much to professional researchers and museum curators as to the fantasies of Hollywood. The "interesting" question about the K/T boundary—at least in the context of popular knowledge about science—is, of course, what happened to the dinosaurs? This question, however, begs another. What are the sources of this widespread and highly stylized puzzle about the history of life?

The most basic answer to this question comes from the scientists themselves: popular science, specifically children's science. Although popular culture is rarely taken seriously in the sociology of science, it is the most powerful explanation of why generations of scientists across a variety of disciplines have continued to puzzle about the fate of the dinosaurs rather than the origins of mammals or the sources of shock metamorphism. Popular knowledge about science is not, that is, a thing apart from professional research. In efforts to capture public imagination and material support, scientific institutions and disciplines have both catered to and transformed the structures of popular knowledge about science. But although interconnected, professional and popular knowledge develop in different ways, in response to varied cognitive and institutional demands, often generating incommensurable taken-for-granted questions and answers. So although the overlap of these questions makes interdisciplinary debate possible, the very different foundations of these questions ensure that the debate will rarely proceed without fundamental disagreements over assumptions and evidence.

So who does care about what happened to the dinosaurs? Most everyone, it seems, although no one seems entirely sure about the origin of this fascination. Most commentators have adopted a folk version of a Freudian analysis (Schowalter 1979). As the humorist Dave Barry explained in his column: "It's a power thing; children like the idea of creatures that were much, much bigger and stronger than mommies and daddies are" (n.d.; see also Gould 1989a). But although tensions between child and parent may approach some level of universality among human beings, children have been able to project these concerns onto ancient reptiles only by virtue of the long-standing diffusion of professional science into popular culture, specifically children's culture. Rather than pop Freudianism, the autobiographical statements of scientists point to the institutions of popular science as the source of their interests in the puzzles of ancient life.

In what has emerged as an almost ritual account of interest in the impact debate, the astrophysicist Richard Muller referred to a childhood book, *The Dinosaur Hunters*:

> The book showed paleontologists digging up rocks, carefully scraping away the mineral around the fossilized bone, taking weeks to remove a single specimen. I told my parents that I wanted to be a paleontologist. But deep down I didn't believe it. Real paleontology looked terribly boring. My most valued possessions were a telescope and a microscope. The world I saw through these seemed particularly beautiful, crystal clear and sharp. From that point on I associated science with beauty. (1985: 34)

This account is by no means exclusive to the impact debate. In explaining his off-hours collaboration with paleontologists, Roland Hagen, a physical chemist at Los Alamos, explained: "When I was a kid I read about the dinosaur expeditions to the Gobi Desert [in the 1920's], and it captured my imagination. So I've gotten quite a kick out of working on a real dinosaur dig" (Mervis 1991: 8). These books, many produced by the expeditionary paleontologist Roy Chapman Andrews as part of his efforts to publicize and secure funding for his projects, helped to define the public perception of paleontology for "anyone whose childhood ended before the 1950's" (Wilford 1985: 138). As a paleontology student in postwar Warsaw, Zofia Kielan-Jawarowska "read the Polish translation of Andrews' book *To the Ends of the Earth* . . . and managed to unearth his reports in the *National Geographic* and a few other popular articles" (1969: 3–4). For the paleontologist Robert Bakker, "It all started very suddenly, in the spring of 1955. I was reading magazines in my grandfather's house in New Jersey, and I found that magical *Life* cover story—'Dinosaurs.' . . . And I made up my mind then and there that I would devote my life to the dinosaurs" (1986: 9). This sort of magazine paleontology was soon supplemented by trips to another great source of popular knowledge about science, the natural history museum, which shaped and nurtured the early interests of many scientists (Cloudsley-Thompson 1978: ii). As Stephen Jay Gould recalled:

> I was a kiddie dinosaur nut in the late 1940's when nobody gave a damn. I fell in love with the great skeletons at the American Museum of Natural History and then, with all the passion of youth, sought collateral material with thoroughness and avidity. I would pounce on any reinforcement of my greatest interest—a Sinclair oil logo or a hokey concrete tyrannosaur bestriding hole 15 at the local miniature golf course like a colossus. (1989a: 14)

Although this passage accurately captures the contrast with the current flood of dinosauria—200 children's books listed at last count (Lessem 1991: 1)—if truly no one had given a damn, it is a safe bet that those great skeletons would not have occupied a fair piece of valuable Manhattan real estate (see Endnote 3.1).

The conventions of museum display were not the product of happenstance, but often reflected an explicit intention to engage popular interest in questions of evolution (Rainger 1988, 1991). Institution-building concerns also influenced the conventions of museum display (see Honan 1990), by valuing big skeletons that would attract both public interest and funding for further expeditions. Though this practice has left paleontologists open to methodological challenges in the current debate (Kerr 1991), it undoubtedly has helped to shape the questions that have been brought to the investigation of terrestrial extinctions. For much of the first half of this century, in fact, the question of dinosaurian extinction appeared to float free of the rest of the discipline. Michael Benton has argued that the period between 1920 and 1970 can be characterized as a "dilettante" phase, in which propositions must be categorized not only as testable or untestable, but also as serious or deliberate jokes: "A large number of the theories, most of which were published in standard scientific journals by scientists who were no doubt expert in their own fields, show a remarkable relaxation of scientific standards. It was as if, at the mere mention of 'dinosaur extinction,' scientists breathed a sigh of relief and felt freed from the straitjacket of normal scientific testing" (1990: 385). In the years since 1970, studies of dinosaurian extinction have entered a "professional phase" marked by testable hypotheses and detailed attention to ecological and physiological factors (Benton 1990: 386), thereby following a path toward the more biological orientation that was charted for much of vertebrate paleontology by the 1930's (Rainger 1988). Still, the more popular framing of dinosaurian extinction persisted both in the institutions of public science and in the background knowledge of scientists who had once visited those museums and avidly read those books.

Although popular knowledge about dinosaurs captured the interest of a wide range of future scientists, the influence of this popular knowledge is not identical for paleontologists and those who abandoned their early fascination for other fields of scientific research. As part of their professional training, would-be paleontologists had to unlearn much of the popular wisdom about the history of life (just as astrophysicists had to forget their scout-camp astronomy), to forget the "great mystery" of what happened to the dinosaurs and, in its place, to master a more complex picture of floral and faunal change.

For these scientific experts, the irony of the impact debates has been the inability of expert knowledge to displace a framing of the question firmly grounded in popular knowledge about science. In 1985, for example, a reporter for *The New York Times* conducted an informal survey at the meetings of the Society of Vertebrate Paleontology and found that the core objection to the impact hypothesis was not to the proposed impact itself but to the claim that it explained an event that had not occurred, namely a sudden extinction at the end of the Cretaceous (Browne 1985). Despite repeated hopes that the questions of impact and dinosaurian extinctions might be separated, despite the fact that the thinness of the fossil record for terrestrial vertebrates makes the fate of the dinosaurs a poor test of the impact theories (Upchurch 1989: 210),

> the fact that the iridium spike was found at the K/T boundary . . . thus immediately appropriated the ever-popular question of the demise of the dinosaurs, rather than of some obscure group of Paleozoic plants. The dinosaur connection ensured that the press would join in force, from serious scientific journals down to and perhaps especially the supermarket-check-out-line tabloids: "Missiles from space killed our dinosaurs!" and so on (something of a mixed blessing). Intriguingly, it turned out to be irrelevant at this stage that a sudden extinction of dinosaurs at the end of the Cretaceous had already been all but ruled out by serious paleontologists and was mostly a popular myth. (Thomson 1988: 59)

This passage suggests a relation between popular science knowledge and specialized research very different from that implied either by a trickle-down model of popularization or by the repeated charges that media popularization has corrupted the debate and caused it to become quite cantankerous, if not downright hostile. Following an incident prompted by a misattributed research parody circulated by a public television producer at the 1991 meetings of the Geological Society of America, for example, one "prominent scientist who supports the asteroid-impact theory, but who refused to discuss the matter if identified, said perceptions of rancor surrounding the scientific debate have 'largely been engendered by press coverage'" (Bigelow 1991). Though this attribution of responsibility may preserve an image of the scientific community as a realm of pure reason, it is also fundamentally misleading. Rather than pointing to "outsiders" as the source of conflict and misunderstanding, the tenacity of the "dinosaur connection" is a prime example of "finding items of popular knowledge from other fields within the depths of these sciences"

(Fleck 1979: 112). The universalistic label of "scientist" obscures the mix of expert science, textbook science, and popular science that informs the thinking of any researcher. Consequently, conflict can arise, because the boundary between expertise and other knowledge about science is unclear. After more than a decade of interdisciplinary debate, in which many paleontologists have pointed to the complexity of the fossil record, Walter Alvarez would still title a public lecture "Apocalypse Then: Massive Impact and the Death of the Dinosaurs" (1991a). Thus claims that attention to paleobiological complexity necessitates a reevaluation of the impact hypothesis have been blocked repeatedly by the deeply held assumption that the sudden death of the dinosaurs is the thing to be explained.

Patterns of selective attention to scientific questions have a long history in American culture. Roy Chapman Andrews, the great organizer of the Gobi expeditions of the 1920's, was plagued by the lack of public interest in broad questions of evolution. His claim that the expeditions would help "to reconstruct the past climate and physical conditions of the great plateau" of central Asia "didn't create a ripple in the newspaper world. Primitive man was what they wanted and anything else bored them exceedingly. In a week, we were known as the 'Missing Link Expedition'" (1943: 174). Ultimately, it was the discovery of dinosaur eggs that seized the popular imagination, but Andrews himself pointed to the diverging structures of interest that characterized popular and professional paleontology: "At the Flaming Cliffs we found seven skulls and parts of skeletons of these Mesozoic mammals. . . . After the dinosaur eggs have been forgotten these little skulls will be remembered by scientists as the crowning single discovery of our paleontological research" (1943: 239). But not by all scientists. For researchers in fields other than paleontology, popular science and textbook science remain major influences on their image of the world and of other disciplines. And contemporary coverage of paleontology has done little either to alter the highly selective structures of popular interest or to raise more general theoretical questions about the history of life.

The celebrity status of dinosaurs, as well as the obscurity of the rest of the history of life, can be easily confirmed by examining the coverage of paleontology within the popular press. The abstracts provided in the annual index of *The New York Times*—an appropriate choice not only for the national status of the newspaper but also for the gatekeeping role of its journalists in the wire-service reporting of scientific research (Clemens 1987; Nelkin 1987: 125)—allow a rough reconstruction of the selection criteria for news related to paleontology. From 1978 to 1989 (inclusive), a total of 230 items were abstracted under the heading "Paleontology." Grouped according to categories of popular interest, rather than the taxonomic categories of the profession, 32 related to the impact debate, 27 to prehistoric animals that flew (including archaeopteryx, dinosaur-bird evolutionary theories, and the model pterosaur sponsored by NASA and the Air and Space Museum), and 70 to other dinosaur-related topics. Taken together, these three categories represent over 56 percent of the total. Slightly more than 20 percent of the total items refer to mammals, but, of these, almost one-third concern mammoths and mastodons, another handful report the discovery of fossil whales, and a number address human causes of extinction during the Quaternary. Clearly, size counts, and association with humans also helps in attracting attention. By comparison, a residual category including reptiles other than dinosaurs, marine life, amphibians, invertebrates, insects, plants, and algae—in other words, the great bulk of life on earth—accounts for little more than 12 percent of the total articles covering paleontology over this period.

Taken together, this distribution points to a popular understanding of the history of life in which dinosaurs and ancient flying creatures, mammoths, and man are the only characters of much significance. Only one abstract referred to the evolutionary origin of mammals, a topic that might be assumed to be of some personal interest to the readers of the *Times*. Obviously, the range of paleontological evidence relevant to an assessment of the impact hypothesis, along with the methods of interpretation, has not been widely available to audiences outside this discipline. And, within the discipline, research has also been shaped by the public-education goals of the great museums that have provided the institutional setting for much of American paleontology. At the American Museum in New York during the early twentieth century, "the emphasis on description, exhibition, and evolution meant that some avenues of investigation were not fully pursued. The interest in large, visible specimens yielded hundreds of studies on dinosaurs, elephants, and horses, but there were few similar studies on other organisms," notably on rodents and other small mammals (Rainger 1991: 8). In the case of paleontology, at least, the institutions of public science have often shaped the structures of both popular and professional attention.

The current state of popular knowledge is also characterized by a lack of attention to a related topic: evolution. Although nineteenth-century audiences in Britain were caught up in controversies over Darwin (Desmond 1982; Ellegard 1990 [1958]) and a few early museum directors in the U.S. highlighted evolutionary questions in their public-education campaigns, current levels of popular attention seem quite low. In the period from 1978 through 1989, abstracts of 236 items appeared in *The New York Times* under this heading, but almost 56 percent (132) of these related to controversies over teaching creationism or the content of science textbooks. This focus on the opposition of creationism and evolutionary theory has tended to minimize attention to debates among evolutionary biologists. The coverage of textbook content also suggests that even without the inclusion of creationist accounts, discussion of evolutionary theory has been watered down to avoid possible political controversies over textbook selection. Nor have the problems encountered in teaching evolutionary theory in the schools been remedied by extensive coverage elsewhere in the mass media. Of the abstracts listed in the *Times* index, 47 (almost 20 percent of the total) were related to human evolution, whereas only 57 referred to more general discussions of theories of or evidence for biological evolution. This averages out to fewer than five per year. By comparison, an average of 110 items per year were abstracted under the heading "Space and Upper Atmosphere" ("Space" as of 1985) during the same period (see Endnote 3.2).

Although an examination of science coverage in *The New York Times* suggests patterned differences in the level of coverage between disciplines and even among research topics, the relevance of this distribution of popular knowledge about science to the activities of professional researchers must be clarified. Some connection can be presumed on the basis of a fairly noncontroversial assumption: scientists are people too. More specifically, with respect to issues outside their disciplinary specialty, active researchers in the United States are not all that different from the rest of the highly educated audience for science coverage published in the *Times*. Although individual researchers may well have had some graduate training outside their field, this becomes increasingly out of date with the passage of time, and, as has been suggested by a recent study of genetics textbooks (Gaster 1990), the science taught in universities may well be ten to fifteen years behind the state of the art in any given field. As David Raup observes elsewhere in this volume, paleontologists were slow to consider the impact hypothesis in part because many had been trained in geology at a time when bombardment of the Earth by asteroids and meteors was almost exclusively associated with the first billion years or so of Earth history. Unless audiences become aware that their background, taken-for-granted knowledge about science is in fact being debated within disciplinary specialties, interdisciplinary debates may well proceed on the basis of assumptions that seem unreasonable by the standards of a specific discipline.

An adequate assessment of this argument would require finding similar patterns of coverage in other media, particularly television, major periodicals, and the staff-written columns of leading scientific journals such as *Science* and *Nature*. Nevertheless, suggestive evidence of the role of the distribution of knowledge about science in forging links between disciplines can be found in two precursors of the impact debate. A decade before the publication of Alvarez et al. (1980), the paleontologist Digby McLaren used his presidential address to put forth an argument about the definition of boundaries within geological columns and to address the question of whether such boundaries should be located within continuous stratigraphic sequences or recognize the possibility that stratigraphic breaks might represent real events with widespread, indeed catastrophic, consequences. He concluded by making a case for one such actually catastrophic boundary, the Frasnian-Famennian of 360 million years or so ago, and began by defining the paleobiological contours of that boundary:

> We have to explain a change in life that extinguished reef organisms, including stromatoporoids and corals, but not coralline algae, and which left deeper and colder-water corals unharmed; and that extinguished many genera and families of trilobites and filter-feeding brachiopods adapted to rocky bottoms and an epiplanktonic environment. Nektonic and planktonic organisms appear to be generally unaffected, in terms of sudden extinction, and, as far as I know, there was no sudden change in land plants or animals. (1970: 811)

McLaren moves from a detailed comparative description of extinction and survival to the identification of a proximate cause of extinction: "Fresh water or turbid water are both fatal to the types of organisms that disappeared at the close of the Frasnian" (1970: 811). On the basis of a close connection between biological consequences and the proposed proximate cause, McLaren then identifies a possible primary cause of this extinction:

> I shall, therefore, land a large or very large meteorite in the Paleozoic Pacific at the close of the Frasnian. Presumably on impact with the ocean surface or at a certain depth below the surface, the missile will explode with an enormous release of energy. . . . The turbulence of the tidal wave and accompanying wind, followed by the gigantic runoff from land would induce a turbid environment far longer than could be survived by bottom-dwelling filter feeders and their larvae. The hypothesis of meteoric impact in the ocean explains equally the nonextinction of many forms of both marine and terrestrial life. (1970: 812)

As a paleontologist, McLaren begins by defining the extinction event through an analysis of patterned differences in biological change, and then invokes a meteorite impact as a cause of extinction that discriminates among different classes of organism and across different environments. This attention to the details of the extinction to be explained distinguishes McLaren's account from the framing of the question by Alvarez et al. (1980), as well as from the popular scenario of the abrupt extinction of a limited set of celebrity species. But for the history of inquiry into the history of life, the difference is largely academic. Despite McLaren's position as an officeholder, which, according to analyses of the mass media should have raised the probability that his argument would become "news" (Gitlin 1980: 153–54), this hypothesis drifted into obscurity until it was resurrected in the context of the Alvarez findings (McGhee 1989: 145). In neither 1969 nor 1970 did *The New York Times Index* or the *Readers' Guide to Periodical Literature* indicate that this hypothesis received attention outside the specialist communities of geology and paleontology. Based on the selection criteria implied by the pattern of paleontological reporting in the *Times*, it is not implausible to suggest that part of the explanation for this lack of attention stems both from the framing of the hypothesis by a technical question of boundary definition and from the fact that what died off at the Frasnian-Famennian were small, shelly marine creatures, rather than organisms with a cult status comparable to that of the dinosaurs.

The role of what got killed—as opposed to how—is underscored by another precursor of the Alvarez impact hypothesis. Harold C. Urey, a Nobel laureate in chemistry, published "Cometary Collisions and Geological Periods" (1973) in the journal *Nature*. But despite the higher status of both the author and the journal in the pecking order of the sciences, the proposed link between extraterrestrial events and terrestrial extinctions received little attention and prompted little subsequent research. If McLaren lacked a compelling victim for his scenario, Urey had motive but lacked both the body and a smoking gun.

These examples should not be taken as a call for us to care more for the little creatures of the ancient oceans, or as a qualification of the role of status in science, but rather as an observation that patterns of interest established outside the disciplinary frameworks of the sciences can have profound consequences in forging links among those disciplines. The long-standing popular fascination with dinosaurs provided a context within which both the general public and research scientists from a wide variety of disciplines became aware of a limited set of highly stylized questions concerning the history of extinction.

Another link between the impact debate and popular knowledge about science formed around the widespread concern over climate change—specifically "nuclear winter" and the "greenhouse effect"—that was evident throughout the press during the 1980's. In discussions of the decimation of rainforests and other ecological disasters, concern over catastrophic climate change merged with the possibility of major manmade extinctions: "The actions of an exploding human population are sundering the ecological webs that support life by setting off a worldwide wave of extinctions comparable to the one in which the dinosaurs perished some 65 million years go" (Stevens 1991: B5; see also Raup 1986: 132). By comparison, there appears to have been comparatively less interest in the question of which species survived the K/T event, or in the implications of the impact hypothesis for evolutionary theory (see Endnote 3.3). As some scholars have complained, the total effects of the proposed climate changes have not been fully stated and investigated (Feldmann 1990: 153). If we know how the dinosaurs were killed, surely that is enough?

The scope and speed with which interest in the impact hypothesis spread cannot be understood apart from the fact that many people already believed that the question of how the dinosaurs died was both answerable and worth answering. Furthermore, distinct advantages flow from addressing questions widely perceived as significant. The effort that some researchers put into gaining acceptance for their work underscores the importance of this widespread public acceptance of the significance of a line of inquiry. In the impact debate, this linkage is clearest in the case of astronomers interested in near-Earth asteroids and impacts. As one advocate of this research program recently explained,

> . . . informed skepticism is probably not what keeps most people from being concerned about the danger of an impact. On the contrary, it is ignorance. Asteroid impacts happen so rarely that it is difficult for us to grasp the risk. What's more, our space age image of Earth may be the enemy of true understanding. We have become accustomed to thinking of Earth as a lonely world, suspended in the vast emptiness of interplanetary space, millions of miles from its nearest neighbors. The reality is quite different. . . .
>
> Could we do anything about it? The first thing we could do is identify the asteroids that might threaten us. That is a simple matter of devoting more astronomers and more telescopes to the search—in other words, it is a matter of more money. . . . Shoemaker and other researchers have long had telescopes on their drawing boards that would be more efficient at scanning the sky for asteroids. But those plans have remained in limbo for lack of funding. (Chapman & Morrison 1991: 40, 43; see also Broad 1991)

Thus popular science influences scientific debate at the level of both cognition and material interests. To the extent that the general scientific community "already knows" that a particular question is interesting or significant, interdisciplinary conversation—if not necessarily research—is more likely to develop. To the extent that the general public, and particularly those nonscientific elites who control resources, already knows that a question is reasonable and important, it will be easier to secure funding for that line of research. Consequently, structures of popular interest can create asymmetries even within a single research effort.

For sociologists of science, one lesson to be drawn from the impact debate is that beyond asking how society "corrupts" science (or, alternatively, how social interests shape science), we should attend to the social processes that influence what questions are asked and when which people believe that they have been answered. If the distribution of popular knowledge about science suggests one reason why the impact hypothesis rapidly attracted a wide circle of interested researchers from a variety of disciplines, it may also help to explain the ways in which these researchers framed their questions and attended to one another's findings.

Reception of the Debate among Disciplines

When *Science* published "Extraterrestrial Cause for the Cretaceous-Tertiary Extinction: Experimental Results and Theoretical Interpretation" in June 1980, a number of audiences were prepared to receive this as an important piece of work. The idea that iridium

TABLE 3.1
Annual Citations of Alvarez et al. (1980), by Journal or Discipline

Journal/discipline	80	81	82	83	84	85	86	87	88	89	90	Total
All citations (131)	8	24	37	36	56	61	64	65	54	55	38	498
Science (1)	2	3	3	9	14	2	10	4	3	3	2	55
Nature (1)	3	5	4	4	10	12	5	6	7	5	1	62
General, U.S. (7)	0	0	2	2	1	0	4	0	0	1	2	12
General, foreign (13)	2	3	3	0	3	3	4	3	6	7	2	36
Astronomy/space (8)	0	1	2	2	4	4	3	7	0	3	2	28
Earth & planet (6)	0	2	5	3	3	5	3	4	5	6	5	41
Physics (4)	0	1	1	1	1	2	1	1	0	0	0	8
Geology/earth science (15)	0	2	3	5	6	8	19	11	12	8	5	79
Geophysics (8)	0	0	0	4	1	2	3	4	4	4	3	25
Geochemistry (5)	0	0	1	2	1	2	1	2	2	3	3	17
Chemistry (8)	0	0	0	1	1	1	4	6	0	2	0	15
Paleontology (9)	1	3	1	1	2	5	3	5	8	5	4	38
Biology/Zoology (11)	0	0	4	0	0	6	0	2	0	2	4	18
Ecology/environment (6)	0	0	1	0	2	1	0	2	1	2	1	10
Nuclear science (5)	0	1	1	1	3	0	0	0	2	0	0	8
Medicine (4)	0	0	0	0	1	2	0	0	0	1	1	5
History/sociology (3)	0	0	1	0	1	0	1	0	0	0	0	3
Misc./unknown (17)	0	3	5	1	2	6	3	8	4	3	3	38

anomalies were a marker of both asteroid impacts (as a geochemical signature) and terrestrial extinctions (by their stratigraphic location) appealed to preexisting interests, both popular and professional, while meeting all the aesthetic criteria associated with "good science" (Clemens 1986). In the spirit of their analysis, Alvarez et al. may be used as a marker of participation in—or at least attention to—the impact debate itself (see Endnote 3.4). Drawing on the *Science Citation Index* for 1980 through 1990, Table 3.1 provides a rough measure of the distribution of interest in the debate both over time and across disciplines (see Endnote 3.5). The most striking finding is probably the sheer number of citations—498 over 11 years—followed by the scope of interest in the 1980 article (although I will refer to "citations," the unit for this analysis is actually "articles that cite at least once"). A full 131 journals, representing disciplines from astrophysics to clinical pathology to herpetology, referred to the research by the Alvarez team. Facilitated by the institutions of popularization and general science, particularly science journalism, which has expanded greatly since the 1960's (Nelkin 1987), news of the Alvarez et al. findings was rapidly conveyed to audiences predisposed to find it interesting and significant.

The pattern of citation also reveals how the debate has been distributed across different publications and, by implication, across different audiences and disciplinary groups. As the totals for *Science* and *Nature* alone indicate, the impact debate has been carried out in front of a general scientific audience: 23.5 percent (117) of all references to Alvarez et al. appeared in these two journals, 33.3 percent (166) in the total set of 23 journals classified as general science. As an arena for scientific debate and development, the institutions of general science have thus received far less attention than have the organizations of particular disciplines and more specialized research groups. Although individual research articles may be indistinguishable from those appearing in specialized journals, the relation of a generalist publication to its audience is quite different. To the extent that a debate develops within such arenas, it is particularly likely to encounter clashes of assumptions and taken-for-granted knowledge that remain implicit in more narrowly circumscribed efforts (see Endnote 3.6).

The remainder of the citations have been assigned to three disciplinary categories (with a residual group) that correspond to the three basic elements of the impact scenario: an asteroid, an impact, an extinction. The first group consists of 20 journals of astronomy, astrophysics, physics, and space and planetary sciences. As a set, this group accounted for 17.7 percent (88) of the citations, with *Earth and Planetary Science Letters* accounting for 23 mentions, or 4.6 percent of the total. Over time, the interest suggested by the level of citations conforms to the general patterns of a shallow curve peaking in the mid- to late 1980's. With respect to the likelihood of an asteroid impact, these scholars seem to have reached a consensus that extraterrestrial objects are indeed likely to hit this planet with some frequency, and debate among astronomers and astrophysicists has focused on alternative mechanisms for the claimed periodicity of such impacts (see Heisler 1990: 104). Research published in earth and planetary science journals, by comparison, has focused on continuing debates over the character of the K/T boundary layer. Discussions are framed as specific tests of the impact hypothesis, but there is no longer consensus on the status of the iridium enrichments as unquestionable signatures of impacts, and, consequently, conclusions differ with the type of evidence under consideration: "The South Atlantic K/T orbital chronology requires an environmental and biological transition whose rapidity is clearly compatible with the asteroid-impact theory of Alvarez et al." (Herbert & D'Hondt 1990: 273); "The white interval below the K/T boundary thus requires a brief episode of unusual bottom-water chemistry, which is compatible with impact-triggered mass extinction" (Lowrie et al. 1990: 311); "Although diffusion might account for part of the extent of this Ir enriched interval, we believe that a causal link with other long-lived KTB markers must also be envisaged" (Rocchia et al. 1990: 217). Not the likelihood of impact in general, but the characteristics of the specific alleged impact, are still at issue.

Reflecting the focus of controversy on the boundary layer itself, those journals associated with geology, geochemistry, and geophysics account for more citations than the other two disciplinary categories. These 36 journals contributed 141 citations, or 28.3 percent of the total. *Geology* alone accounts for 38 mentions, or 7.6 percent of the total. As a set, these journals show a similar pattern of citation over time, although the peak in the mid- to late 1980's is far more pronounced. The level of consensus, however, is lower than is the case in the astronomers' camp. Individual scientists continue to promote alternative explanations—notably, the claim that much of the geochemical and paleontological evidence can be attributed to a period of increased volcanism (e.g., Officer 1990)—and both popular and professional articles continue to acknowledge the presence of competing explanations for the presence of shocked quartz and geochemical anomalies (Beatty 1991: 40; Doukhan et al. 1990). A related, and similarly unresolved, debate has developed around the proposal that impacts might be a cause of paleomagnetic reversals (Merrill & McFadden 1990; Schneider & Kent 1990). As one lengthy review article concluded,

> Even after nearly a decade of intense study of all aspects of the K/T boundary phenomena, a clear consensus about the fundamental details of the proposed impact event is lacking. These disagreements reflect the conflicting lines of evidence regarding the composition and number of the impacting bodies, and the general character (marine or continental) of the target area(s). (Cisowski 1990: 111)

In the absence of consensus, positions on debates within the geosciences have influenced the description of the biological events to be explained. Impactors have tended to promote a fairly clean framing of the paleobiological consequences of whatever happened at the K/T. In some cases, biology barely figures in the statement of the question: "An array of stratigraphic, mineralogical, chemical and isotopic evidence has accumulated in support of the theory that the Cretaceous Period was terminated by an impact of a large . . . asteroid or comet" (Hildebrand & Boynton 1990: 843). Other arguments invoke a definition of the consequences of impact congruent with the popular image of sudden, dinosaur-centered, mass extinction: "The event terminating the Cretaceous Period and the Mesozoic era caused massive extinctions of flora and fauna worldwide." Recognizing the links between popular science and professional research, the author acknowledges that "The cause of the mass extinctions at the Cretaceous/Tertiary (K/T) boundary has been a matter of controversy among geologists for a long time, mainly because of the apparent rapidity of the extinctions and because they include the demise of the dinosaurs. . ." (Bohor 1990: 359). Evidence that appears contrary to the frame of catastrophic event and catastrophic consequences is minimized by invoking subsequent processes: "The fine structure of the tails [of the iridium enrichment at Gubbio, Italy] is probably due in part to lateral reworking, diffusion, burrowing, and perhaps Milankovitch cyclicity" (Alvarez, Asaro & Montanari 1990: 1700; for a different argument, see Rocchia et al. 1990). Complex patterns of biological change are attributed to the multiple effects of an impact: impact winter, followed by greenhouse warming, acid rain, and worldwide wildfires (Alvarez & Asaro 1990: 81–82).

Proponents of volcanism, by comparison, have been more likely to emphasize differentiated patterns of extinction at the K/T: "Both the terrestrial and marine extinction records show that we are dealing with a complex sequence of extinctions which extend over a measurable period of geologic time. It is difficult to explain such a record by a single causative event of *either* extraterrestrial or internal origin" (Officer 1990: 405–6). As the major battle line has shifted from impact vs. gradualism to impact vs. volcanism, however, differences in the framing of paleobiological events have lost their determinative influence over conclusions. As one paleontologist acknowledged at the end of a review of selective patterns of extinction, "It should be admitted, however, that the vertebrate fossil record is not of much help in choosing between the various hypotheses outlined above. The scenario of extinction by food chain disruption . . . is compatible with both impact and volcanic hypotheses. . ." (Buffetaut 1990: 344). After a decade of apparently interdisciplinary debate, once again the question of dinosaurian extinction (and of the K/T more generally) appears to be floating free of the biological sciences.

The final set of journals is most closely allied with the biological or extinction component of the impact debate. Even though popular interest in the extinction of the dinosaurs and their contemporaries accounts for much of the high profile of the debate, paleontologists, zoologists, and evolutionary biologists have often been absent from the debate. The 30 journals cited Alvarez et al. 73 times, accounting for only 14.7 percent of the total. This figure is lower than that for the 20 journals in the astronomical group, even though the level of consensus over the impact hypothesis among paleontologists has been lower and the level of controversy correspondingly higher (see Endnote 3.7).

Among those professionally associated with the dependent variable—the paleontologists, paleoecologists, and evolutionary biologists—the verdict appears to be both Scottish and Victorian: not proven, but nice people don't speak about such things. One of the more striking results of the analysis of citation patterns is the low frequency with which the 1980 Alvarez et al. article has appeared in the published record of the various biological disciplines—as distinct from general journals such as *Science* and *Nature*. To the extent that the debate has been addressed, the response has been mixed. On the one hand, there appears to be a general sense that the extent and character of biological extinction had been underappreciated in the recent history of these disciplines, and that the impact debates have forced a needed redirection of paleontological inquiry: "The new findings attracted many sciences to consider the fossil record, a reborn paleontology included. . . . The resurrection of interest in catastrophe opens new vistas for the understanding of global change" (Martin 1990: 187). But the character of the catastrophe at the K/T remains very much a contested issue.

On the basis of the published record alone, those with a professional claim to speak on the biological character of the K/T have not yet reached a consensus, and many appear disinclined to speak to the question at all. There has been only one major instance in which paleontological evidence appears to have caused changes in the impact hypothesis—stepwise extinctions have prompted revisions from single impact to comet showers. Otherwise, the modal reaction to claims of complexity in the fossil record has been to dismiss paleontological evidence as artifactual or statistically insignificant (Alvarez 1983), with the consequence that the initial framing of the extinction as geologically instantaneous and generally catastrophic is defended (see Endnote 3.8). Critics of the impact hypothesis, by comparison, increasingly converge on a framing of the K/T extinctions as a complex and internally differentiated phenomenon: "Extinction of diverse organisms at the end of the Cretaceous has always been puzzling, because the organisms inhabited very different ecological settings and, perhaps even more perplexing, contemporaneous biotic associates in similar ecological settings survived the extinction interval unscathed" (Feldmann 1990: 151); "With regard to the macroinvertebrates the pattern of change for different groups is a complicated one, with both gradual decline and sudden extinction being well documented in some cases" (Hallam & Perch-Nielsen 1990: 347; see also Buffetaut 1990). Field research has uncovered both unexpected patterns of survival at the K/T (such as dinosaurs at high latitudes particularly sensitive to climatic shifts, Brouwers et al. 1987) and very large impacts with few biological consequences (Aubry et al. 1990). Taken together, however, the rhetorical strategies informing much of the biological discussion have the effect—intended or not—of complicating the framing of the biological events at the K/T and thereby downplaying the relevance of paleobiological evidence to the adjudication of the two major causal scenarios: impact and volcanism.

These disciplinary differences in both the focus of attention and the level of consensus highlight a curious disarticulation at the heart of this allegedly interdisciplinary debate. For all three broad groups of participants, the framing of the problem reflects both prior disciplinary allegiances that define what is problematic (e.g., the dynamics of impact, the character of the boundary layer, or patterns of extinction and survival) and the reliance on broader, more popular knowledge about science to fill in the background of these questions. In these juxtapositions of professional and popular knowledge about science, researchers have sought to protect their areas of expertise while simultaneously appealing to more general understandings of other aspects of the science at issue.

The relation of scientific research to popular knowledge about science resembles neither the old model of pristine isolation nor images of intellectual corruption by popularization. If, as Steve Woolgar and Dorothy Pawluch have argued, problems are defined by a process of "ontological gerrymandering" (1985), in which one aspect of the world is problematized, then intellectual trespassing will create a conflict between two definitions of a question: the version of disciplinary specialists and the version provided by the taken-for-granted image of the world derived from general knowledge. This trespassing also involves asymmetries produced by the conventions of scientific training.

As Walter Alvarez explained in a manifesto for intellectual trespassing as a "gentle art": "As students, we are all trained in the primary or basic sciences—mathematics, physics, and chemistry. However, the secondary sciences—geology, paleontology, biology—are studied almost exclusively by practitioners of those sciences. . . . This puts a one-way valve in the communications system, and . . . good communications are the prime consideration and the prime difficulty in doing good interdisciplinary science" (1991b: 30). When combined with the status order of the disciplines and the comparatively low level of attention to their concerns in more general media coverage, however, the result is often that practitioners of the primary sciences dismiss the specialized training of others.

Acting on the defensive, many paleontologists have focused on questions of dinosaur behavior, metabolism, and paleoecology that developed during the 1970's (Benton 1990)—questions that drew the field closer to evolutionary biology and away from its more traditional alliance with geology. Given this fine-grained ecological approach, the most "interesting" questions are not those of mass extinction but those bearing on differential survival and evolution. For astrophysicists, geologists, and geochemists, however, the link with a compelling question such as the death of dinosaurs brought the promise of a new source of publicity, celebrity, and, perhaps, even greater funding (e.g. Chapman & Morrison 1991).

But the stakes in the impact debate are cognitive as well as material and political. Much of the variation seen in the enthusiasm for the impact hypothesis may be attributed to the basic scenario itself: a simple, elegant model of causation meets complex, incomplete evidence of consequences. Much of the heat generated by this debate may be attributed to a clash of scientific styles (Clemens 1986: 438–42), styles that inform not only research practices but public expectations of science and the ideological strategies pursued by scientists in promoting their work (Hilgartner 1990). In a now notorious interview with *The New York Times*, Luis Alvarez explained: "I don't like to say bad things about paleontologists, but they're really not very good scientists. They're more like stamp collectors" (Browne 1988: 19). Robert Jastrow, a physicist/astronomer/geologist, responded by accusing Alvarez of "pulling rank on the paleontologists" by claiming status for a particular model of scientific inquiry (1983: 152). Stephen Jay Gould pointed to this same "stereotype of scientific method" in explaining the lack of attention paid to the stunning record of the Cambrian "explosion of life" in the Burgess Shale:

> The impact theory has everything for public acclaim—white coats, numbers, Nobel renown, and location at the top of the ladder of status. The Burgess redescriptions, on the other hand, struck many observers as one funny thing after another—just descriptions of some previously unappreciated, odd animals from early in life's history. . . .
>
> The common epithet linking historical explanation with stamp collecting represents the classic arrogance of a field that does not understand the historians' attention to comparison among detailed particulars, all different. This taxonomic activity is not equivalent to licking hinges and placing bits of colored paper in preassigned places in a book. The historical scientist focuses on detailed particulars—one funny thing after another—because their coordination and comparison permits us, by consilience of induction, to explain the past with as much confidence (if the evidence is good) as Luie Alvarez could ever muster for his asteroid by chemical measurement. (1989b: 281)

This clash between the methods of natural history and the stereotypical model of science epitomized by the laboratory experiment has drawn attention away from problems of reconciling different kinds of evidence with the model of extinction by impact. Any resolution of the current disarticulated state of the debate (see Endnote 3.9) must await the collection of systematic and comparable data for assessing both the impact component of the Alvarez hypothesis and its extinction component. But the existence of an interdisciplinary debate, despite fundamental professional disagreements over what is to be explained, documents how taken-for-granted questions outside of disciplinary expertise can shape the dynamics of basic research.

In the meantime, however, the disarticulated character of the debate has had consequences in its own right. Combined with differing levels of participation and varying channels between the involved disciplines and broader audiences both within and without the scientific community, the disarticulation of the debate among research scientists has led to a cacophonous chorus of consensus—a chorus in which the character of the consensus still depends on what and whom one reads. And the available reading material is determined, in turn, by the conventions and selective attention of the institutions of scientific communication, both professional and popular.

Multiple Audiences and the Reception of the Impact Debate

So how did the dinosaurs die? The impact hypothesis provided a clean answer to this time-honored question, a satisfyingly concrete image of a large rock crushing the ancient reptiles and, perhaps, a few of the pesky mammals eating their eggs. But how is such a local dying to be linked to mass extinction, encompassing different regions and many species? The initial Alvarez et al. piece linked impact to extinction by way of catastrophic climatic change: "Impact of a large earth-crossing asteroid would inject about 60 times the object's mass into the atmosphere as pulverized rock; a fraction of this dust would stay in the stratosphere for several years and be distributed worldwide. The resulting darkness would suppress photosynthesis, and the expected biological consequences match quite closely the extinction observed in the paleontological record" (1980: 1095). This causal mechanism received further support, both direct and metaphorical, from the "nuclear winter" simulations published by Turco et al. (1983) and the extrapolations concerning the biological effects of nuclear war (Ehrlich et al. 1983). By suggesting the generalizability of climate-mediated extinction—and underscoring the identification of the fate of mankind with that of the giant reptiles who once ruled the earth—these findings heightened the salience of the Alvarez hypothesis for both scientific and general audiences.

At a most basic level, this resonance between popular concerns and scientific research has been constitutive of the impact debate itself. Over the past decade, the unusual role of the science press has been frequently noted, sometimes as a matter of interest, sometimes as a threat of major proportions. Midway through the 1980's, one survey found that almost 30 percent of a sample of paleontologists in five countries first read of the Alvarez hypothesis in "scientific commentaries" rather than in journal articles. The level of this commentary reflected, in turn, the fact that the hypothesis appeared as the lead article in *Science*, a key gatekeeping journal in the system of scientific communication. As the authors concluded, this cross-national comparison "clearly indicates how important is the role of *Science* in placing a scientific problem in the focus of interest of the North American scientific community, though considerably less so in Europe" (Hoffman & Nitecki 1985: 886).

This selective attention was evident in the popular media as well as in the more in-house organs of the scientific community. Again, the *Readers' Guide to Periodical Literature* provides an estimate of this interest (see Table 3.2). For a set of thirteen periodicals over a twelve-year period, 143 articles either mentioned the impact debate in the title or could be linked to the debate by the identity of the subject of an article or its author. The fact that the impact debate was featured ten times, including at least one cover story, in the two major news-

TABLE 3.2
Discussions of the Impact Debate in the Popular Press, by Periodical

Periodical	78	79	80	81	82	83	84	85	86	87	88	89	Total
Astronomy	0	0	2	3	3	0	1	1	1	0	1	1	13
Discover	0	0	0	0	0	0	2	2	1	3	1	0	9
Earth Sciences	0	0	0	0	1	1	1	0	1	2	0	0	6
Natural History	0	0	1	0	1	0	5	1	0	0	0	0	8
New Yorker	0	0	1	0	0	0	0	0	0	1	0	0	2
Newsweek	0	0	0	0	0	0	1	1	0	0	2	0	4
Omni	0	0	0	0	0	0	1	0	2	0	0	1	4
Science Digest	1	1	0	3	2	2	1	3	2	0	0	0	15
Science News	1	2	3	2	2	3	7	12	8	7	5	1	53
Scientific American	0	0	0	0	3	2	3	2	0	2	1	0	13
Sky & Telescope	0	0	1	0	2	0	1	0	0	1	1	0	6
Technology Review	0	0	0	1	1	0	0	2	0	0	0	0	4
Time	0	1	0	0	0	1	0	3	1	0	0	0	6
TOTAL	2	4	8	9	15	9	23	27	16	16	11	3	143

NOTE: Articles were selected on the basis of a reference to the impact debate in the title or of the identification of subject/author as a participant in the debate.

weeklies—*Time* and *Newsweek*—where competition for coverage and constraints of space are particularly severe, underscores the widespread nature of interest in the alleged catastrophic extinction of prehistoric life. This high level of interest, in turn, has had consequences for the development of knowledge about science on a number of levels: in the unfolding of the impact debate itself, in the reshaping of a number of the disciplines involved in the debate, and in the development of popular knowledge about science. To the extent that various audiences needed to be primed to regard Alvarez et al. as an important scientific contribution, the popular scientific press and general news media had already begun this job. Even before the publication of Alvarez et al. (1980) in *Science*, *Science Digest* had published two articles on preliminary research reports, *Science News* had published three, and *Time* magazine—not usually considered a central pillar of the scientific community—had published one.

As the debate has unfolded, coverage in the scientific and popular press has also influenced the timing and heatedness of its development. The most striking instance of such mass-media-mediated scientific communication involved the dissemination of Raup and Sepkoski's statistical study that reported periodic extinctions in the marine fossil record to astrophysicists who then proposed a series of extraterrestrial mechanisms to account for this periodicity. As Dave Raup has observed:

Normally, such challenges go unheeded, I suppose because communication between disciplines is not very good. Each field is a highly complex culture pursuing interesting problems of its own. Furthermore, none of us has the breadth of training necessary to appreciate questions posed in most other fields—especially if we are talking about fields as distant and different as paleontology and astrophysics. (1986: 131)

In this case, however, structures of interest in science that were external to any specific discipline promoted discourse among disciplines:

Within a few weeks of Jack's [Sepkoski's] Flagstaff presentation, three good treatments of the extinction research appeared in the scientific and popular press. Roger Lewin of *Science* wrote it up as part of a general report on the Flagstaff meeting. Cheryl Simon of *Science News* followed with an excellent account based on Lewin's piece and some interviews. And George Alexander of the *Los Angeles Times* wrote a clear and authoritative article. . . . Astrophysicists were so intrigued that a number of them started serious work on the problem. This ultimately produced Nemesis. (Raup 1986: 132)

This episode also produced a wave of controversy over the proprieties of scientific communication. The visibility of popular accounts notwithstanding, many took the rapidity of the astrophysicists' response as a sign of some conspiracy among individual researchers to circumvent the legitimate gatekeepers of scientific publication. The editor of *Nature,* John Maddox, criticized the "widespread circulation of preprints. . . . The most obvious complaint against the system is that it is discriminatory, excluding from the circle of those in the know people who happen not to be on the authors' mailing list" (1984: 685). But whether or not such a network of individuals was at work in this instance, the organizational and habitual selection criteria of the mass media are sufficient to explain the rapid diffusion of *certain* kinds of findings and hypotheses.

But if the embeddedness of this debate in broader structures of popular interest has sometimes facilitated the development of research, it has also tended to oversimplify arguments and, at times, to foreclose further discussion. In part, this is a consequence of the simplification that occurs as research reports are translated for ever broader audiences (Fleck 1979: 112; Nelkin 1987: 120). Yet this simplification interacts with the disarticulated character of the impact debate to produce the cacophony of consensus described above. As paleontologists continue to treat the K/T extinction as a complex phenomenon, more popular pieces in periodicals oriented toward astronomers present the biological nature of the K/T extinction as settled. As early as 1979, *Astronomy* explained that "Something mysterious happened about 65 million years ago. Geologically, that time is known as a transition period between the Cretaceous and Tertiary periods, and it is marked by the relatively sudden disappearance of half the different kinds of life on Earth. The most famous to vanish were the dinosaurs, giant reptiles that had dominated the Earth for millions of years" ("Death of the Dinosaurs" 1979: 64). A decade later, this biologically clear-cut framing of the extinction question had not changed. In a 1991 cover story with the declarative title "Cataclysmic Collection: The Event That Changed Life on Earth," *Sky and Telescope* described a possible crater in the Yucatan that might be "the Holy Grail of impact geology—the long-sought site of the blast that triggered the eradication of most life on Earth 65 million years ago" (Beatty 1991: 38). In sharp contrast to McLaren's (1970) framing of the question, in which patterns of survival were a critical tool in assessing the nature of the catastrophic event to be explained, over the course of the 1980's, evidence of biological complexity has had little influence on the development of the impact debate.

Not surprisingly, the level of consensus among astronomers and astrophysicists has led to a parallel discrediting of their evidence by critics of the impact hypothesis. J. Robbins quotes the paleontologist Robert Bakker describing his work on the Jurassic-Cretaceous boundary:

> "The asteroid partisans have this great theory that explains why everything on Earth should die at the K-T," he says. "But everything on Earth doesn't die. At the J-K, the K-T, and the other extinctions the pattern appears to be the same: large and medium-size animals are eliminated, and small animals—the overwhelming majority—are free to range out and evolve. . . ."
>
> "People are so wedded to the idea of an asteroid or a climatic cataclysm," Bakker says, "they refuse to entertain anything else. It's tough to change minds when beliefs have become more a religion than a scientific hypothesis." (Robbins 1991: 59)

Such exchanges prompt two observations. First, contrary to commentators who have stressed the opposition of uniformitarian and catastrophic frameworks in explaining the poor reception of the impact hypothesis within much of the paleontological community (e.g. Feldmann 1990), the tension between complexity and simplicity seems equally salient. Although impactors within the paleontological community may argue that apparent complexity is merely artifactual, for many of the other disciplines involved in the debate, biological complexity and evolutionary processes have not really been issues. Nor should this be particularly surprising. Given the low level of popular coverage of research related to evolution and the exceedingly skewed character of reporting on paleontology, the gap between the questions of disciplinary specialists and those taken-for-granted mysteries conveyed as part of popular knowledge about science was particularly large with respect to the fossil record.

Second, the enduring differences of opinion among sets of disciplinary specialists suggest the limited potential of broader structures of interest to transform the practices and organization of scientific research. The durability of discipline-centered models of scientific activity stems from their recognition that disciplines do serve as the primary—albeit not the only, and certainly not an insulated—matrix for research. To the extent that disciplines constitute a variety of partisanship, any interdisciplinary debate is likely to degenerate into name-calling:

> Debates that cease to break new ground begin to spin their wheels as the old ground is simply fought-over again and again. For the study of mass extinctions, the heady days of major discoveries that dramatically revolutionize the field may be past (unless Nemesis is found), and what is needed now is a lot of hard work, gathering data under the auspices of a coherent agenda. If such an agenda is not forthcoming, then the study of mass extinctions may ultimately disintegrate into a bunch of provincial squabbles. (Miller 1990: 383–84)

Whether or not this will occur depends in large part on the successful institutionalization of an interdisciplinary research program. The high profile of the debate, along with the considerable resources of certain groups involved in this interdisciplinary research, argues in favor of a successful outcome. Other developments, however, such as the increasing migration of paleontology from departments of geology toward biology, suggest that even greater gaps among the constituent elements of the impact debate may develop in the future.

So what might be the consequences of this partial, often muddled and disarticulated, diffusion of knowledge about the history of life on Earth and the forces that shape it? The development of the impact debates has been shaped by the distribution of knowledge about science across a variety of disciplinary specialties and more general publics. By resonating with preexisting interests and an established aesthetic of scientific explanation, the impact debate has attained and maintained a strikingly high profile in both the scientific community and the public imagination, throughout the 1980's. But this high level of visibility has the potential to change the conditions that made the impact debate such a "charismatic" issue in the first place.

What impact have the impact debates had in this respect? To address this question, I spent a Sunday afternoon looking at 21 books related to dinosaurs in the highly regarded children's section of a popular bookstore in an academic neighborhood in Chicago (see Endnote 3.10). Of these, many told stories where the characters could have easily been kittens or rabbits or lions: "While vacationing in Africa, the Lazardo family finds and brings back to America a friendly dinosaur that becomes the talk of the town." The book jacket informs the reader that "Dinosaur Bob blows a mean trumpet, dances, and plays baseball. But most important, Dinosaur Bob is a pal" (Joyce 1988; see also Otto 1991). Other stories contained little information apart from the fetishism of species names characteristic of many five-year-olds these days: "No ordinary pet will do for Alex. But when Alex brings Fred, a pet Massapondylus, home from the Dino-Store, he gets a whole lot more than he bargained for" (Oram & Kitamura, 1990). Another book promises "Seventeen whimsical poems featuring allosaurus, stegosaurus, tyrannosaurus, and other dancing dinosaurs" (Yolen 1990).

Almost half the books did, however, present factual information concerning dinosaurs. But of these, few made a strong claim about the fate of the beasts. Often, the question was simply never raised (Barton 1989; Durrell 1989; Most 1987; Schlein 1991; Watson 1990). In some cases, extinction was mentioned, but not explained: "Dinosaurs lived almost everywhere on Earth. They lived for millions of years. Then they died out. No one is sure why they became extinct. But they did. There hasn't been a dinosaur around for 65 million years" (Brandenburg 1988: 10; see also Brandenburg 1985, 4; Lauber & Henderson 1991). Others trotted out the old chestnuts of explanations:

> Brachiosaurus was truly immense
> its vacuous mind was uncluttered by sense,
> it hadn't the need to be clever and wise,
> no beast dared bother a being its size.
>
> Stegosaurus blundered calmly through the prehistoric scene,
> never causing any other creature woe,
> its brain was somewhat smaller than the average nectarine—
> Stegosaurus vanished many years ago. (Prelutsky 1988: 7, 11)

A number of books took the option of listing a series of possible causes:

There were dinosaurs living on earth for millions and millions of years. Then something happened. All the dinosaurs died out. In a very short time—maybe only months—they were all gone.

Nobody knows what killed the dinosaurs. Some scientists think a huge fiery meteor crashed into the earth, filling the air with black dust. Then the dinosaurs couldn't breathe. [Note that the specific causal mechanism of extinction has been altered.] Perhaps a fiery volcano covered the earth with thick ash, killing the plants. Then the dinosaurs couldn't eat. Was there a change in the weather so that it became too cold for the dinosaurs? Did a new kind of animal appear that ate up all the dinosaur eggs? Maybe the dinosaurs began to eat a new kind of plant and it was poisonous. Maybe you'll become a paleontologist. Maybe you will find the answer to this great dinosaur mystery! (Ingoglia 1991: 11; see also Benton 1987: 72–73)

Clearly, the children of the 1980's have been receiving a mixed message from their storybooks, although coverage in the mass media and television appears to have given the edge to the impact hypothesis in terms of taken-for-granted popular knowledge about science.

To the extent that the impressive mastery of species names required by kindergarten culture translates into an appreciation of biological complexity, the simple framing of the extinction question may be undermined. But, at the same time, the astounding popularization of dinosaurs may lead them to lose their status as a favorite mystery of the unusual bookish child. Familiarity may breed contempt, after all, as two cartoon dinosaurs discussed in an installment of "Zippy":

A: "Th' dinosaur thing's dying down. . . ."
B: "Nah . . . It's a perennial!! . . . Like bugs . . . or Fudd. ."
A: "We were big in Books, Toys and Hollywood for a while!"
B: "Hey, don't forget the deal at ABC!"
A: "I'm Worried! . . Look what happened to Dick Tracy . . & I hear th' Simpsons heat is cooling. . ."
B: "Maybe you're right. . . Hey, isn't that a tar pit up ahead?" (Griffith 1991)

Knowledge about Science and Scientific Knowledge

Our usual image of the sciences is of a congeries of institutionally separate disciplines, each governed by a particular set of practices, professional norms, and cognitive orientations. But, as the impact debates graphically demonstrate, popular culture can serve as a matrix which fosters connections among disciplines that otherwise protect their institutional and intellectual autonomy. The broad distribution of popular knowledge about science, in turn, is the joint product of the efforts of professional researchers and developments within a variety of "culture industries." As the contours of that popular knowledge shift—with fashion, with politics, or whatever—the possibilities for new and potentially transformative connections among disciplines are also redefined.

Thus the seemingly irreverent claim that the impact debates have been shaped by children's culture and the conventions of the popular press is intended, not to denigrate the quality of the science in question, but to argue against an uncritical delimitation of the object of analysis. Simply because scientists sometimes claim—for reasons political or otherwise (Hilgartner 1990)—to leave their extra-professional lives at the laboratory door, analysts of scientific debate should not ignore the highly improbable psychology implied by such a claim. Rather than assuming a sharp line between professional scientific expertise and popular knowledge about science, we should remember that such distinctions can only be maintained by intensive "boundary-work" (Gieryn 1983) of a cognitive sort that is far from always successful.

4

Impacts and Extinctions: Science or Dogma?

Digby J. McLaren

The forces ranged against empiricism in the earth sciences are formidable. We tend to doubt the evidence of our senses unless it agrees with our ideas of how things work. Once we can fit the evidence into our model we can then look for additional observations that support it, and we always find them, without difficulty. The general validity of induction in our empirical science continues to be questioned, largely by relying on principles derived from valid deduction. The two systems, however, cannot logically be linked in this manner, and, in spite of Bertrand Russell, the sun will rise tomorrow.

Hypotheses are central to the advance of science. They are the stepping-stones that lead toward truth, which may never be reached in its entirety. But they are hesitant steps that sometimes advance toward their goal, whereas others may be a step sideways, i.e. toward irrelevance, or backwards, i.e. wrong. We all know this in theory, and we also know that many of those who have made major advances in our own field of inquiry have done so by proposing, discussing, and happily discarding hypotheses when they don't work. Hypotheses often lead to the building of models that may be valuable as a basis for calculations, predictions, or further investigation. But hypotheses or their models are also dangerous, because it is abundantly clear that we have a tendency to fall in love with them, and, like all true lovers, we become blind to their faults and blemishes.

In the earth sciences there are few known or assumed principles that allow us the luxury of deduction—or reasoning from generalities to particulars. The reverse is true. We work in a science that has developed by observation and, to a lesser degree, experiment. In this paper I shall refer to several major empirical general theories that have been proposed in the past in the earth sciences, all of which were rejected on the grounds that they described effects without causes and were at variance with the conventional wisdom of the time. Many of the models used to attack the empiricists were so strongly entrenched that they had, in fact, already been assumed as principles, and thus formed the unquestioned base for deductive conclusions. These range in time from Hutton's classic *Theory of the Earth* to the subject of this symposium, impacts and extinctions, about which controversy still rages, and they furnish good examples of the strength of models in opposition to empiricism. If there is a single coherent lesson to be learned from this discussion, it is that if a deduction is questioned by conclusions from empirical evidence, then the presuppositions, or the model itself, may be flawed.

James Hutton: Theory of the Earth, 1788

Hutton's amazing synthesis of field observations gave the first coherent theory of how geology works, and it is still correct in principle. He wrote of "a cyclic progression of changes so ancient as to obscure any vestige of a beginning and no prospect of an end." Time was so incomprehensibly vast as to defy calculation, and its exact duration, therefore, was of little consequence. The model used to attack his theory was nothing less than the book of Genesis. Richard Kirwan stated: "How fatal the suspicion of the high antiquity of the globe has been to the credit of Mosaic history, and consequently to religion and morality" (Gillespie 1959). Hutton also offended the Neptunist school of geology, led by Werner of Freiberg, which denied the igneous nature of basalts, rocks he claimed were precipitated from water. He was able to show from field evidence within his own home city of Edinburgh that basalt had intruded sedimentary rocks in a liquid state as a result of heat.

Religion, Neptunism, and the age of the Earth continued to be linked, and indeed they blocked acceptance of Hutton's ideas for some time to come. Empiricists, such as William Smith, ignored theory and went on to develop the techniques of stratigraphic correlation and geologic mapping based on observation. Smith did not speculate on why fossils could be used for correlation, except to say that "the earth is formed as well as governed . . . according to regular and immutable laws, which are discoverable by human industry and observation." Although he ascribed such regularity to a Creator, there was no question but that we must determine what these laws are by our own efforts, and no conflict is involved (Smith 1815).

Louis Jean Agassiz: The Glacial Period, 1840

The History of the Geological Society of London (Woodward 1908) contains a good account of the reception of Agassiz's paper "On glaciers, and the evidence of their having once existed in Scotland, Ireland and England," which was read in November 1840. This was followed by a paper by William Buckland, who acknowledged that he had found compelling evidence in England and Scotland that matched phenomena he had observed in Switzerland, and therefore withdrew his previous opposition to the theory. Roderick I. Murchison, however, argued against the theory, essentially because the mountains of Scotland were considerably lower than those of Switzerland, and that therefore there was not sufficient cause for the glacier buildup cited by Agassiz and Buckland. He would stay with "our old ideas" and would examine the subject under the old name of Diluvium.

Thus Agassiz's original theory was attacked by the familiar argument that there was no obvious cause. Charles Lyell fought back: "If we do not allow the action of glaciers, how shall we account for these appearances?" William Whewell countered with: "Now it is not within our reach at present to refer each set of phenomena in geology to its adequate cause, but that is no reason why we should receive any theory that is offered to account for it." The physical evidence, combined with Agassiz's brilliant and clear descriptions of his theory in all its complexity, quickly convinced the geological fraternity, except for the few who preferred their own explanations for Diluvium and Drift, or, as it used to be called, "Extraneous Rubbish." We should note, however, that the theory, in its acceptance, was not considered any less likely to be right, in spite of the lack of a convincing cause.

William Thomson (Lord Kelvin): The Age of the Earth, 1854

Following Helmholtz's work on the meteoric hypothesis for generating the heat of the sun, Thomson, in 1854, suggested 20 million years for the age of the Solar System, which he based upon the rate of cooling of the Sun and the Earth. For the next 40 years his estimates varied, and were commonly less than 100 million years. His last suggestion, in 1881, was that the limits were between 20 and 50 million years (Thomson 1881). Any uneasiness on the part of geologists that the time available for all of geological history was surprisingly short was quickly overcome, as Thomson's prestige was huge, and most scrambled ingloriously to fit their interpretations of the past to these figures. The theory was purely deductive, and opposition was, essentially, silenced before it began. To the end of his life, William Thomson, finally Lord Kelvin, continued his opposition to efforts to extend geological time, but he lived to hear Rutherford explain the significance of radioactivity.

The story of Thomson and the age of the Earth is a remarkable example of the power of theory, or model-building, overcoming for half a century an already huge body of empirical evidence. It demonstrates that there is an innate prejudice in favor of deduction from existing models, or even axioms, over induction from observed facts. There was, however, one opponent to Thomson who was willing to take him on, on the grounds of geological evidence—Charles Darwin.

Charles Darwin: The Origin of Species, 1859

Two issues of relevance to our argument concerning induction vs. deduction arise from Darwin's classic work. The first concerns the age of the Earth, and the second, the theory of evolution.

In Chapter IX of *On the Origin*, Darwin discusses rates of sedimentation, rates of denudation, and imperfections in the geological record. Together with Hutton and Lyell he is impressed by "how incomprehensibly vast have been the past periods of time." Within his discussion, he proposes a simple thought experiment for the denudation of the anticline of the Weald in Kent, England. The strata involved range in age from low in the Cretaceous to mid-Tertiary. Darwin's estimate of the time taken for erosion to remove the 300 or more meters of sediments from the anticline was about 300 million years, with an uncertainty of about 200 million each way. He arrived at this figure by estimating the speed of coastal erosion and the activity of the rivers of the Weald. The object of the "experiment" is not so much a serious attempt to make a reasonably close estimate, but rather to suggest an order of magnitude for a known geological event that plainly represents a very short time interval relevant to the total geological time scale. "I have made these few remarks because it is highly important for us to gain some notion, however imperfect, of the lapse of years." Fully aware of Thomson's views on the age of the Earth, he was attempting to quantify the evidence for the slowness of geological processes.

Presumably because of Thomson's unquestioned authority, opponents of Darwin's speculation made little attempt to demonstrate that erosion must take place at a pace many orders faster than he had suggested, if all geological history was to be fitted into Thomson's time scale. Darwin's attempt to justify the vastness of geological time was universally opposed by the physicists, who saw it as another effort to support uniformitarianism. Darwin advised his fellow naturalists to be slow in admitting the conclusions of astronomers, and to be cautious in trusting mathematicians. He observed that workers trained exclusively in mathematics and physics seemed peculiarly incapable of understanding him (Hull 1973). Nevertheless, he was so ridiculed for his Wealden "experiment" that it was removed from the third and subsequent editions of *On the Origin.* Another victory for deduction.

Erasmus Darwin, Charles's grandfather, was an enthusiastic evolutionist, and seems to have had views on evolution similar to those subsequently elaborated by Lamarck. He was not alone in suggesting that all life must be considered to be interrelated. Evolution was in the air, but it is not possible or necessary to summarize the events leading up to the publication of *On the Origin* in 1859, prefaced by the two separate papers read and published the year before by Darwin and Wallace. The brilliance of this work, and the manner in which the data are presented, make it a classic example of a theory arrived at or induced from interpretation of empirical evidence. The way that Darwin considered himself forced to the theory of natural selection almost reluctantly is a recurring theme in his writings during the long gestation period of *On the Origin,* which he called an "Abstract" of his theory. He expected opposition from nonscientists, especially theologians, "but he had not anticipated the vehemence with which even the most respected scientists and philosophers of his day would denounce his efforts as being not properly 'scientific'" (Hull 1973).

His principal sin, implicit in his empiricism, was to call into question the ruling orthodoxy in both science and theology. The outrage was compounded when his readers found that they were quite capable of following his argument, but saw that it threatened their own beliefs. Darwin's friend and mentor, the well-known geologist Adam Sedgwick, fully appreciated the importance of his work, but believed it repudiated all reasoning from final causes. "I have read your book with more pain than pleasure. Parts of it I admired greatly, parts I laughed at till my sides were almost sore; other parts I read with absolute sorrow, because I think them utterly false and grievously mischievous." Darwin replied: "You could not have paid me a more honourable compliment than in expressing freely your strong disapprobation of my book. I fully expected it." And: "I cannot think a false theory would explain so many classes of facts, as the theory seems to me to do." It was not further facts that were brought forward by his opponents to confound him, but cherished beliefs, models, and deductions from many disciplines.

Alfred Wegener: Continental Drift, 1912

Although obvious in principle, the idea of continental drift offended many cherished models, and was one of the most staunchly opposed theories of any. Several editions of Wegener's book were published between 1912 and 1924, the last being an English translation, under the title *The Origin of Continents and Oceans*. Wegener was not the first to suggest that the continents have changed their relative positions over time, but he builds a convincing case for such movement. He was, however, in trouble from the start, because he suggested a rate for the spreading of the North Atlantic that proved to be geodetically testable and was shown to be wrong, and the theory could therefore be set aside. This is a good example of the "baby with the bath-water syndrome." Theorists also condemned it on the grounds that it was physically impossible, ignoring further empirical evidence for movement. By the 1920's, many geologists in the Southern Hemisphere had realized that it was impossible to describe the geological history of the southern continents without invoking some adjustment to their relative positions. Du Toit's classic publications (1927, 1937), as well as many studies of the Carboniferous glaciations in the Southern Hemisphere, required a Gondwana continent.

As a student in England before the 1939–1945 war, I was aware of the evidence and arguments for and against continental movement. Furthermore, in 1944 I read the first edition of Arthur Holmes's *Principles*, and after the war heard Harold Jeffreys and Edward Bullard debate the issue. It was clear that continental drift was to be taken seriously; there were powerful protagonists on either side of the debate. But it was not until I emigrated to the New World in 1948 that I discovered, not just that no one was prepared to discuss continental drift, but that no one knew anything about the arguments that made it an obviously important hypothesis. At graduate school I achieved a certain notoriety as an unrepentant heretic. This episode in the history of geology, in which purely authoritarian edicts were based on existing models, in effect denied students the right to judge an interesting and exciting idea for themselves.

There is irony in the fact that it was an English scientist, Harold Jeffreys, a distinguished geophysicist, who was one of the leaders of the proscription against continental drift, and this continued well past the time of the plate-tectonics revolution. Jeffreys, in a manner similar to that of William Thomson and the age of the Earth, declared that continental drift was impossible, and that was an end to it. Geophysically, apparently, the argument was unanswerable, and therefore had to be accepted, in spite of the volume of data and interpretation—largely from geological and paleontological field evidence—that existed in the world literature, and that demanded, at least, discussion of the evidence.

The similarity between the use of authority by Thomson and that by Jeffreys to impede progress in the empirical science of geology is further brought out in an interesting phenomenon that is found in Burchfield's book on Lord Kelvin (William Thomson) (1975), in which Thomson is shown to be right and Darwin wrong, because Darwin's model did not fit the conventional wisdom at the time. Similarly, the recent obituary on Harold Jeffreys published by the Royal Society of London makes the point that his opposition to the "old form" of continental drift is valid, but that this does not apply to plate tectonics (Cook 1990). Yet the "wrong" evidence still stands, and continues to play an important role in past continental reconstructions. This is yet another example of the denial of evidence for an event or events on the grounds that the cause is not known.

Ironically, in 1963, Lawrence Morley, and then independently and almost simultaneously Fred Vine and Drummond Matthews, formulated virtually identical hypotheses to explain the provocative, zebra-striped magnetic pattern of the eastern Pacific seafloor mapped by Arthur Raff and Ronald Mason. Their hypotheses—which melded the ideas of magnetic-field reversals and seafloor-spreading theory—

were ridiculed until they were confirmed overnight in 1966 by the Eltanin 19 magnetic-anomaly profile; that profile demonstrated bilaterally symmetrical magnetic blocks whose widths correlated with the time intervals of the geomagnetic polarity-reversal time scale that had been recently formulated by Cox, Doell, Dalrymple, and McDougall. That proof of the Morley and Vine-Matthews hypotheses confirmed seafloor spreading and continental drift, instantly focused attention on Tuzo Wilson's transform-fault paper that had contained the germ seed of plate-tectonics theory, and thus triggered the modern revolution in earth science (Glen 1982).

Impacts and Extinctions

We continue the discussion on the contrast between empiricism and deduction with the controversy over possible causes of mass extinctions, and whether some of them were triggered by the advent of a bolide. One would have expected the protagonists in the debate to inform themselves, as far as possible, on the nature of the evidence and/or opinion being discussed within the several disciplines involved. This, indeed, has happened, in many cases, but, astonishingly perhaps, there are those who argue only from the limited view of their own discipline, and appear to be unaware of the nature of the evidence. The evidence for impacts that must be considered involves:

1. The demonstration of asteroid or cometary bodies in Earth-crossing orbits (Apollo objects);
2. The ubiquity of impact scars on all observed solid bodies in the Solar System, including Earth, scars that are inversely correlated in frequency with tectonic and/or volcanic activity;
3. The calculation, based on three independent methods, of the rate at which some of these objects will strike this planet, and the relation of that frequency to their size;
4. The calculation of the energy released by a body's entering the atmosphere and striking the land or ocean, which is dependent on the velocity and size of the body;
5. Both the instantaneous and the longer-term effects such an impact will have on the atmosphere, hydrosphere, lithosphere, and biosphere;
6. The physical and chemical traces left by such events, both at the point of impact and globally, in the geological record;
7. And, perhaps the most difficult problem of all, the biological

effects that might be expected as the result of an impact and the consequent physical and chemical changes to Earth and its atmosphere.

This is not the place to discuss all the possible mechanisms that might cause sudden extinctions, but one important factor in the bolide-impact hypothesis should be emphasized. The physical changes brought about by a large impact will be of such a magnitude that the equilibrium of all systems on the planetary surface will be disrupted. The rates at which the equilibria are restored will be highly variable, and will determine what organisms survive or become extinct. The immediate mechanisms of extinction are many and may appear unrelated, and this has led to contradictory hypotheses for causes of coeval extinctions observed globally. Postulates have involved physical and biological systems including, for instance, anoxic ocean turnover or volcanicity, both of which are claimed as ultimate causes of extinctions.

We face a rather unusual situation in our empirical science. The examples discussed from the past show that ideas in geology have been held back by physicists and others who have denied effects simply because there appeared to be no possible cause. Yet there were always some geologists who maintained that empirical evidence led to inductions that differed from the current physical model. Paradoxically today, we have astrophysicists offering a cause that might have some importance for both geology and evolutionary theory, while some of us deny its effects. In paleontology in particular, the nature of some of the evidence cited against possible effects from bolide bombardment precludes finding a sudden event in the geological record. For instance, plots showing numbers of extinctions of genera or families (N) divided by the duration of a stage (T) vary in significance (N/T) over a range of 10 to 60 percent, and some of the larger figures are claimed to signal a mass extinction (Raup & Sepkoski 1986). A biomass disappearance observed at many independent global localities, however, may be highly significant, and this significance becomes infinite if the event can be shown to have been instantaneous, essentially irrespective of the number of taxa affected (McLaren 1988).

Many paleontologists, consciously or unconsciously, evidently find it difficult to relinquish a uniformitarian approach to evolution, and that stance leads them to rather extreme statements in the defense of gradualism. I should make it clear that it is not just a matter of not "believing" in bolides, major volcanic episodes, or other catastrophes that might cause extinctions above the background rate.

Many of the specific pieces of evidence for or against various postulates of causes of extinction are indeed controversial, and must be openly discussed. But there is a strong tendency to limit the discussion of evidence to one discipline, operating within the assumptions of the ruling model in that discipline.

Let us examine a few examples in which, because of the strength of a ruling model, an interdisciplinary approach is rejected in seeking evidence for mass extinctions. Taken out of context, perhaps, this is not entirely fair, but they are remarkably similar to some of the examples from earlier times discussed above, of empiricists switching to deduction when their models have become well established.

Thus, "I have not mentioned as such the end-Cretaceous bolide. I simply do not believe in it. Unseen bolides dropping into an unseen sea are not for me" (Holland 1989).

"Environmental changes on this planet as recorded by the facies should be thoroughly explored before invoking the *deus ex machina* of strange happenings in outer space. . . . It is intuitively more satisfying to seek causes from amongst those phenomena which are comparatively familiar to our experience" (Hallam 1981).

Dave Raup has given some good examples in his book *The Nemesis Affair* (1986). He quotes Robert T. Bakker, a noted vertebrate paleontologist, from an article in *The New York Times* (October 29, 1985), commenting on some of his fellow scientists in other disciplines: "The arrogance of those people is simply unbelievable. They know next to nothing about how real animals evolve, live, and become extinct. But despite their ignorance, the geochemists feel that all you have to do is crank up some fancy machine and you've revolutionized science."

Raup finds this statement more than a little appalling, and I am inclined to agree with him. He also gives an example of a nonscientist jumping on the bandwagon, again from the *Times* in 1985, referring to the idea of extinction by impact: "Astronomers should leave to astrologers the task of seeking the cause of earthly events in the stars." As I remarked earlier in this paper, the forces ranged against empiricism in earth science are formidable, and, I might add, come from many quarters.

Causes may begin in space, but effects are very much down-to-Earth and, in fact, represent a normal and continuing phenomenon in the evolution of the Solar System, including our planet and its biota (McLaren & Goodfellow 1990). In general, no claim is made or has been made that impacts cause all extinctions, or that any one extinction has been caused only by an impact. Like many other phenomena in geology, such causes may be multiple and resolved only with difficulty. But surely the burden of proof now rests with those who do not accept the existence of this violent phenomenon. Such a proof would require a denial of the existence of asteroids in Earth-crossing orbits and their occasional arrival on Earth, or the demonstration that such energetic events have had no detectable effect on crustal evolution, including the development of life. Furthermore, it must be recognized that failure to conform to existing hypotheses or models is not sufficient justification to deny the relevance of alternative hypotheses induced from new empirical observation.

The earth sciences continue in a state of flux that probably began with the freeing of continental drift from proscription by dogma at the beginning of the plate-tectonics revolution. Reasoning from strongly held opinions as a result of promoting plausible models is, as we have seen, all too common in the history of our science, and is encouraged by restricting scientific argument to single or related disciplines. There has seldom been a better example of the need for knowledge synthesis than the debate on impacts and extinctions. Looking at problems of planetary as well as biological evolution demands the consideration of many disciplines and varying philosophical approaches. There are ample opportunities for building hypotheses and models that will interact with others and lead to broader interdisciplinary syntheses.

5

What I Did with My Research Career: Or How Research on Biodiversity Yielded Data on Extinction

J. John Sepkoski, Jr.

During the five-year heyday of arguments about periodicity of extinctions, from 1984 to 1989, I was often questioned about my data on fossil animals and their times of extinction. My favorite naive question, most often asked by reporters, was "Did you assemble the data to demonstrate periodicity?" The answer was an emphatic No: Dave Raup and I were totally surprised to discover the periodic pattern of extinction. But more generally, the data were not even compiled to investigate mass extinction, and only serendipitously contained information useful in the extinction controversies launched by L. W. Alvarez et al. (1980).

In this contribution, I present an autobiographical history of the construction of my familial data base up to 1984. I discuss my research activities, why I was pursuing them, and what my attitudes toward the data were, especially with respect to their uses and accuracy. This history was put together from old notebooks, correspondence, and publications. There is no diary, so I relied a lot on memory (and people close to me will attest to how shoddy that is). Still, I hope I have not been overly revisionistic in this commentary.

Data bases of various kinds and forms have played important roles in the debates over the nature and causes of mass extinction: lists of worldwide iridium abundances at the K/T boundary (e.g. W. Alvarez et al. 1982), catalogues of terrestrial crater ages (e.g. Grieve 1982), compilations of Phanerozoic geochemical anomalies (e.g. Holser in prep.), and compendia of times of extinction of fossil organisms. The most frequently cited example in the last category is my *Compendium of Fossil Marine Families* (Sepkoski 1982a), recently published in its second edition (Sepkoski 1992). This compendium was an attempt to summarize known times of origination and extinction of all taxonomic families of animals known from marine strata throughout the world. It lists approximately 3,500 families recognized by paleontologists and represented in the fossil record by one or more specimens. The "times" of origination and extinction are actually 81 stratigraphic intervals ranging from 1.6 to about 20 m.y. long. These intervals are mostly internationally recognized stratigraphic stages, but also include some series as well as nine informal divisions of the Cambrian System.

Data in the *Compendium* have been used to locate extinction events in the fossil record, to measure the magnitudes of these events, and to argue for a 26-million-year periodicity in extinction. Periodicity is an empirical hypothesis derived from statistical time-series analysis of the data conducted by David Raup and me during 1983 (see Raup & Sepkoski 1984). It was a bold hypothesis, and it generated considerable debate concerning the mechanisms for such periodicity, the nature of the statistical analyses performed, and the quality of the data (see reviews by Raup 1986, 1991; Sepkoski 1989).

How to Assemble a Global Taxonomic Data Base

There is an easy way and a difficult way by which to compile taxonomic data bases on fossil organisms. The easy one is simply to use data gleaned directly from a standard source, such as the *Treatise on Invertebrate Paleontology* (Moore et al. 1953–84), *The Fossil Record* (Harland et al. 1967), or *Vertebrate Paleontology* (Romer 1966; Carroll 1987). Data are abstracted for the taxonomic groups of interest, the listed times of first and last appearances are copied directly, and the resulting compilation is used to explore for pattern or to test some hypothesis.

The difficult method is to ignore previous compilations and go directly to the primary literature and assemble a "first-generation" data base. (The even more difficult method is to do it all from personal observations, as Seilacher [1974] did for trace fossils.) The paleontologic literature is vast, it is presented in a babel of languages, and it has been published mostly over the last two centuries. Much of it is description of fossil specimens from particular localities and/or stratigraphic intervals, with assignments (or reassignments) of taxonomic names. This is the ultimate basis for all paleontologic data bases, but first-generation data bases derived from the time-intensive review of all or most relevant literature are usually the most precise and comprehensive. Magnificent examples for Proterozoic and Cambrian organisms have recently been published in J. W. Schopf & C. Klein (1992).

The approach I have used in my work is a hybrid of the two. I began with listings from the standard sources and then updated the classifications, expanded the taxonomic coverage, and improved the stratigraphic precision with primary sources. The best sources are monographic revisions by acknowledged experts who review all species within a superfamily, suborder, etc.

The whole process is iterative, beginning with a skeletal data base of uneven accuracy and precision and then continuously improving it in bits and pieces. The task has involved monitoring the current taxonomic literature and talking to experts on particular fossil groups to learn about the obscure literature (especially from Eastern Europe and Asia) and about the reliability of various workers. It has also involved the accumulation of background knowledge about the class and ordinal classifications of all fossil animals; about the international time scale and how local stratigraphic units correlate; and about where there are problems in the correlations, in the classification of groups, and in the work of some taxonomists. This knowledge is not trivial, and I have had difficulty assigning compilation tasks to students. As a result, I have done most of the work in assembling my data bases. Of course, I too began as a naive and fumbling student, in the early 1970's.

History of the Familial Compendium

The history of the data that sparked the debate over periodicity in extinction goes back to my time as a graduate student at Harvard University from 1970 to 1974. There I learned the elements of global stratigraphy from Bernhard Kummel, of macroevolution from Stephen Gould, and of demography and island biogeographic theory from E. O. Wilson. I divide this history into four phases: getting started, compiling familial data, publishing these data, and discovering periodicity.

Phase 1: Getting Started. In the early 1970's, Steve Gould was working with Dave Raup, Tom Schopf, and Dan Simberloff on stochastic models of the evolution of diversity (see Raup et al. 1973).

Steve was interested in "clade" shape: how diversity within taxonomic groups fluctuated through geologic time and whether there were any consistent patterns in their waxing and waning (see Gould et al. 1977, 1987). The study focused on diversification, and mass extinction played little role. Steve Gould and Dave Raup had devised indices for measuring clade shape in the stochastic simulations but needed data on real fossil taxa for comparison. In the fall semester of 1973, Steve employed me to compile data on all orders within classes and as many families within orders and genera within families as I could obtain. That was the beginning.

The task was to compile ranges of fossil taxa in terms of the 67 stratigraphic intervals used in *The Fossil Record*. I took the project one step further. I was working on a dissertation on Upper Cambrian paleoecology and had become interested in the great radiation of animals at the Precambrian/Cambrian boundary. The stratigraphic scheme in *The Fossil Record*, as well as in the *Treatise*, used only the three series (Lower, Middle, and Upper) of the Cambrian. But newer literature had shown that the Lower Cambrian might be split into four stages, as in the Siberian sections, providing far greater precision for investigating the explosive radiation of animals. I used sources such as Rozanov (1967) and Zhuravleva (1970) to locate all orders to Lower Cambrian stages. This gave me the first inkling that I could go beyond standard sources and improve taxonomic data. Otherwise, the task involved cross-referencing information in *The Fossil Record* and the *Treatise*, which gave me elementary experience in the world of conflicting classifications and changing stratigraphic assignments.

In 1974, shortly before I left Harvard, Steve discovered that most of my compilation work had already been done. He acquired several copies of Kukalova'-Peck (1973), which listed all animal orders to the time scale of *The Fossil Record* and also illustrated the hypothesized phylogenetic relationships as understood at the time. I took my copy and immediately began annotating the differences in classification and stratigraphy that I had discovered, as well as adding the subdivisions of the Lower Cambrian. I continued this annotation for the next five years, differentiating marine and terrestrial orders, updating classifications and stratigraphic ranges as they appeared in the literature, and adding information about the Ediacaran and uppermost Vendian orders; eventually I subdivided the entire Cambrian into nine stages. This heavily marked book became the data base for my study of the diversity of marine orders (Sepkoski 1976a, 1978), in which I argued that diversification across the Precambrian/Cambrian boundary was exponential and subsequent diversification was basically logistic. Again, mass extinction played little role in the study, and in fact orders were used in part because they ". . . tend to smooth out, much like running averages, many relatively minor temporal fluctuations in diversity [-like extinction events]" (Sepkoski 1978: 224).

Phase 2: Compiling Familial Data. In the summer of 1974, I left Harvard with my doctorate unfinished to become an instructor at the University of Rochester. There, I teamed up with Dave Raup, who continually offered encouragement, often much needed, for my research. By the time I arrived, I was more interested in diversification than in Cambrian paleoecology, as a result of work with Steve and reading seminal publications like Valentine (1969, 1973). Furthermore, my dissertation was not going well. The original proposal had been to investigate the roots of community evolution as synthesized by Bretsky (1968, 1969). He had shown that there was considerable stability in the general composition and environmental distribution of marine benthic communities through the Paleozoic Era. But his synthesis began in the late Ordovician and ignored the first 100 million years of Phanerozoic evolution. I was attempting to drop 50 million years below Bretsky's starting point and extend the synthesis down into the Upper Cambrian. But a summer of fieldwork in 1973 showed me that Bretsky's generalized "communities" did not exist in the Cambrian: the faunas were entirely different and the communities could not be traced up into the Ordovician—250 million years of stability, but no roots 50 million years earlier! I combed the literature to learn about this great faunal change and found little. I became convinced that the route to understanding would be from the top down, beginning with an examination of global diversity patterns. In the meantime, I settled on writing a sedimentological dissertation, which I did not complete until 1977, as other interests took hold.

During my first few years at Rochester, I began two notebooks on fossil animal families, attempting to garner more detail on diversity patterns. The first notebook was reserved for Cambrian and Vendian families. It contained mimeographed forms for taxonomic names and stratigraphic ranges, and plenty of room for notes. Cambrian and Ediacaran (latest Precambrian) animals were becoming hot in the 1970's, and I was attempting to keep track of new discoveries, rapidly changing classifications, and extensions of stratigraphic ranges. (I compiled another notebook on the Cambrian stratigraphy of the world—

mostly photocopied correlation charts—so that I could educate myself on local sections and fit these into my nine-stage framework.)

The second notebook was a listing of all Phanerozoic marine animal families and their ranges. I cannot recall exactly when I began this, but it was probably in the fall of 1975. I remember that transcribing family names and ranges from the *Treatise* and *The Fossil Record* was extraordinarily tedious, so I did it in front of TV, half watching football games. The transcription must have been reasonably complete by the spring of 1976, because I mention some of the data in an abstract for the annual meeting of the Geological Society of America that year (Sepkoski 1976a).

By the spring of 1977, I must have been sufficiently confident in the data, especially for the Cambrian, to generate Paleozoic diversity curves. I submitted another abstract for the GSA annual meeting (Sepkoski 1977) in which I argued that Early Cambrian diversification was exponential, but that there was equilibrium through the Middle and Late Cambrian. This then broke down during the Ordovician when another radiation ensued and more than tripled diversity, followed by a longer equilibrium. Thus, I had discovered the problem encountered in my dissertation: there were two great radiations in the early Paleozoic, not one as implied by earlier workers.

Also around this time I must have been becoming more aware of mass extinctions. One cannot fail to miss the end-Permian mass extinction, even in very sloppy familial data. Furthermore, I was interested in extending island biogeographic ideas to model extinction events, following the lead of T. J. M. Schopf (1974) and D. S. Simberloff (1974) (see Sepkoski 1976b). I was also aware of significant drops in diversity in the late Ordovician, Devonian, Triassic, and Cretaceous, times that Newell (1963, 1967) had recognized as containing mass extinctions. But I made an interesting discovery, published in Sepkoski (1979, 1982b). Newell had identified the end of the Cambrian as a time of massive extinction, but he had used familial data with only series-level resolution. When I split the Upper Cambrian into three stages, I found high extinction throughout and not a unitary pulse at the end, corroborating observations made by Cutbill and Funnell (1967) and Valentine (1969). I knew from Palmer (1965a, b) and Stitt (1971, 1975) that there were three biomere events in the Late Cambrian, but these must have been more minor than the five later mass extinctions, since they did not depress diversity to anywhere near the same degree.

Two events occurred during the 1976–77 academic year that fueled my data collection. First, Curt Teichert retired from the University of Kansas and become professor emeritus at Rochester. He brought with him an incredible knowledge of the fossil record and an almost equally huge library of monographs and reprints; he also brought galley proofs for the crinoid and Ediacaran parts of the *Treatise on Invertebrate Paleontology*. Second, I defended my dissertation in June 1977, freeing me of an albatross. I immediately began mining Teichert's library, which included many difficult-to-acquire European papers. Although I concentrated on Cambrian and Vendian taxa, I began revising large parts of the family notebook, incorporating more modern classifications, adding new families, and cleansing imprecise stratigraphic data.

Before leaving Rochester in the summer of 1978, I must have had growing confidence in the diversity patterns generated by the data. Simon Conway Morris, then at Open University, may have helped this. He visited Rochester that spring and took an interest in my data efforts. Simon has a voracious appetite for the literature, and he provided me with terrific advice on the classification and ranges of odd animals, not only from the Burgess Shale but from throughout the early Paleozoic. I remember sitting on my lawn in Rochester, paging through the Cambrian notebook with Simon; I now find his handwritten literature references scattered over various facing pages in the notebook. Later, I would benefit from the deep knowledge of another bibliophile: David Jablonski became interested in my compilations as soon as he learned of them, and he would send me notes about this, that, and the other family as he moved from the University of California system to Arizona and finally to Chicago.

In the spring of 1978, I counted numbers of families within classes and orders and punched them onto computer cards to generate long strips (on fanfold paper) of Phanerozoic "clade diversity" diagrams. This was done on a dean's account, since none of my work to date had received external funding. The computer account expired at the end of the academic year, and in order to drain it on the last day, I decided to repeat the factor analysis of the fossil record performed by Flessa and Imbrie (1973) on data from *The Fossil Record*. Half a dozen huge factor analyses drained the account quite adequately, but I did not have time to scrutinize the several inches of output until 1979. The results formed the basis of my three-fauna categorization of the marine fossil record (Sepkoski 1981).

Phase 3: Publishing the Data. Transfer to the University of Chicago did not change my data-gathering operations. The University and the Field Museum of Natural History had much better libraries than did Rochester (although nothing so compact and organized as Teichert's). I continued scouring new and old literature for data, hoping to improve my resolution on diversity patterns. This took more time than I should have devoted, and I did not have particularly productive years when first at Chicago.

But a funny thing happened in the late spring of 1980. My senior colleague, Tom Schopf, walked into my office and informed me (incorrectly, it turned out) that I would be up for a tenure review during the coming fall. I panicked and immediately began thinking about what work I could churn out quickly. The data base I had been compiling was an obvious candidate, but it took me several weeks to think of this. I knew the accuracy of the data was limited and many ranges were still imprecise. I was not enthusiastic about having taxonomic experts find my errors (which actually did not happen in print until Patterson & Smith 1987); and I wasn't sure, on the other hand, if anyone would be interested in a listing of familial data.

I spoke to Dave Raup (then at the Field Museum), and he encouraged me to pursue publication. I also discussed it with Leigh Van Valen (*Evolutionary Monographs*) and John Bolt (*Fieldiana: Geology*), who were both interested. So, I spent the summer in the library further cleaning the data with as much speed as possible. In the evenings, without the blessing of football, I created a new handwritten notebook, now being called the "Compendium," with a stereotyped four-letter code for stages. This was typed as manuscript by Jean Pasdeloup (no personal computers, yet).

Somewhere in this process, Tom discovered his error in timing, and my tenure review was properly postponed for a year. But too much effort had been invested in the project by now. In December 1980, I mailed portions of the manuscript to half a dozen taxonomic experts for criticism and help. In one cover letter I wrote, "The rationale for this publication is that currently available comparable data sets, such as those in the *Treatise* and *The Fossil Record*, are not entirely comprehensive, have many stratigraphic and taxonomic inaccuracies, and are no longer up to date." Also around this time, I was sending copies of the data to some colleagues, including Ted Foin (at U.C. Davis), Rebecca German (then a student at Harvard), and Jim Valentine (then at U.C. Santa Barbara). In the letter to Rebecca German in

May 1981, I wrote, "Enjoy the data but don't take them too seriously. In the year since I had the list typed I have had to revise the information for nearly 15 percent of the families." This summed up my confidence: good for diversity studies, but originations and extinctions were still subject to a lot of change.

In 1981 I acquired some NSF funding to publish the *Compendium* and continue my diversity studies. I submitted the manuscript to *Fieldiana: Geology*, but it was rejected in the summer of 1981 because it was too unusual and because the journal had too large a backlog. At the time, I was working with Peter Sheehan of the Milwaukee Public Museum on a project concerning community replacement during the Ordovician radiations (my dissertation proposal at last; see Sepkoski & Sheehan 1983). We were conversing on the phone about something, but I got around to complaining about *Fieldiana*'s action. Pete suddenly needed to get off the phone but promised to ring back in 15 minutes. He did. He had spoken to the editor of their *Contributions to Biology and Geology* and had convinced her to publish the manuscript. She did, although she ended up less than pleased with all the corrections I kept making.

The manuscript was submitted in October 1981. It was accepted after review in December with only one major suggestion for revision: that all families within orders be alphabetically ordered. I spent several nights with razor blade and tape rearranging the wretched manuscript lines and wondering why I had not typed the thing into a computer file. A photocopy was mailed back in February 1982 with a number of handwritten corrections based on new literature. Galleys were delivered in July, and I made 554 alterations based on yet more literature. Finally, in November, the *Compendium* was published, and I immediately stuck a list of 82 corrections into the front cover. (I continued doing this annually until 1988: people who I knew had copies or who requested updates would be mailed an expanding list of changes and additions. This led to various citations in the literature that read something like "Sepkoski [1982] with corrections through 1986.")

Phase 4: Discovering Periodicity. By the time the compendium manuscript was submitted, mass extinction had become the hot topic in paleontology. In fact, in the letter of submission, I wrote that the data could constitute a basis for studies of "mass extinctions, evolutionary rates, and patterns of diversification." Diversity—the very reason the data base existed—had been relegated to last place.

I had attended the Snowbird Conference on impacts and extinctions in October 1981 as an uninvited skeptic, mostly wishing to hear paleontologists speak about mass extinction so that I could learn more about their effect on diversity patterns. I learned little about that, but I had my paleontologic worldview changed. I learned (or, more aptly, realized) that large-body impacts were frequent over geologic time and therefore, in a sense, were uniformitarian; I learned that the mortal effects of impact could be specified in some detail, something other hypotheses of mass extinction could not do; and I learned that this data set I had been compiling had taught me a lot about mass extinction. There was a question from the floor after one talk about how many mass extinctions there had been and how big they were. (Norman Newell was not in attendance.) From my seat, I presented a three-category ranking of extinction events, based upon the familial data. Later, I volunteered to write this up for the conference proceedings (see Sepkoski 1982b). The data now were being used for more than just diversity studies.

Around this time, Dave Raup was sent the manuscript of the *Compendium* for review. He, of course, was well aware of my efforts to generate the data base but had actually never worked with it. Now he had a copy, and he immediately began studying it. The first thing he did was to produce an extraordinary graph, plotting extinction rates against time for each of the 81 stages of the Phanerozoic. Previously, I had always computed extinction rates over stratigraphic series (except for the Cambrian), because I was worried about the finer-level accuracy of the data. But with fresh vision, Dave had produced a stage-level graph that clearly showed the five big mass extinctions of the Phanerozoic as well as a secular decline in background extinction in between. I helped Dave clean up the graph a bit, identifying what data points reflected extraordinary fossil deposits such as the Burgess Shale, and noting where there were minor extinction events. But the paper (Raup & Sepkoski 1982) was his.

A little later, I began work on my manuscript for the Snowbird proceedings. I wished to review not only the five big mass extinctions but also the smaller events that workers had been finding of late (e.g. the Cambrian biomere events, the Toarcian event, and the Late Eocene event). As I studied the literature, I discovered that the familial data, when assessed at stage level, had little peaks of extinction at each of the post-Cambrian minor events. Hallam's (1977) earliest Toarcian (Early Jurassic) event was there (although registered in the preceding Pliensbachian Stage), and there was a comparable peak in the Tithonian (Late Jurassic), where he had also noted, almost in passing, a small extinction event. Likewise, the Upper Eocene appeared as a minor extinction peak in the data. The magnitudes of these peaks were in what I had previously assumed to be the noise level of the data, but each in the Mesozoic-Cenozoic portion corresponded to a documented extinction event, except for one—the Cenomanian in the mid-Cretaceous. No extinction event had yet been described there in the literature. In the spring of 1982, I phoned Erle Kauffman at the University of Colorado, who had been at the Snowbird meeting, to inquire whether anything was going on in the Cenomanian. My memory of the conversation is that he was as surprised as I: indeed, his group had recently identified a significant extinction event at the end of the Cenomanian, and how did I know about it? That conversation greatly changed my attitude toward the sloppy data I had in press: I was making predictions in the level I had been considering as noise, and they were correct.

Other people were taking interest in the data base, too. Ted Foin wanted to test Van Valen's (1973) "Law of Extinction." In 1980, I sent him a copy of the first manuscript of the data, and by October 1982 Ted had it computerized with all of the changes to date. Prior to this time, I had been tabulating originations and extinctions of the 3,500 families by hand, about one evening's work. For me, it would have taken several evenings' work to get Ted's computer tape up and running on the mainframe. But Raup did it quickly and began exploring the data via CalComp graphics to discover anything else that might be discovered about extinction, and especially the small extinction events.

Early in the winter of 1983, Dave invented what he called "pseudocohort" survivorship curves. Standard cohort curves consider only the individuals (taxonomic families in this case) that originate in some interval of time; they then trace them forward to determine how the cohort decays. A "pseudocohort" consists of all of the taxa present in a given geologic interval, whether they originated there or before; the survivorship curve then plots how they all decay through subsequent geologic time intervals. Dave showed me graphical output for one of these, and I commented that it would be interesting to compute curves for all stages of the Phanerozoic and plot them together. A few days later, Dave had a magnificent strip of CalComp paper that looked like a seismic reflection profile; there were even offsets like vertical faults where each pseudocohort lost diversity at major or minor extinction events.

"Do you see it?" Dave asked me. I stared but didn't. "They're regularly spaced in time."

My response: "Oh shit, Fischer and Arthur were right." I had just concluded, in Sepkoski (1982b), that Fischer and Arthur's (1977) 32-million-year extinction cycle was wrong, and here were my data exhibiting periodicity.

Dave spent the rest of the winter and all spring attempting to kill the periodicity in the familial data. He couldn't. I presented the preliminary results in August 1983 at the "Dynamics of Extinction" symposium in Flagstaff, Arizona (see Sepkoski & Raup 1986); I was also introduced in my hotel room there to a continuing guilty pleasure, MTV. Dave and I published the formal results on periodicity in Raup & Sepkoski (1984). Much of the surrounding issues and ensuing hoopla are chronicled in Dave's book, *The Nemesis Affair* (Raup 1986).

I need to add a footnote to this fourth phase of data compilation: in 1983 my confidence in the details of the familial data was hardly supreme. The periodic time series in Raup and Sepkoski (1984) contains several very small local maxima that are below the magnitudes of the Pliensbachian and Cenomanian peaks of familial extinction. The detailed paleontologic literature up to 1983 contained no information to suggest that these smallest local maxima might correspond to extinction events. However, in order to remain objective and not introduce any subjective cutoffs, Dave and I called all peaks, big and small alike, "mass extinctions." I suspected that some of these peaks were noise rather than signal, even at the Flagstaff meeting. When asked by a participant how I might obtain better precision, I answered that I would have to put together a data base on fossil genera.

That sealed my fate. Since late 1983, I have been iterating again through the standard and primary sources, compiling a data base that is an order of magnitude larger than the families, listing marine animal genera and their first and last fossil occurrences, now at the resolution of stratigraphic substages. Analyses of these data to date corroborate results from the smaller familial data base: there are five big mass extinctions, and events through the Mesozoic and Cenozoic remain very periodic (Raup & Sepkoski 1986; Sepkoski 1986, 1989, 1990); furthermore, the extinction peaks appear much more distinct than they did among families. Someday, this generic data base will be published, maybe in electronic form, when I gain more confidence in the accuracy of the data, and, perhaps, when someone implies that my career is dependent upon their publication.

Conclusion

This has been a story about how a data base assembled as a side interest to a dissertation project became a focus in controversies about how extinction works. Most of the narrative has had little to do with extinction, because that was not an exciting issue before 1980. Mass extinctions were evident in the data, but without interesting hypotheses to test and argue, they could be ignored as annoying perturbations in the history of diversity. It was hypotheses about diversification that provided the energy for tedious transcription of data and long hours in the library. But that was not the only motivation. Paleontologic literature is interesting simply for the phenomenology it holds. Fossils are fascinating in and of themselves, as is the history of life that they tell.

Acknowledgments

David Raup has encouraged and counseled me in my work since 1973, and I thank him dearly. Christine Janis's insightful comments helped the manuscript; I met her briefly at the beginning of the first phase, and met her again at the end of the fourth phase. Research described here, when funded, received support from NSF grant DEB 81-08890 and NASA grants NAG 2-237, NAG 2-282, and NAGW-1693.

6

The Extinction Debates: A View from the Trenches

David M. Raup

The extinction debates of the past decade have been a wrenching experience for nearly all paleontologists. The 1980 paper by Alvarez et al., arguing that comet or asteroid impact caused the end-Cretaceous mass extinction, was met with nearly universal derision by the paleontological community. Many specific objections were offered, but, I submit, the driving force was dominantly intuitive: the Alvarez proposal was simply not within the accepted coordinate system governing inferences used to understand Earth history. My own attitude in 1980, and I suspect that of most of my colleagues, was that "Alvarez *must* be wrong, and it should be easy to show that he is wrong. After all, he is a physicist unfamiliar with the data of geology and paleontology and with the complexities of the extinction phenomenon."

Many reviews of the evidence for and against the impact theory have been published, and even though much remains to be learned, a consensus appears to be developing in the broader scientific community in favor of the Alvarez hypothesis, virtually in its original form. The consensus is far from complete, however. Although a majority of practicing paleontologists have come to accept that an impact event probably occurred at the end of the Cretaceous, most have not yet accepted its causal link to mass extinction. The opinion of the paleontologists is important, of course, because no group in the scientific community is closer to the actual data of extinction. Do paleontologists know something the rest of the scientific community has missed, or are they blinded by tradition? As a contribution to answering this question, my purpose here is to enumerate and analyze some of the factors that led to that initial, intuitive reaction against the impact theory.

Failure of Communication Between Disciplines

When I was training to be a paleontologist, in the 1950's, I was taught that most meteorite impacts on Earth were confined to what was known as the "early bombardment," the relatively short period of accumulation of debris left over from the formation of the Solar System. The bombardment was assumed to have ended well before the origin of life and certainly before the big mass extinctions. Owing to sedimentation and erosion, the craters left by the early bombardment were covered or obliterated. The craters on the Moon and the other terrestrial planets were still visible only because of the lack of these destructive processes. The following passage is from Strahler's excellent, and widely used text, *The Earth Sciences* (1963):

> It is reasonable to suppose that, at an earlier time in the history of the earth and moon, meteoroids in the solar system were far more numerous and much larger, so that meteorite impacts were much more frequent and of much greater intensity. Although the earth must have sustained a similar bombardment [to that of the moon], the terrestrial processes of erosion and deposition (not acting on the moon because it had no atmosphere) would have long since obliterated most of the earth's craters. (p. 84)

A few impact craters were recognized on Earth, notably the 1.2-km-diameter Meteor Crater in Arizona, but these were seen as flukes, having no importance in Earth history and having no bearing on the evolution of life.

Given this background, it is no wonder that so many paleontologists found the notion of a large Cretaceous impact, one alleged to have produced a crater 150 km in diameter, incredible and against all geologic experience. The notion was viewed as a betrayal of the present-is-the-key-to-the-past dictum and, therefore, a *deus ex machina* argument not worthy of serious consideration.

I think this is a clear case of failure of communication between scientific disciplines. Unbeknownst to most paleontologists (and a surprising number of geologists), major strides were made in the 1960's and 1970's in the field of meteoritics, a small subdiscipline often more closely associated with the space sciences and cosmochemistry than with geology. Among the advances were (1) the recognition of high-pressure forms of the mineral quartz (e.g., stishovite and coesite) as sure indicators of meteorite impact, (2) satellite photography that rendered easier the recognition of topographic structures of possible impact origin, (3) radiometric dating of samples returned from the Moon that indicated crater ages much younger than the "early bombardment," and (4) the discovery of large numbers of present-day asteroids in Earth-crossing orbits.

With these new tools, more than 100 confirmed impact features on Earth were identified, and a small industry developed in search of more. More important, the data from these impacts were combined with new information from other sources to make rigorous estimates of the impact flux for the younger periods of Earth history. It became clear that although the impact rate had indeed fallen off after an early bombardment, it remained substantial to the present day. But Shoemaker (1984) has estimated that impact rates during the Phanerozoic (the most recent 600 m.y.) were actually higher than the average for the last 3,300 m.y. of Earth history, higher by a factor of between two and six! In other words, the rate has been climbing since the drop following the early bombardment.

In a 1982 review, Wetherill and Shoemaker made the following comment: "Although the physical encounter with the earth of these objects can properly be termed 'catastrophic,' in terms of the magnitude of the effects they produce, they are at the same time 'uniformitarian' in that *they represent the extension of presently observed geologic processes to earlier geologic time*" (p. 1; emphasis added). Thus, we have a situation where one group of scientists, the meteoriticists, sees large-body impact as a normal geologic phenomenon—something to be expected throughout Earth history—but another group, the paleontologists, is confounded by what appears to be an ad hoc theory about a nonexistent phenomenon.

Therefore, a simple failure of communication between disciplines was responsible for at least some of the negative reaction by paleontologists to the Alvarez proposal. How much influence this had on the extinction debates of the 1980's is unknowable, because many of the paleontologists were probably not consciously aware of why they found the impact theory so incredible.

Are Extinctions Episodic or Gradual?

The two patron saints of paleontology, Charles Darwin and Charles Lyell, taught that changes on Earth occur gradually, as accumulations of countless small increments. Both Darwin and Lyell applied this principle also to biological changes, including extinction.

In *On the Origin* (1859), Darwin noted that ". . . species and groups of species gradually disappear, one after another, first from one spot, then from another, and finally from the world. . . . There is reason to believe that the complete extinction of the species of a group is generally a slower process than their production . . ." (pp. 317–18).

A particularly clear statement of Lyell's position is contained in his treatment of faunal turnover at the end of the Maestrichtian stage of the Cretaceous, now known as the K/T mass extinction. The following is from his *Principles of Geology* (vol. 3, 1833):

> *Chasm between the Eocene and Maestricht formations.*—There appears, then, to be a greater chasm between the organic remains of the Eocene and Maestricht beds, than between the Eocene and Recent strata. . . . It is not improbable that a greater interval of time may be indicated by this greater dissimilarity of fossil remains. . . . so we may, perhaps, hereafter detect an equal, or even greater series, intermediate between the Maestricht beds and the Eocene strata. (p. 328)

Lyell argued repeatedly in *Principles of Geology* that extinction is a continuous process. Indeed, he subdivided the Tertiary Period on the basis of the proportions of species found in fossil assemblages that are still living today: one-thirtieth in the Eocene, one-fifth in the Miocene, one-third in the older Pliocene, and nine-tenths in the newer Pliocene. He emphasized that these were arbitrary boundaries drawn for convenience in a continuum of change, and included the following caution in *Principles of Geology*:

> . . . we are apprehensive lest zoological periods in geology, like artificial divisions in other branches of natural history, should acquire too much importance, from being supposed to be founded on some great interruptions in the regular series of events in the organic world, whereas, . . . we ought to regard them as invented for the convenience of systematic arrangement, always expecting to discover intermediate gradations between the boundary lines we have first drawn. (vol. 3, p. 57)

Thus, Lyell ruled out, on principle, what we now call mass extinctions. In the first passage quoted, and elsewhere, he predicted that further exploration of the geologic record would reveal intermediate fossil assemblages to bridge the apparent gaps. And he covered the possibility of failure in this endeavor by noting that there must be intervals of Earth history that lack record. With this caveat, he was able to argue for continuous change despite cases that appeared to be discontinuous. It is, to me, a classic case of interpretation driven by theory.

Darwin echoed Lyell's reasoning, as exemplified by the following passage from *On the Origin*:

> With respect to the apparently sudden extermination of whole families or orders, as of Trilobites at the close of the palaeozoic period and of Ammonites at the close of the secondary period, we must remember what has already been said on the probable wide intervals of time between our consecutive formations; and in these intervals there may have been much slower extermination. (pp. 321–22)

It is ironic that so many paleontologists still subscribe to the paradigm of extinction as a continuous process, despite the fact that so many major and minor boundaries in the geologic time scale are defined by the relatively sudden disappearance of large numbers of species—just as the difference in fossil content was used to distinguish Maestricht and Eocene beds in Lyell's time. It is no accident that two of the largest mass extinctions are used to define boundaries between the three major eras of the Phanerozoic (Paleozoic, Mesozoic, and Cenozoic), and that lesser pulses of extinction mark most of the secondary boundaries of the time scale.

The negative reaction to the Alvarez theory was certainly driven, at least in part, by the legacy of Lyell and Darwin. Often-heard statements like "the extinction event was too gradual and protracted to have been caused by an instant of mass killing" and "mass extinctions are often artifacts of gaps in the record" reflect this legacy.

The principle of gradualism developed by Lyell and Darwin has been applied to all aspects of the evolutionary process, not merely extinction. In the words of Roland Brinkmann (1929), in his classic analysis of the Jurassic ammonite *Kosmoceras* [translated from the original German by L. R. Fischer, 1981]: "Evolution proceeds continuously, and all jumps are deceptions caused by gaps in the record" (p. 242).

Suggestions of an Anti-Darwinian Flavor

Charles Darwin, too, argued that the extinction of a species is primarily the result of that species' failure to compete successfully with other species. Two quotes from *On the Origin* (1859) illustrate this view: "The inhabitants of each successive period . . . have beaten their predecessors in the race for life, and are, insofar, higher in the scale of nature (p. 345). "If . . . the eocene inhabitants . . . were put into competition with the existing inhabitants, . . . the eocene fauna or flora would certainly be beaten and exterminated; as would a secondary [Mesozoic] fauna by an eocene, and a palaeozoic fauna by a secondary fauna" (p. 337). Although Darwin acknowledged that purely physical forces could cause extinction, it is clear that he saw the struggles between species as paramount. And this accords easily with his formulation of natural selection, which, although acting within rather than between species, was also based on interactions that were primarily biological.

The Darwinian view of extinction has been challenged in recent years by a number of paleontologists frustrated by the lack of evidence for the inferiority of the victims of extinction. Did the dinosaurs deserve to die and the mammals deserve to survive at the end of the Cretaceous? Although many biological failings of the dinosaurs could be concocted after the fact, an equal number could be found for the mammals. In no case was it possible to say with certainty which group was the more likely to survive, save for the fact that one survived and the other did not.

Although some members of the paleontological community have found this challenge to Darwin's view convincing, especially for the most dramatic mass extinctions, my impression is that the majority of practitioners have held quite closely to the notion that extinctions ought to be explicable in terms of the inferiority of the victims, even though specific causal factors may be elusive.

Notwithstanding this uncertainty over interspecies competition as a cause of extinction, the Alvarez proposal flew in the face of the conventional Darwinian view. His proposed mechanism was purely physical, and the environmental effects of a large impact would be so profound that species interactions could play little or no role in determining survival. Survival would simply be a matter of chance, depending on whether a given species—or type of organism or habitat—happened to be susceptible to the rare physical stresses triggered by the meteorite impact.

Again, it is impossible to say how much of the negative reaction to Alvarez was due to a feeling that the impact theory was an attack on Darwin. This argument has been used only rarely in debate, but I suggest that it was the basis for many of the arguments that have been used. For example, the Alvarez proposal has been criticized repeatedly for being too simple to be viable in the complex world of biology.

This strange reversal of Occam's razor may in fact be a proxy for arguing that the Alvarez proposal fails because it ignores the myriad complexities of species interactions and community structure. One of the most common arguments made by paleontologists is that Alvarez, as a physicist, did not understand just how complex extinction is.

Conclusion

My thesis in this discussion is that the negative reaction of most paleontologists to the impact theory is deeply intuitive, stemming from a host of preconceptions, which may or may not be misconceptions, but which rarely come to the surface as explicit arguments. Rather than quoting Darwin or Lyell, the paleontological opponents of the impact theory have relied on presenting long lists of specific objections (often reminiscent of the creationists' arguments against evolution) that must be satisfied before the theory can be considered seriously. Some of the objections are valid and are important contributions to the debate, but, I submit, the majority are trivial. This mode of argument has imposed an almost impossibly high standard of proof. To win their case, the advocates of large-body impact must remove all uncertainty over a wide range of topics. That the burden of proof should be with the new (and revolutionary) theory is reasonable, but the standards that have been demanded are far higher than is normal in science, and far higher than is practiced by most of its participants.

7

Hazards from Space: Comets in History and Science

S. V. M. Clube

By concerning itself and NASA with the hazards that space offers to Earth-bound civilization, the U.S. Congress unwittingly revived the issue at the heart of theological debate these last two thousand years, namely, the fundamental question raised by Plato and his successors down the ages whether the "revolutions" of an invisible circulation in space sometimes affect the Earth. Thus a large circum-Solar stream of cometary-asteroidal material intersecting the Earth's orbit, previously only suspected in modern times, has now been confirmed in sub-Jovian space. Streams of this kind, containing both active and inactive cometary debris, are a natural consequence of the breakup, under the influence of the Sun's radiation field, of very large comets of heterogeneous composition. They are important by virtue of the occasional major disintegrations responsible for the longer-lived (kilometer-plus) debris that join the inner Solar System population of asteroids in planet-crossing orbits, fundamentally affecting the survival of species; they are also important by virtue of the more frequent minor disintegrations responsible for the shorter-lived (sub-kilometer) debris that give rise to unexpected surges in the fireball flux, principally affecting the survival of civilization.

These surges were systematically recorded in China, where they were evidently regarded as signifying a potential threat to our planet. In Europe also, these surges were a source of concern, for they marked the occasions of extreme protestation against prevailing cosmological orthodoxy. This orthodoxy declares the cosmological environment generally safe on centennial-to-millennial time scales, a supposed fact that was a matter for theological justification in the past (assuming that a divine Savior acts on our behalf) before it became a matter for scientific verification. But as we learn more about that large cometary-asteroidal stream, it has become clear that the fireball surges are not necessarily without serious global consequences.

In the decades since the baton of western civilization was passed across the Atlantic, there has been an unparalleled advance in the West's material prosperity, coupled with two world wars and one major depression. The Darwinian thinking that led humankind to perceive in nature a uniformitarian process has not been unaffected by these changes, and we are more inclined these days to accept a general advance punctuated by catastrophe. Twelve years ago indeed, Luis Alvarez rose above the scientific debate and chose to warn the American people as well as the world at large that catastrophists were probably right after all, and that not even our cosmological environment can be regarded as safe. There are differences of scale, of course. The intervals between major biological extinctions exceed the million-year intervals between kilometer-plus asteroid impacts by a factor almost as great as that by which the latter exceeds the time scale of western civilization. By the same token, civilizations have fallen many times in the past, often for reasons that are not well understood, and there have been other occasions during the long history of western civilization when political sources close to the fountain of power have concerned themselves to some effect with the potentialities of our more immediate cosmological environment. Usually these potentialities were the subject of intense theological debate, but with the space age and the historical record both fostering a degree of concern about the state of our cosmological environment, it is hardly a surprise that Congress should now require NASA to assess the hazard to civilization from near-Earth space. We cannot be sure in the present instance what the outcome of these enquiries may be, and there is as yet little certainty regarding the nature of the public reaction or the future response of our political masters, should it be discovered, or even postulated, that such a cataclysm will soon be upon us. The purpose of the present article is to describe how our ideas about comets and fireballs have evolved through history and science, and to place the cosmological hazard to civilization in perspective.

Turning Points in the Past

In England, for example, some three and a half centuries ago, Cromwell rode to power on the crest of a millenarian wave fathered by what Norman Cohn in his *The Pursuit of the Millennium* (Palodin, 1970) called "supernatural illumination," only to discover that the "destroying angel" was not about to materialize. Thus the ranting parliamentarians were made to look foolish, and "it was only a matter of time before the reins of power passed to others of a more sophisticated persuasion, whose vision of the world embodied the kind of enlightenment that has sustained western civilization ever since. Enlightenment of course builds on the providential view and treats the Cosmos as a harmless backdrop to human affairs, a view of the world which Academe now often regards as its business to uphold and to which the counter-reformed Church and State are only too glad to subscribe" (S. V. M. Clube 1992, *Celestial Mechanics and Dynamical Astronomy* 54: 179–83). It was Newton, in fact, who at the very foundation of modern science took an Aristotelian or providential view of comets, hinting at their capacity to replenish planets and stars, and it was Darwin who took over during the nineteenth century, when the replenishment was attributed solely to harmless meteors—the more obvious cometary debris. The ground was thus prepared for the new evolutionary mainsprings of catastrophe and revolution—survival of the fittest and dictatorship of the proletariat—and where formerly the Church on behalf of the State had defused divine revelations, the responsibility for an orderly Cosmos was now taken over by Science. It was Huxley who outmaneuvered Bishop Wilberforce during their famous 1860 Oxford debate, and who went on to be the chief architect of the environmental policy that emerged. This policy, as the baton of western civilization crossed the Atlantic, would require that repeated cosmic stress—the key to any divine interference in terrestrial affairs—be deliberately programmed out of mainstream Christian theology and natural science, arguably the two most influential contributions of western civilization to the control and well-being of humanity.

The public Newton, however, is well known now to have been economical with the truth. He eschewed the relationship between supposed divine relations, intense millenarian fears, and social revolution, commonplace though that relationship was until the middle of the nineteenth century. Attention was thus deflected from the large meteors that provided Taurid illuminations, as we now know from Chinese astronomical records these last two thousand years (see Figure 7.1). In China, of course, the expectation was a cleansing by fire, the action of broom-stars with the ability to sweep the Earth clean and start a new age. In sixteenth-century Europe also, the perception was much the same: of comets with their occasional streams of larger debris, implicitly as great as the bodies that produced the Tunguska explosion on the Earth in 1908 and the Giordano Bruno crater on the Moon in 1186, both now commonly attributed to the Taurid stream. Indeed, the modern evidence is of a very substantial though largely invisible cometary-cum-asteroidal Taurid stream (Table 7.1) that is evidently the source of significant disintegrations in space every few hundred years, producing a variety of divine revelations (e.g. large meteors) and, no doubt, most of the zodiacal dust as well. The very probable explanation of all this is the continuous, hierarchical breakup of an exceptionally large comet in sub-Jovian, near-Earth space these last 20,000 years, requiring a degree of purblindness from the providentially minded to its recent role as an established, fundamental, and very real aspect of God's creation. It seems, however, that the space age's providentially minded, programmed as they are against cosmic stress, are obliged to discount global catastrophes due to celestial inputs on time scales as short as this. Thus the remnants of an essentially Anglo-Saxon tradition are still with us, and the message reaching NASA is that we should resist the challenge of the catastrophic Beast.

This is not the message of Chinese records and European history, however, for the decadal-to-centennial intensifications of the sporadic-cum-Taurid meteor flux clearly coincide with the principal revolutionary periods associated with the emergence of millenarianism, Protestantism, and fundamentalism (Figure 7.1). As such, they provide a secure basis for Cromwell's (but not Marx's) understanding of "revolution," revealing a supposedly predestined effect of the circulation of heavenly bodies. At the same time, they represent interruptions on the plateau of dominance occupied by the Roman ministry—the plateau on which the latter, deciding against "second causes" and treating divine interventions as revelations or signs, eventually found itself joined by the increasingly materialistic viewpoints of the counter-reformed Church and secularism, which did and did not admit the possibility also of a "primary cause." In other words, we see now an Anglo-Saxon tradition that perpetuates a long-established providential view of interruptions, or punctuations,

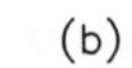
(b)

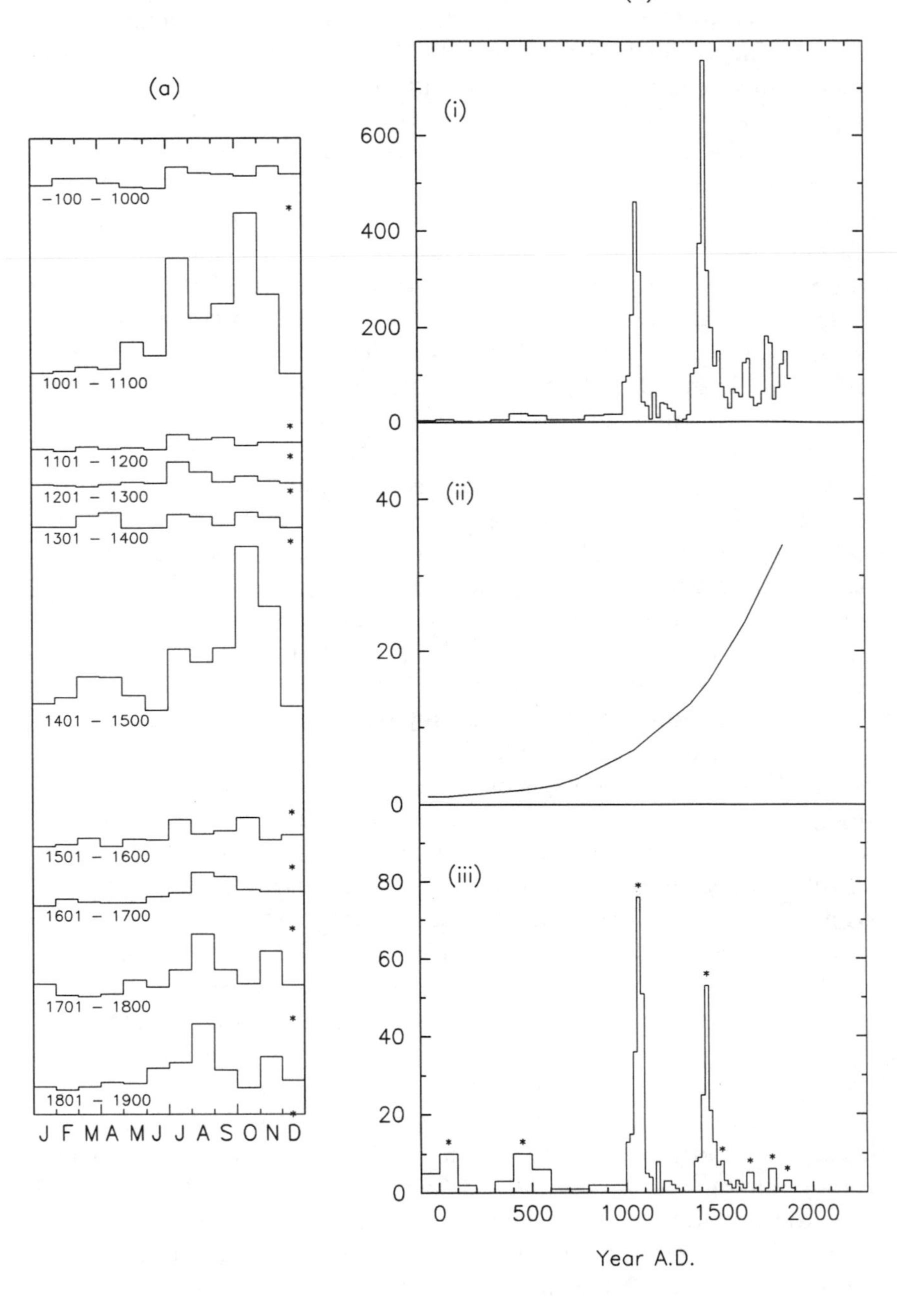
(a)
-100 – 1000
1001 – 1100
1101 – 1200
1201 – 1300
1301 – 1400
1401 – 1500
1501 – 1600
1601 – 1700
1701 – 1800
1801 – 1900
J F M A M J J A S O N D
(i)
(ii)
(iii)
600
400
200
0
40
20
0
80
60
40
20
0
0
500
1000
1500
2000
Year A.D.

TABLE 7.1

Comets and probable asteroids now associated with the Taurid Complex; and their similar orbital parameters, implying a likely common origin. Assuming a roughly uniform detection efficiency for Earth-crossing asteroids, these discoveries imply a total Taurid Complex population of kilometer-sized (and larger) objects of 50–100.

Comets and probable asteroids	Semi-major axis of orbit (in astronomical units)	Eccentricity	Inclination	Longitude of perihelion (π)
1991 GO	1.96	0.66	10	113
(4341) Poseidon	1.84	0.68	12	123
(4197) 1982 TA	2.30	0.77	12	129
1991 TB2	2.40	0.84	9	132
1984 KB	2.22	0.76	5	146
(4183) Cuno	1.98	0.64	7	170
(2201) Oljato	2.18	0.71	3	172
(5143) 1991 VL	1.83	0.77	9	177
1991 BA	2.24	0.68	2	189
P/Encke	2.20	0.85	12	160

Figure 7.1 Combined monthly (a) and annual (b) variations of the large-meteor (fireball) flux during the last two millennia, based on Chinese records. These records have been partially calibrated in accordance with three subdivisions of panel (b), thus giving (i) the recorded total flux, (ii) the smoothly interpolated inter-peak flux, and (iii) the recorded total flux rectified and normalized on the basis of (ii), assuming that the true inter-peak flux is constant. This calibration is at best only approximate, and takes no account of a possible limitation to only the very largest fireballs before about 1000 A.D.

Note the persistent July-August and October-November peaks in the millennial and centennial fluxes, which are consistent with the Taurid-stream association. Note also the major enhancements of the annual flux coinciding with periods of supposed divine revelation and militant, God-fearing fundamentalism, which were responsible for significant revolutionary epochs in Europe, namely, the West's split with Islam, the Great Schism, the Reformation, the English Revolution, the American War of Independence and the French Revolution, and the recent years of European revolutions. These epochs punctuate the "plateaux" during which the various providential views of the Cosmos have been preferred (see text): those of the Catholic Church since ca. 1300, those of the counter-reformed Church since ca. 1600, and those of secularism since ca. 1850. Note also the corrected Dark Ages enhancement, assuming an approximately uniform interpeak flux.

characterized by a sustained sporadic-cum-Taurid meteor flux with repeated spectacular events. So far as civilization is concerned, however, it seems that the inferred danger is more cerebral than Pavlovian, requiring a mature perception of the underlying reality. For example, there is clearly a close correspondence between late-twentieth-century events (the "Cold War," with its sustained ICBM threat and prior examples of A-bomb disaster, set alongside highly principled, lesser disputes) and these past upheavals (each with its enhanced meteor flux and its authorized view of Armageddon, set alongside holy [civil] wars). The Chinese records, in other words, are straightforward evidence that the reactivated Taurids have always had a perceived association with multi-Tunguska bombardment. The question we have to address, however, is not whether the reactivated Taurids *always* induced multi-Tunguska bombardment, but whether they have done so *sometimes*, thus justifying the past perception that such events may occur again.

The Roman Response

During the fourth century, there was a dawning realization, close to the fountain of power in the embryonic civilization, that the then-prevailing view of the potentialities of the more immediate cosmological environment was probably incorrect. Initially, as the defects became clear, the captains of society and cosmology made every effort to ensure that the existing beliefs were preserved. The writings of the "mathematici" and the astrologers were destroyed, and both cadres were silenced on pain of death. In the event, these efforts were in vain, and the Roman Empire duly declined and fell.

Managed and led from Rome by a patrician society that was both proud and practical, the embryonic western civilization collapsed, it is often said, because it was too attached to its pagan gods. This is a conveniently simplistic view that altogether misses the point. Thus, with its Empire built, its central administration in place, and its army everywhere within reach, the patrician society of Rome evidently continued for centuries to hold the Empire together and to deal successfully with the unexpected pressures that arose through deserted lands and among displaced peoples. Eventually these pressures were too great, and there is no question that the management did effectively cease to function by the mid-fifth century, when, for example, a section of the British people migrated to northern Gaul—displaced, so it is said, by the effects of "the fire of righteous vengeance." These facts do not mean, however, that the Roman civilization, during its period of success, was not extremely businesslike and essentially secular in its approach and outlook. In fact, it seems more than likely that the upper echelons of Roman society did successfully absorb the atheistic teachings of the more intellectual Greeks and were happy to go along with the trappings of an ancient religion for the sake of maintaining an orderly civilization. We can therefore be rather certain that the embryonic western civilization basically accepted the planets as suitable vehicles for the names of ancient gods whilst believing in practice that these vehicles had no very serious capacity to interfere in terrestrial affairs. Indeed, although it is certainly true that the planetary theory of Ptolemy and the cosmological theory of Aristotle—the former was merely a part of the latter—were both seriously in error, we can be confident that in this particular respect, namely, the planetary influence on human affairs, the attitude of the hard men behind western civilization was not so very different in the fourth century, at this formative stage, from what it is today.

What dawned on the embryonic western civilization at this time, therefore, was the realization that a religious sect, known as Christians, though thought at first to be of no great importance, was privy to a deeper and, in some sense, more valid cosmological view. The secular component of this cosmological view was the one eventually expounded with considerable success by Augustine, a component that accorded rather closely with the teachings of Proclus, a near contemporary, head of the neo-Platonic school in Athens. According to this cosmological theory, the space around the Earth contained a variety of demons, or fallen angels, and it was the Christians in particular who anticipated a catastrophe for the human race toward the end of the fourth century, a catastrophe borne of the angelic host, and one that would mark the "second coming." In the event, any catastrophe that did occur was widely dispersed and not altogether of the kind that unnerved the Roman administration during the anticipation; furthermore, there was little sign of the "second coming." Rather, there was a period of intensified pressure through the deserted lands and displaced peoples, rapidly descending toward the prolonged Dark Ages. Augustine managed in due course to put a more hopeful gloss on affairs, suggesting that the anticipated catastrophe was already in the past, though the actual events did little or nothing to rid Church or society of the new cosmological conviction that near-Earth space was filled with demons.

Indeed the conviction intensified, and there was a great deal of concern subsequently to understand and predict the role of demons in terrestrial affairs. Astrology (both judicial and horoscopic) has its roots, of course, in the fear of demons (visible and invisible), but the subject loses its meaning whenever society restores Aristotelian or providential views of the Cosmos. Aristotle, it will be recalled, held a particularly providential view of comets and meteors, treating them as purely atmospheric phenomena. On this occasion, the restoration came first with Islam, and it was not until the thirteenth century that western civilization, led by the Catholic Church, allied itself again with Aristotelian principle. Subsequently, of course, the Protestant Church resisted the new alliance, but with no eventual success. The point here is that the eventual removal of demons from space, the apparent key to everything that was then considered wrong with cosmology, is now widely and very deeply regarded as one of the major intellectual achievements of western civilization during the centuries leading up to the Enlightenment. The result is a wholesale trivialization of objects that our distant forefathers regarded as intensely real. And yet here we are, some 300 years into the period of the Enlightenment, in the presence of a near-Earth space that is once again being filled with demons! The demons are now the invisible planet crossers and crumbling Taurids, and like their medieval counterparts, they are greatly feared.

As before, the captains of society and cosmology are doing their best to ensure that pre-existing beliefs are preserved. As before, the "mathematici" and the astrophysicists, successors to the astrologers of old, are silenced on pain of everything, so far, but death. The politicians have wind of something in the air; the U.S. Congress has invited NASA to investigate; some professionals have now said their piece; but the outcome is still very far from clear.

Twentieth-Century Demonology

Physically akin to mainbelt asteroids, dynamically akin to short-period comets, and dimensionally akin to both, being approximately 0.1 to 10 kilometers in size, the demons of the twentieth century produce impact craters on planets and their satellites. This is by no means a remote astronomical threat of the kind that is fashionable among scientific hedonists, such as the red-giant phase of the sun, or a supernova explosion, or the universe collapsing in on itself to produce another supposed Big Bang; *rather, it is a hazard that is immediately at hand*. In other words, the probabilities imply an event tomorrow, next year, next century, or the next millennium; the demons are in effect the greatest scourge of the environment and the greatest threat to life presently known to humanity. It is hardly any wonder that they are greatly feared, most especially by those who are directly concerned with the evidence.

Being mostly invisible, these demons act without warning, and we cannot predict when the next catastrophe will occur. It is generally accepted, however, that they threaten to take out areas the size of a small nation (as a demon 0.1 km in diameter would) through a continent (1 km) to virtually the whole surface of the globe (10 km), depending on their size. On the basis of studies of lunar craters, it is known that the successive catastrophic experiences on Earth relating to these levels of hazard are separated on average by a few centuries (0.1 km) through a million years (1 km) to several hundred million years (10 km), the progression being straightforwardly due to the relative numbers of demons of various sizes in near-Earth space, which increase steeply with decreasing size. Even the shortest of these time scales, so far as civilization is concerned, is rather too short for comfort.

Assuming that it is desirable to continue subscribing to the providential view, mainstream cosmologists have managed to handle this situation with some degree of skill. The problem, of course, is that the lunar-based averages do not actually mean anything unless we know whence the twentieth-century demons come. It is possible that the demons are displaced from the Main Asteroid Belt and arrive at the Earth fairly regularly, in which case the true and average rates of arrival are the same, and there is one demon every few centuries. Since three-quarters of the globe's surface is ocean, it is then possible to arrive at the theologically inspired scientific conclusion that the threat to civilization is about one small demon per millennium, and they can be safely ignored. This is the conclusion that mainstream cosmologists would like to see accepted by government agencies around the world. This is also the conclusion, however, that accords with the Anglo-Saxon tradition, thus deflecting attention from the demons that come from crumbling Taurids.

The theologically inspired conclusion is of course as irresponsible as it is speculative. It is also possible, even likely, that many demons are the products of very large comets like Chiron, which have asteroidal or heat-processed cores, in which case they are present in substantial numbers only when large comets settle in short-period orbits and disintegrate under the influence of the Sun. The mean arrival rate of demons at the Earth for the duration of these active spells is then substantially increased over the average, amounting to about one demon every few years. The bombardment scheme that fits this particular picture is a burst of demons from time to time: say, nothing at all for several centuries followed by 100 in a short space of time; or nothing at all for a millennium followed by 1,000 in a short space of time. It is this bombardment scheme we should be considering if in fact a very large comet produced the crumbling Taurids.

Of these differing schemes, involving the Main Asteroid Belt or an occasional large comet, it is obvious which we have most reason to fear. But in order to preserve the illusions of an outmoded tradition, just as in Rome centuries ago, the public *and* our political masters are persistently fed only the theologically inspired specification of the celestial hazard to civilization. It may be that NASA, in turn, will assume that specification of the theologically inspired hazard is the one Congress will wish to hear.

Cosmic Winters

The physical reality we have most reason to fear, then, is a pack of demons traveling along with a generally inactive but still disintegrating cometary core; the whole thing is mostly invisible, though it could light up if the core itself fragments, revealing a not completely devolatilized section of the original comet. The hazard that civilization faces in this demonic host is the sudden incidence of multiple impact events. Of these events, there are two kinds. The first is massive and regular, occurring at those epochs, millennia apart, determined by orbital precession, when the path of the demonic host runs very close to the orbit of the Earth. The second, less massive and less regular, occurring at other epochs, just centuries apart, is due to random host encounters in space, when major fragmentations of the core cause a temporary enhancement of host debris, possibly even including a short-lived but bright comet.

Either way, multiple impact events are like a nuclear war, and even a failed nuclear war would be frightening enough. Single large explosions will take out many large areas and countries, but there is a global influence as well, owing to raging wildfires and dust lifted into the stratosphere, producing a celestial version of nuclear winter for which the term "cosmic winter" is not inappropriate. It is known of course that the terrestrial record includes many instances in the recent and very distant past of sudden, otherwise unexplained, deteriorations in Earth's climate, lasting for decades to centuries, and developing on occasion into a full-blown ice-age. The presence of occasional very large comets close to the Sun implies, therefore, a long-term pattern of evolution on Earth in which glaciations are interleaved with interglacials, and the interglacials are studded with cosmic winters and dark ages. We thus expect the late Pleistocene glacial, followed now by the early stages of the Holocene interglacial, to be attended by a long-lived demonic host. We also expect that such dark ages occurred during the *past* several thousand years, not to mention the future!

A long-lived demonic host may in fact say something important about cometary orbits. Thus it is known now that most short-period comets switch orbits rather frequently, as a result of close planetary encounters, and there is a suspicion that many longer-lived capturees from the Oort Cloud, which remain close to the Sun, may only be present through some combination of orbital resonances and commensurabilities. Resonances are said to occur whenever the orbital period of a Solar System body oscillates close to a mean value that is in a simple integer ratio to the period of a planet. So far as capturees are concerned, then, resonances can have the effect of allowing the planets to be generally avoided, at least for a while. A longer-lived demonic host among the terrestrial planets may therefore oscillate about a Jovian resonance while at the same time filling the resonance with much of its fragmentation debris. A telltale assembly of this kind in the inner Solar System would certainly be interesting not only for what it would tell us about the past evolution of a very large comet, but also for the fact that it would continue to generate further debris and dust as the host periodically passes through the resonance center. One expects, then, that the terrestrial and solar equatorial atmospheres would be simultaneously and rather significantly perturbed, cooling down and warming up, respectively, depending on the increased backscatter from such dust. The possible consequence is an underlying solar-terrestrial relationship involving long-period variations in global warming and in the cosmic-ray flux, both reflecting the motion of the demonic host. Because the cosmic-ray flux controls the distribution of isotopic species in the terrestrial atmosphere, the Earth happens to possess a climate-proxy record (that is, certain elements and organic structures and much else vary with climate, and such "proxies" leave a telltale record of the ephemeral climate change). Thus the demonic host may also have a significant say in the ^{14}C content of arboreal cellulose and in the ^{10}Be content of atmospheric aerosols, trapped in polar ice.

The situation, then, is that there may well be millions of twentieth-century demons skulking in near-Earth space that have so far evaded detection. In addition, there may be a host of demons with the ability to creep up on us unawares and cause immense devastation, though the host may also continually reveal its presence through its rather gentler influence on the atmosphere and the climate. Thus where, for the sake of an outmoded tradition, it may seem that global warming either does not occur or is an industrial side-effect, thereby encouraging billions of dollars toward useless cures, it may be that a host of fallen angels is at play, and that the theological inclinations of influential scientists should be more closely examined!

The Cosmological Can of Worms

The Space Age has brought several unexpected additions to the crumbling Taurid stream, showing that there is much we still need to learn. Thus in addition to the several comets and asteroids so far discovered, and their great swath of cosmic debris (meteors of various sizes together with cosmic dust), we are also aware now of a huge swarm of fragmentation debris at the 7:2 Jovian mean-motion resonance (period = 3.39 years), which the Earth-Moon system appears to have penetrated in June 1975 and November 1981, and possibly at other epochs as well. These particular observations would allow the demonic host to be at the positions of Comet Encke in 1786, when it was first detected, and of the so-called Encke dust-trail in 1983, when it was detected by the Infrared Astronomical Satellite IRAS.

IRAS functioned for only a year, and it has unfortunately not been possible to study the trail since. However, the observations evidently allow an invisible core object that underwent a chance fragmentation just before 1786, producing both Comet Encke and the increased flow of Taurid meteors recorded by Chinese astronomers during the last few decades of the eighteenth century. The invisible demonic host would then be oscillating in the resonance with a half-period of around 200 years, exactly in phase with the present round of global warming and with the observed climate-proxy cycles. The remarkable thing about this demonic host is that its orbital precession would also have had it intersecting the Earth's orbit around 400 and 600 A.D., more or less in line with the European Dark Ages and in accordance with sustained bombardment during the decline of the Roman Empire, culminating with the latter's collapse at the end of the fourth century. Not all contemporaries were of course blind to these celestial events in their midst. Earlier fragmentations, reinforcing this bombardment, may well have been professionally observed. The star of Bethlehem, for example, apparently coincided with the observations by a civilized people on the other side of the world of a notable broom-star considered capable of marking the beginning of a new age. Locally, the cosmologists claimed the re-emergence of a primary deity and predicted the apocalyptic return of its celestial horse-drawn carriage 400 years hence. The evidence can lead us to believe that the fourth- and twentieth-century demons, like the incipient new cosmologies of these times, are in fact very much the same.

One has to bear in mind, of course, that a small demon gratuitously struck the site of Tunguska in Siberia in June 30, 1908, wiping out some 2,600 square kilometers. One has also to bear in mind that most of our ancestors, at least those extant more than 300 years ago, were not really under any illusions concerning the possible existence of a demonic host. Europeans were extremely fearful of occasional comets, which were commonly associated with dense packs of demons, heralding the day of judgment. Most normal comets, which are small, are not usually a danger, and the problem may always have been to recognize a particular, presumably intermittently bright, comet *along with its invisible demonic host*. The Chinese, who were more advanced than the Europeans and generally more knowledgeable in these things, seem to have been particularly concerned about the arrival of broom-stars. Their astronomers were certainly directed to keep a lookout for celestial hazards, as were the Babylonian astronomers between two and five thousand years ago. It is hard to believe that such thinking would have arisen so frequently without repeated prodding from the celestial environment. Historically speaking, in fact, any tendency on the part of western civilization during the last 300 years to argue from the scientific consensus that global catastrophes due to the demonic host are improbable is plainly anomalous. Indeed, the findings of twentieth-century demonology would even suggest now that the recent scientific consensus has been seriously mistaken. This should not occasion too much surprise—the scientific establishment is particularly inclined, in fundamental matters, to make a virtue of conservatism, and has often been known to be wrong.

It may be, then, that the theologically inspired establishment will now have to face a period of readjustment. There are demons out there that we still have to trace. There were also demons out there in the past, and we should listen to what our more perceptive ancestors had to say. To understand the considered views of our ancestors, one can do worse than to go back to the accounts of the likes of Proclus, from whom the great scientific revolutionaries of the European Renaissance, versed in the ways of Aristotle, gradually learned to question the planetary spheres and take on elliptical orbits. Proclus, it seems, was aware of very ancient secular knowledge, and pictured such orbits bearing a host of fallen angels, known to be derived from archangels, themselves the offspring of an original and omnipotent source. For Augustine, this varied community of angels was the "civitas dei." In the Platonist account, to which Proclus subscribed, the source was originally responsible for two intersecting rings of material in the sky, "heaven" and "earth." The former split further to produce the more important deities, which were supposedly bunched together, therefore, Taurid-like, in a collection of inclined elliptical orbits, while the latter, known as the demiurge, was both guardian of day and night, like the zodiacal light, and responsible for creating all things in the ecliptic, including our planet and every form of life thereon.

Plato was very insistent on the precessional motion of "heaven," counter to the normal orbital motion, though much to the bafflement of commentators since, who have been generally unaware of anything in the sky to which the ancient knowledge might reasonably refer. In addition, while describing these phenomena in the *Timaeus*, Plato referred to both the intersection of "heaven" with the ecliptic (the cross) and the return (with catastrophic consequences over long periods of time) of invisible bodies in space. The combined cosmic and terrestrial significance of the intersection would evidently suggest a very potent symbol during the centuries of orbital approach by the demonic host, and under circumstances where the devastating consequences of the approach were increasingly observed and appreciated, it is understandable that the cross should have been perceived as such by Plato, by a little-known sect of Essenian astrologers with a direct line to Babylon and the East, by Constantine, and then by western civilization as a whole.

The Turning Point to Come

To sum up, then, the population of planet-crossing asteroids and the past cratering record are not the only aspects of the inner Solar System environment that matter. We also have to concern ourselves with the *origin* of the planet-crossers, and where it was once supposed that the Main Asteroid Belt must be the primary source, the evidence now points rather firmly to an additional source in the form of giant comets, or Chirons, from the Oort Cloud that are deflected by Jupiter into sub-Jovian space, where they proceed to break up rapidly and intermittently under the influence of the Sun. Lesser events, producing smaller debris, are part of history, while greater events, producing larger debris, contribute to the cratering record. The celestial threats to civilization and species are thus two aspects of the same story. Indeed, by studying the "disintegrating giant comet" environment of the present, we learn about the "disintegrating giant comet" environments of the past, one of which may well have been responsible for the main features of the Cretaceous-Tertiary event: a mass extinction due to the rapid descent of a large, entrained piece of debris and a mass speciation due to the much-slowed descent of entrained, much smaller, pieces of debris. These latter are vouchsafed by the amino-acid deposits above and below the Cretaceous-Tertiary boundary, processed material apparently from the core of the giant comet, leading us to suppose that the building blocks of life may be cometary as well. Natural selection is not, then, the likely source of novelty in evolution: rather, it is the filter through which celestially weakened and celestially modified species must be drawn. Even the centennial-to-millennial inputs are examples of this filter, and the hazard to civilization from near-Earth space may have to be accepted as having a biological component as well.

The guardians of knowledge are often responsible for its pace of change. Readers not familiar with the constraints on science imposed by political correctness may be surprised that a line of investigation concerned with supernatural forces can penetrate (and undermine) so many citadels of knowledge without public attention. This, however, is in the nature of things, as those who deal with deep traumas of the collective mind are well aware. Indeed, there is a kind of justification for official inertia and preserving the illusion of cosmic safety, since new findings are often by no means what they first appear. So, while the demonic host continues to be invisible (to date it has not been directly observed), there will be suspicions of a cosmological cover-up. Nevertheless, the fact that NASA is now assessing the hazard to civilization from near-Earth space will probably serve to counter the worst excesses of cosmologists fixated by the wider problems of the universe, who persistently trivialize the historically important role of comets.

Acknowledgments

Profitable discussions with Dr. D. J. Asher and Dr. D. I. Steel are gratefully acknowledged. Original sources are not cited in this essay, but appropriate references may be accessed through the bibliographic note, which lists some recent publications.

Bibliographic Note

In Appendix I of *The Sirius Mystery* (Sidgwick & Jackson, Cambridge, England, 1976), R. K. G. Temple has drawn attention to the only unbiased assessment in English of the Platonic tradition relating to "invisible circulations," which is due to Thomas Taylor, writing around 1800. Monthly and annual variations in the fireball flux recorded by Chinese astronomers during the last 2,000 years are summarized in I. Hasegawa (1992) in *Celestial Mechanics and Dynamical Astronomy* 54: 129. A discussion of the implications of these findings in terms of the evolution of the Taurid meteoroid complex is provided by D. J. Asher & S. V. M. Clube (1993) in the *Quarterly Journal of the Royal Astronomical Society*. The significance of active meteoroidal streams in relation to the population of near-Earth asteroids has been discussed in D. I. Steel (1992) in *Nature* 354: 265. These developments are based extensively on earlier work relating to the evolution of giant comets discussed in S. V. M. Clube & W. M. Napier, "Fundamental Role of Giant Comets in Earth History," *Celestial Mechanics and Dynamical Astronomy* 54: 179–93; in S. V. M. Clube & W. M. Napier, *The Cosmic Winter* (Basil Blackwell, Oxford, 1990); and in M. E. Bailey,

S. V. M. Clube & W. M. Napier, *The Origin of Comets* (Pergamon, Oxford, 1990). The idea that the population of near-Earth asteroids is largely of *cometary* origin, thereby explaining many longer-term aspects of the terrestrial record, was developed before the seminal 1980 discoveries at the K/T boundary: see W. M. Napier & S. V. M. Clube (1979), *Nature* 282: 455.

8

The Liturgy of Science: Chaos, Number, and the Meaning of Evolution

Herbert R. Shaw

I have earlier expressed my views (Shaw 1987a, 1988a) on the role of *nonlinear dynamics* and *theories of chaos* in interpretations of co-periodicities, or resonances, between timings of impacts, biological extinctions, geological boundaries, and other terrestrial and extraterrestrial phenomena. Little more has happened since then to warrant refinements or revisions with regard to the extinction problem per se (see McLaren 1970, 1982, 1986; Alvarez et al. 1980; Raup & Sepkoski 1984; Silver & Schultz 1982; Alvarez 1987; McLaren & Goodfellow 1990; Alvarez & Asaro 1990; Courtillot 1990; Glen 1990; Sharpton & Ward 1990; Officer et al. 1992). Evidence favoring specific correlations has been greatly strengthened in some instances, especially at the K/T boundary, but community opinion in the earth sciences remains divided (but not equally; see Glen, this volume) between exogenous and endogenous controlling factors in extinction dynamics. In the meantime, I have continued to explore concepts of nonlinear evolution in both the organic and inorganic realms and find that neo-Darwinian evolution—which some allege is as simple as a randomized clock mechanism—holds a subtle, but rich, nonlinear-dynamic store. In my more recent work I have described the circumstances by which orbital motions in the Solar System have conditioned the general character of impact-extinction phenomena. These are expressed in a book currently in production at Stanford University Press (Shaw 1994).

The idea that the K/T and other extinction boundaries have been created by impacts presupposes that there is a direct one-to-one relationship between the history of impacts on Earth and the boundaries of the geological time scale. The time-scale boundaries are punctuations marked by both the direct and the indirect effects of impacts in Earth history. Assuming that that coincidence has been (or can be) proved for all boundaries, it is taken to be prima facie evidence of *nonlinear resonances*. In simple terms, resonance implies synchronization; nonlinear resonance is synchronization that is not linearly calibrated. It is as simple as that. Imagine a clock face (on a two-dimensional globe) that is marked off (by impacts) in unequal parts, but those unequal subdivisions of the circle still constitute a reproducible standard of measurement (the geologic time scale is such a nonlinear clock; Shaw 1987a, 1991). The impact-cratering record of any other body in the Solar System also constitutes such a nonlinear clock—and all of these Solar System clocks should thus match when correlated, if they are calibrated against the nonlinear evolution of planetary dynamics. If the Raup-Sepkoski view of impact-caused extinctions is, as they purport, periodic (in the harmonic sense), then their postulated calibration scale (26-million-year period) is a kind of linear clock (or simple measuring stick). But if all impact phenomena in the Solar System are nonlinearly synchronized, and if all are correlated with Earth's cratering and extinction records, than all, too, are synchronized with a common dynamical celestial reference frame (Shaw 1994). The whole constitutes a dynamic wedding of processes across the interface between the terrestrial and extraterrestrial realms, and all such processes can be calibrated against the same nonlinear-dynamically calibrated system of clocks (Shaw 1987a, 1988a, 1994).

Such a grand premise mandates the search for the precise matching of the nonlinear clocks (and meter sticks) of biostratigraphy, geochemistry, geochronology, and the cratering record. But even if the fruits of this hopeful search were not realized, that finding would only admit of greater play (still fuzzier calibration of the nonlinear clocks, which, as we will shortly learn, equates to wider chaotic bandwidths) in the coupling phenomena than are implied by the unqualified term *nonlinear resonances*, which I amplify below.

Put in the vernacular of modern beer commercials, it can be inferred that what goes around in the Solar System comes around in Earth processes—and the gusto of this dangerous liaison is undiminished over the course of the history of life on Earth. For instance, the internal "orbital" motions of vortical (whirlpool-like) structures in the Earth's core—analogous to the motions of Jupiter's Great Red Spot and of other prominent Jovian vortex structures—are characterized by internal flows, sudden jerks, and excursions within the Earth. Such behaviors characterize the history of the geomagnetic field relative to other Earth processes; all are of a piece, dynamically speaking, with analogous orbital motions in the Solar System. The orbital motions of planets, satellites, and other objects around the Earth have effected analogous jerks and excursions (punctuated rhythms) that are written in the Earth by the cratering record of impacting meteoroids and the effects of grazing encounters, such as the Tunguska event (a presumed airburst of a bolide on 30 June, 1908, named after its epicenter near the Tunguska River in Siberia; cf. Clube & Napier 1982, pp. 140ff; Chapman & Morrison 1989, pp. 3–5 and 277–79; Shaw 1994, note 1.2). The effects of impact dynamics in Earth's history have been coupled with Earth's thermodynamic engine both directly and indirectly through *impact-induced* and/or *impact-correlated* magmatism, tectonism, and mantle convection. All of these effects and processes act in concert with the dynamics of the oceans, atmosphere, and core to mediate (*tune*) the magnetostratigraphic and biostratigraphic records (*codes* or *scores*) that are imprinted in the solid Earth, and by which the orbital synchronizations of impact dynamics can be demonstrated (Shaw 1994).

This information-feedback loop indicates that in the deciphering of the geologic column we are dealing with a system of nonlinear-dynamical *coupled oscillations* (our nonuniformly marked-off clocks, treated as measuring sticks). Examples of how such oscillations are linked can be seen in the coupling between cyclic Solar System processes (e.g., planetary, cometary, and asteroidal orbits) and cyclic terrestrial processes (e.g., earthquake cycles and glacial cycles). This form of coupled oscillation is expressed in the form of resonances in time and space (cf. Shaw 1987a, 1988a, 1991, 1994; Shaw & Chouet 1988, 1989, 1991). The term *nonlinear resonances*, used in a loose sense, may include everything from the metaphor of the precisely but unevenly calibrated clock face to the chaotic regime in which those constructs and processes described above emerge from a conceptual dichotomy conveyed by the term *deterministic chaos*, wherein intervals of both sharp and "fuzzy" calibration coexist. Deterministic chaos (essentially synonymous with the term *chaos* within the context of nonlinear dynamics) is a type of behavior that is both exact and inexact at the same time. Its exact component (determinism) is given by a recipe or formula (an algorithm) for generating the chaotic structure. The inexact component arises because the chaotic structure produced by the exact algorithm is not identical in successive cycles of recursion (or replication). Successive operations of the recursive algorithm (or replicative scheme) produce differences in the detailed paths of evolving parts of the system because of what is called *sensitive dependence on initial conditions*: small differences in any arbitrary "initial" state of the dynamical system result in large differences in later stages of the system (equivalently, repeated starts from the same, arbitrarily precise, "initial" states do not produce identical structures). As an example, nonlinear dynamics describes how a small change in a single variable, such as a fraction of a degree of temperature rise locally in a weather system, could produce a global change out of all proportion to the magnitude of the minuscule initial change. That such a small change in initial conditions can yield in due course such a huge change in distant conditions is possible because initially adjacent trajectories diverge from each other exponentially. The work of science since Newton's day has proceeded with the tacit belief that a flight of geese over Berkeley could have no significant effect on the formation of clouds over Boston. Science practice has mandated a belief that vanishingly small variables can, and even *must*, be overlooked in order that we might progress in deciphering the complex real world, but chaos science demonstrates that the behavior of complex dynamical systems requires an entirely new and different analytical approach—one that takes into account the effects of sensitive dependence and nonlinear measures (clocks, meter sticks, etc.) in the replicative cycles of universal evolution.

The Earth's stratigraphic record is interpreted by Shaw (1994)—on the basis of several types of demonstrable nonlinear-dynamical phenomena and lines of argument—to comprise a series of small to large catastrophes, distributed in a rootlike (dendritic) hierarchy, in the sense of Ager (1984), that are directly analogous to nonlinear-dynamical *crises* (Shaw 1994, Chap. 5). Such a distribution is hierarchical in a sense much like the distribution of earthquakes, in that it is described by *fractal* scaling of spatiotemporal frequencies ranging from many small events to few large events, hence ranging from the smallest ecological scale to the global scale (and related sequences tend to be distributed logarithmically in both time and space). *Fractals* are geometric objects that show similar patterns of complexity at different scales in space and time, like the dendritic branching pattern of a river system as it moves from brooks to streams to rivers. Objects in which scale ratios are expressed by fractal geometry, and which appear similar at different scales, are said to possess *self-similarity*. Ager's concept of hierarchical biostratigraphic catastrophes is simply an expression of the self-similar properties of the geologic column.

In this essay I refer to the hierarchical distribution of scale-dependent catastrophes as a system of *impact-related, nonlinearly resonant boundary crises* (INRBC); extinction is only one among a suite of crises that characterize the boundary. *Crises* are self-induced anomalies (transients) within systems of coupled oscillators. In such a system the oscillator of interest (e.g., the belt of asteroids in the Solar System) is operating in the chaotic regime; i.e., crises are universally characteristic of coupled, dissipative nonlinear-dynamical systems, and are said to be *universal*, because they exhibit characteristic scaling coefficients that apply to any type of system, regardless of its physical and/or chemical makeup, as discussed in Baker & Gollub (1990, p. 81) and in Shaw (1994). Examples of universal dynamical structures are, loosely speaking, such things as the branching systems of both streams and organic vascular and pulmonary systems, and the orbital hierarchies of satellites and planets, in which the same general class of proportionality constants (universality parameters) is preserved. The concept of chaotic crises mentioned above is closely related to the phenomenon of *critical self-organization*, the latter term usually referring to dynamical systems with many degrees of freedom (see Bak et al. 1988), as discussed in Shaw (1994, Chap. 5). Self-organization, in turn, can be viewed as a natural analogue of the generating functions (algorithms) that produces these universal systems of self-similar structures. The term *critical* reflects the unstable, precariously balanced state of a dynamic system that is perched precisely between nonlinear periodicity and chaos. An analogy is a stream channel in which the flow of water oscillates between two characteristic states: the stream flows smoothly with intermittent jitters at many different scales that reflect the near-criticality of its state, but then suddenly and intermittently is driven into fully developed turbulence by the slightest additional disturbance. Any complex system with numerous intermittent processes, such as the entire Earth system itself, can be said to be critically self-organized; so, too, the Solar System, the Galaxy, and the universe at large (Shaw 1994).

But applying this systematic point of view to the impact-extinction debates seemed to surprise some and to be largely resisted or rejected by others. Such a reception was not surprising: appeal to authority and adherence to orthodoxy are as solidly entrenched today as always. To some, authority is represented by the infallibility of probability and statistics; to others, by the infallibility, in general, of Newtonian dynamics and, in particular, of Newton's laws of motion, as applied to celestial mechanics. But to yet others there is a still greater infallibility, cobbled together arbitrarily from Newtonian and probabilistic viewpoints and run by computers like a Monte Carlo gambling casino (hence the name *Monte Carlo simulation*). All the machines and games in the casino are set to give, on average, a certain percentage of profit to the house. The net result of such an operation (which includes the machines, players, house employees, and computers) is a critically self-organized process that guarantees the house a profit in the long run, but achieving that self-organized state, and the guarantee, requires the efforts of a great number of players over an extended period.

Nonlinear dynamics constitutes a framework both synthetic and analytic, a framework that allows us to view familiar processes and their products from wholly new vantage points, at some remove from the paradigms of established science. Ironically, however, the "new" vantage points turn out to be—in truth—older than the paradigms, in their correspondences with our own physical and biological makeup, dynamically speaking (e.g., Skarda & Freeman 1987; Freeman 1991, 1992). The language of nonlinear dynamics is often felt not to be understandable to persons either in or out of science, because the resulting visions of familiar things are often so alien to the traditions of science, and therefore objectionable, that resistance to them is natural. It is natural to reject any mechanism that aberrates our touchstones and the sacrosanct tenets of the scientific method—the *liturgy of science*. Resistance to nonlinear-dynamics analysis has been the rule since its advent, and only in the last half-decade, with repeated successful applications in various fields from meteorology to cardiology to geology, has its reception begun to warm. Novel ideas meet resistance as a function of the extent to which they threaten existing ideas, and new ideas that lack analogues in extant systems of knowledge produce threatening visions within the communities charged with their appraisal, leading either to the vilification of their authors, or, even worse, to their being simply ignored. The history of science is replete with such cases, and it will be interesting to compare their receptions to the fate of nonlinear dynamics. We shall also try to identify the qualities and components of a new idea that provoke rejection, and segregate the objective and the rational from the visceral and the subjective in the reception process.

The opposition of certain mathematical physicists to continental drift, seafloor spreading, and the new global tectonics during this century exemplifies a historically recurrent mode of dissent that seems a prologue to a yet larger mode, indeed what one might call a paradigm of ideological metamorphosis. Such patterns of dissent emerge like instances of a genetic memory template, throughout the history of science. I choose two examples in support of this point. William Thomson (Lord Kelvin) was the most famous predecessor of one of the doyens in the fight against continental drift, Sir Harold Jeffreys—both being descendants of that magisterial lineage of mathematicians and physicists that takes Sir Isaac Newton as its paragon. Kelvin, for example, adopted the same spirit of dissent in initially opposing the overthrow of the "political" regime of the *Caloric Theory of Heat* in favor of the new regime epitomized by Joule's demonstration of the *Mechanical Equivalent of Heat* (see Sandfort 1962). The interdisciplinary comparison of Kelvin's stance with that of Jeffreys shows that the seemingly divergent evolutions of thinking seen in theories of heat and motion, on the one hand, and in geological-paleontological theories, on the other, have turned out to be but alternative expressions of thoughts that both then and now have been on a convergent, if not collision, course.

Rheology, a discipline that treats the deformation and flow of materials, is centrally concerned with questions of heat and work. It is a field in which I have been long engaged, in an effort to better understand the phenomenon of melting and transport of molten materials in the Earth. The generic term for naturally occurring molten material in the Earth is *magma*, which includes *lava* as the volcanic form (e.g. Shaw 1965, 1969; Shaw et al. 1968). In a recent review of nonlinear dynamics as applied to magmatic phenomena (Shaw 1991), I pointed to a historical bifurcation between the disciplines of mechanics and "thermics," and to their potential reunion in an improved understanding of magmatic phenomena in the Earth and Solar System (cf. Shaw 1983a)—noting that the word *magma* stems from the Greek for *to knead*, a nonlinear dissipative process of repeated stretching and folding. In Shaw (1991) I illustrated this point with a series of experiments that I originally performed in 1968, experiments that I had deliberately cast in the eighteenth- to mid-nineteenth century modes of Benjamin Thompson (Count Rumford) and James Prescott Joule, but which—in order simultaneously to demonstrate two kinds of paradigmatic shift—I recast in the mold of nonlinear dynamics (Shaw 1991, Figs. 8.1–8.9). In reviewing the application of those experiments to magmatic phenomena in 1991, I had acquired the added insights and *language* of nonlinear dynamics and theories of chaos, and found that I could apply them to the interpretation of the observed kinematic patterns of the 1968 experiments as a form of *dissipative structure* called a *strange attractor*. A strange attractor is a complex (fractal) form of an attractor, and an attractor is simply a locus toward which components (atoms, molecules, planets, galaxies, or dots on a computer screen) of a system are attracted in space and time as they move toward a balanced state. The experimental system of 1968 was, in fact, a *thermodynamically irreversible pumped system operating far from equilibrium*, the sine qua non of all complex self-organized systems, hence of all living systems (see Prigogine et al. 1972). The notion of a strange attractor—a term first coined by Ruelle and Takens (1971)—can be likened, in simplest form, to patterns such as those seen in the rising streams, swirls, and toroidal vortices of cigarette smoke, or in the analogous structures seen in the zonal streaming motions and vortices in the atmosphere of Jupiter, such as The Great Red Spot (e.g. Ruelle 1980; Shaw 1987a; Gleick 1987).

These experiences with experimental rheology, thermodynamics, and nonlinear dynamics convinced me that there has been a marked ideological vacuum in mathematical physics since its inception. It came dramatically into focus for me while reading the history of heat engines by Sandfort (1962), particularly for the period between the late eighteenth century and the mid-nineteenth century. The vacuum I came to recognize lay in the schism between, on the one hand, the experimental simplicity and mathematical "purity" of theories in mechanics and, on the other hand, the experimental difficulty and messy mathematics of studies related to the traditionally imprecise and ambiguous theories of thermal phenomena. Therefore, even when thermodynamics became an established discipline, following the acceptance of Joule's work around 1850, mechanicists continued to go their own historic way, while thermodynamics evolved into a new discipline that took on a separate-but-equal thrust of its own. In between the two were, and are, the rich phenomenologies and fertile theoretical fields of *dissipative processes* and *theories of attractors* that are only now beginning to be seen in the "new" light of nonlinear dynamics (e.g. Prigogine & Stengers 1984; Gleick 1987; Shaw 1987a, 1991, 1994).

When I joined my observations concerning dynamical phenomena with my geological studies of global tectonics, I found therein the very same bifurcation of thought—hence the convergence of ideologically divergent concepts in science I mentioned at the beginning of this historic aside.

It was, in fact, my struggles to reconcile the theoretical and experiential aspects of the rheology of magmatic processes with the larger framework of geodynamics and thermodynamics that originally led me to suspect that the schism I allude to represented a massive blind spot not only in geological thought but in what were, at the time, the "present-day" disciplines of physics and applied mathematics as well (cf. Gruntfest 1963; Shaw 1969; Gruntfest & Shaw 1974). Irving J. Gruntfest helped me to realize that my training in the physical sciences had somehow blinded me to the phenomena, for example, of thermomechanical feedback in rheology. Putting the matter differently, I came to concur with May, who in 1976 pointed out the deficiencies in the traditional methods of education in physics that were illuminated by the insights of nonlinear dynamics. Later I recognized that my thoughts concerning the roles of ideological vacuums and patterns of dissent in the history of scientific thought fell into a natural resonance with those of William Glen (1982), which he presented through meticulous, detailed descriptions of the way in which the now-paradigmatic concepts of continental drift and plate tectonics were resisted during the early and mid-twentieth century. The more subtle and more pervasive historical bifurcation and vacuum of which I speak still persists in the disciplines of mechanics, rheology, and the physics of the Earth and planets.

I chose the two protagonists alluded to above, William Thomson (Lord Kelvin) and Sir Harold Jeffreys, to emphasize the parallels that I see between the misapprehensions of the "exact" sciences and those of the "inexact" sciences. My reference to Kelvin's initial reaction to Joule's experimental demonstration in 1847 of the energetic equivalency of heat and work is found in historical sketches by Sandfort (1962, pp. 76–80). Joule, in turn, and in effect, had "cleaned up" and made scientifically presentable the rheologically more profound experimental studies originally brought to the fore by Rumford in 1798 (see Sandfort 1962, p. xviii; Shaw 1991). Rumford's visceral descriptions of what took place in his cannon-boring mill in Bavaria seem to me to have marked the dramatic point of departure that can now be seen to lead directly to the present-day collision—and ultimately, I hope, the effective merger—between the principles of mechanics and the principles of dissipative processes.

I have found the study of dissipative structures in rheology useful in hinting at the nature of potential geometric and kinematic differences between dissipative and conservative structures in celestial dynamics and cosmology. The many existing studies of nonlinear yet conservative *Hamiltonian systems* (simplistic nonlinear-systems theories that descend directly from classical mechanics, and hence lack dissipation and any evolutionary capability; an equivalent statement is that *attractors* do not exist in such systems; cf. Ruelle 1980; Marcus 1988). Though rich in revealing regimes of nonlinear periodicities and chaos, the theories of conservative, nondissipative systems (Hamiltonian systems) ultimately are misleading by reason of their tacit reinforcement—or reenactment—of the age-old, but still not worn-out, closed-system point of view in cosmology (cf. Helleman 1980; Jensen 1987, 1992). There is no evolution of regions of attraction in the Hamiltonian type of "conservative" systems, and like an attempt to learn the topology of the Earth by studying maps of a flat world, it is impossible to learn the nature of dissipative structures in cosmology from such syntheses (this usage of "conservative" is placed in quotes because it refers to conservation of area, volume, and so on, in classical *phase space*. [*Phase space* is that space in which one plots the motions of the dynamical system. A *phase portrait* was originally just a plot of the position vs. the velocity of an object in phase space; the line that results from connecting the sequence of points in time and space (the phase space) is called a phase-space *trajectory*. In general, phase space can be thought of as any space in which the key relationships that best characterize a system of motions are plotted—hence the phase-space plot can be thought of as the system's signature, in the sense of a fingerprint or DNA print, etc.]

By contrast with the conservative approach to celestial dynamics, the nonlinear resonances predicted from my recent investigations of the coupling between terrestrial and extraterrestrial dynamics are based on the existence of dissipative structures and attractors in the Solar System (Shaw 1983, 1987a, 1988a, 1991, 1994). I came to recognize that the notion of a Hamiltonian, nondissipative system—a frictionless world—is merely an idealization, and that particular idealization is what we have called the Solar System, a system composed of an immutable set of planetary orbits. A system that is Hamiltonian or frictionless does not contain attractors. But the distribution of the planets, their satellites, the comets, and the belts of "minor" planets, such as the asteroids and other objects postulated by some workers (e.g. Stern 1991) to be planetesimals still surviving in regions near and beyond Neptune are *not* free of friction and thus represent a *system of attractors*. I intend this phrase to indicate a system that is not only largely chaotic (in a broad sense, as discussed below) but consists of subsystems, such as a planet and its satellites, the Asteroid Belt, and so on, that correspond at their own characteristic scales of motion to the same, or at least similar, chaotic descriptions. The dynamical properties of the system as a whole, and/or its parts, are in some respects either analogous to or identical with the concept of a strange attractor, in which the evolution of the trajectories of motion are *fractal* in character (Ruelle & Takens 1971; Ruelle 1980; Shaw 1994). The term *fractal*, recall, refers to properties of *self-similar invariance*, meaning, in a simplified sense, that the spatial and/or temporal properties that characterize the system at a large scale are mimicked in kind at smaller scales, or vice versa. In caricature, then, this is equivalent to saying that all satellitic systems—the planets being "satellitic" to the Sun—look alike and act alike, in a scale-model sense, regardless of size. Although this obviously is not true in special cases, it is close to being true as a rule of thumb. Such structures are not necessarily evident by inspection over short periods of time because many cycles of the motion must be plotted with great precision if we are to uncover the nature of fractal chaotic motions (Ovenden 1975; Ruelle 1980; Shaw 1987a, 1991, 1994).

In my new studies (Shaw 1994), the records of mass impacts on Earth, and on the Moon and planets in general, are envisioned as constituting patterns generated by a nonlinear and figurative "hourglass" of actual and potential impactors flying about wildly throughout the Solar System, but in such a way as to be highly organized (*critically self-organized*) in diverse groupings of orbital characteristics, whether heliocentric (Sun-centered), geocentric (Earth-centered), or otherwise localized relative to "random" models. A random model is maximally complex, meaning totally lacking in any coordinating, hence simplifying, structures of any kind relative to the number of bits of information *that would be required by an algorithm* to describe the system as a whole, either as it is observed or as the observer *chooses* to observe it (a sometimes subtle, and sometimes not so subtle, distinction, the latter *usually* implying a blind faith in the conventions of probability and statistics; Shaw 1994, Introduction). Such models of algorithmic complexity tell us only that we cannot tell in a given case, for a given expenditure of effort, whether the complexity of some aspect of nature really reflects a class of entirely uncoordinated and uncomputable structures (equivalent to an undecidable proposition in mathematics; cf. Post 1965; Ruelle 1991), or whether we have just been unwilling to search for an algorithm that is capable of reducing the maximally complex description to a simpler, hence more redundant, one. The discovery of the genetic code in biology, and of the geomagnetic time scale in geology, were examples of dramatic *and unexpected* reductions in the algorithmic complexities of natural phenomena, because suddenly less information than before was required to describe a given biological structure or a given geological history, and some structures and histories that had never been described at all could now be given quantitative expression.

Arrant presumption may be—and often is—all that denies us the possibility that analogous codes exist within geological and cosmological structures (see the preliminary searches for, and indications of, such codes in Shaw 1987a, b, 1988a, b, 1991, 1994). On the contrary, the nonlinear dynamics of dissipative systems assert that any natural system, subsystem, or supersystem is *bound to be* highly organized relative to the maximum complexity of states the system would be capable of if there were no interactions, and no feedback, among its *actual* states (see the discussions of *correlation dimensions* and *degrees of freedom* in Shaw 1994, Chap. 5, Note 5.3; cf. Chouet & Shaw 1991). The complexity of the genetic code, though miraculously simple to those persons who were steeped in the preceding paradigm of molecular genetics (see Judson 1979; Schrödinger 1992), apparently is capable of even greater simplicities based on *linguistic principles* (Searls 1992), an approach that Shaw (1987a, b, 1988a) concluded should apply, given comparable amounts of information, to complex natural systems in general.

The hourglass mentioned above symbolizes an open-system nonlinear process of critical self-organization (Bak et al. 1988; Chen & Bak 1989; Shaw & Chouet 1988, 1989, 1991; Shaw 1994, Preface, Preface Note P.2, Introduction Note I.2) that extends, in principle, beyond the Solar System into the interstellar reaches of the Cosmos (implicating interstellar processes, Galactic and intergalactic processes, processes of supergalactic clustering, and so on. In the Solar System, critical-state processes provide an intermittently continuous source of showers of asteroids, comets, and other near-Earth objects (the recently much-discussed NEOs; cf. Steel 1991). A fractionally very small, but poorly known, intermittent population of these objects eventually comes to Earth, along both ballistic and resonant geocentric trajectories, creating intermittent systems of small natural Earth satellites—which may correspond to "large body" impactors—that interact resonantly with longitudinal mass anomalies in the Earth (e.g., Allan 1967a, b, 1971; Shaw 1994, Chap. 6). This general process writes (by *impact printing*), on the surfaces of all planetary and satellitic objects in the Solar System, spatiotemporal patterns of impact craters and other impact-related scars that record the nonlinear thermomechanical histories of these smitten objects. As a result, a kind of impulsive "ratcheting" of the surface and interior motions is effected by impact-tectonism and rheological feedback, representing a form of orbital-ballistic telegraphy that punctuates planetary surfaces and interiors somewhat like the action of a nonlinear laser printer. And the printed patterns, in turn, are transcribed through the media of paleontology, isotope geology, and geochronology into the languages of biostratigraphy, magnetostratigraphy, and magmatic stratigraphies (volcanic and plutonic), becoming what we call the *Geologic Column* and the *Geologic Time Scale* (see Harland et al. 1990). In actuality, this record might be better described as a multidimensional calibration chart, or code, that memorizes the spatiotemporal patterning of nonlinear geological-cosmological processes.

My own notice of the predisposition toward dissent by Jeffreys during the early- to mid-twentieth-century developments in geodynamics (cf. Glen 1982) came from his authoritative pronouncements concerning the workings of the Earth tides, a stance that was virtually necessitated by his position (or vice versa) concerning the inefficacy of rheological phenomena within the Earth to support the migration of its "solid" parts over long distances (cf. Shaw 1970; Shaw et al. 1971). The thermodynamic *condition* of "solidity" often is equated with the rheological *property* of "rigidity" even today, depending sometimes on the outcome favored (e.g., one person might argue for the "fixity" of rigid tectonic plates while another argues for the "rafting" of portions of rigid plates halfway around the globe, both using the same terminologies). An aspect of Jeffreys' writings that struck me powerfully was a form of bicameral advocacy that posed a contradiction: he was influential in imposing a still-effective conviction among geophysicists that virtually all (read absolutely all) tidal energy is dissipated, as in Joule's experiments, in the oceans; but at the same time, he had to conclude that the Moon and other planetary objects lacking oceans—or any evidence of the action of fluid dissipation (other than the, to him, little known details of dissipative phenomena associated with the dynamics of coupled magmatism, tectonism, and mass impacts)—must have evolved to their present states of orbital and rotational motions by unidentified mechanisms of internal dissipation (cf. Shaw et al. 1971, p. 871). Had Jeffreys taken such evidence a step further, he would have had to recognize the integral nature of *coupled orbital-rotational dissipative oscillations* in the evolution of planetary bodies as the direct manifestation of indirect couplings between their internal dynamics. This is the same logical step that had been traditionally eschewed by mathematicians and physicists in an insistence on the apartheid-like separation between the pure disciplines historically associated with the mechanics of solid bodies and the calorics of reactive substances.

It seems to me that neither Jeffreys nor scientists in general (with cases of overt enmities excepted) have malevolently exercised authoritarian power toward partisan ends. Therefore, the effect realized must be lodged somehow in conditioned practices to which the practitioner eventually becomes oblivious. Such an effectively unconscious mechanism of behavior seems tantamount to a form of constrained mentation or cognition that precludes any alternative view—and, in that sense, is analogous to a form of gestalt response that has been described by Glen (1991, and this volume) as commonly occurring in and strongly conditioned by one's theory choice and disciplinary affiliation. Since such gestalt responses appear to be the rule, it may be more beneficial to view them as the manifestation of a phenomenon that affects all scientists in one way or another.

Another example of such bifurcated thinking or gestalt response can be found among scientists and lay persons alike, as for example in thinking of the mechanical and thermal parts of heat engines (e.g. the internal combustion engine) as separate and distinct entities. It is therefore not surprising that geologists tend to think of the mechanical and thermal parts of the Earth as separable phenomena, or that cosmologists tend to think of the mechanical and thermal parts of the cosmic engine in the same way (e.g. celestial mechanics vs. thermonuclear processes and plasma dynamics; cf. Alfvén & Arrhenius 1975; Alfvén 1981; Zirin 1988). It seems natural to suppose that the roots of bias in science likely lie in the same psychological sources that house the unconscious bigotries obliviously practiced outside of science.

This area of investigation is likely to be a fertile field for psychological applications of nonlinear dynamics, especially with respect to the evolution of "fixed-point behaviors," "bifurcational states," and other types of attractor-like patterning (e.g. Abraham 1990).

Carrying the above sort of comparative analysis further, it would seem that neither Kepler nor Newton was entirely forthcoming in revealing the motivations for their science. All scientists fear that their hidden motivations, if discovered, surely must detract from their credibility, impugn their credentials, and erode their security. Historians have ferreted out a good deal of "the dirt" about Kepler and Newton (cf. Hall & Hall 1962; Beer & Beer 1975; Beer 1979)—perhaps more about Kepler than about Newton because of Kepler's open enthusiasms for diverse subjects—but I seriously doubt that we know much of what they really thought about the rigorous disciplines that have been promulgated in their names. It is ironic—but is as true in science as in religion—that Kepler's and Newton's self-appointed disciples have often used the authority of the very principles their masters passed on to them to squelch the kinds of curiosity and ranges of inquiry that characterized the creative environment within which both Kepler and Newton had thrived, and without which they would not have become known to us at all (cf. Hall & Hall 1962; Cohen 1975; and see also Endnote 8.1, this volume).

It is no secret that one of Kepler's greatest hopes—perhaps taking his cue from Pythagoras—was to decipher the playing of "the music of the spheres" (see testimonials in Beer & Beer 1975). Analogously, it would seem that Newton found inspiration in, and mimicked, the rigors of liturgical formalisms in his formulations of a *universal law* of gravitation, and in the litany of correlative statements called, in familiar Anglican style, *The* Laws of Motion (see Hall & Hall 1962, pp. 77–79; Cohen 1975, Note 16). Most of us have been trained to accept these laws as a gospel less forgiving than the Old Testament (cf. two of the best and most forthright texts in, respectively, physics and astrodynamics: Halliday & Resnick 1966, pp. 80–82; Bate et al. 1971, pp. 3–5.)

Putting ourselves in Newton's place (in itself a sacrilege to many of his unsolicited disciples), we might have thought as follows about the problem of interpreting the combined observations of Tycho Brahe, Galileo, and Kepler:

What might we write down as the essential premise underlying our understanding of the material world, the premise that will give expression to the principle of the extension of quality, but not substance, that is essential to the nature of reality in the spiritual world? And, given such a question, how might we state a system of law that would provide for the uniform application of that principle, just as the laws of the Church were intended to ensure the uniform practice of religious observances proper to the nature of a spiritual hierarchy?

Intuitively, a major clue might have been the idea that there should exist some agency of action that remains unaffected in kind and potential during the course and consequences of that action as it is manifested in the movement of the Heavens. In other words, this would be a *universal* agency as all-powerful in its ministrations to the physical world as is the spiritual power of *the Church* in its ministrations to the spiritual world. Let this physical analogy with God be called *gravity*, and let a corollary liturgical property of the universe be called *inertia*. Thus there is established a body of physical law with all the attributes that in religious terms are called theism, atheism, and agnosticism. The first two are explicit and implicit, respectively, relative to the law of gravitational attraction, whereas agnosticism is analogous to a necessary null state called *inertial motion*. That is to say, loosely paraphrasing in the language of the laws of motion the nature of agnosticism as the term is commonly employed in reference to the "fence-sitter," rather than as it would be employed in the more mathematical context of unknowability,

A body set in motion, which includes "rest" because there is no datum for the absence of motion, continues unmodified in that motion except as it may come to be acted on by God or by the anti-God.

The idea of such a principle underpinned both a system of thought and the documentations of Newton's predecessors—Tycho Brahe, Galileo, Kepler, and a host of others we never hear about, especially those who had a theological influence on Newton's education. Such ideas were cast in the form of rigid statements couched in the language of the mathematical formulae and proportionalities that fitted them. These formulae, in turn, employed the minimal number of variables required to satisfy the universal premise. These variables, as symbols of the faith, were accorded names appropriate to their measures, and were expressed relative to the common notions of length and time (the idea of mass being dependent on the formulae). The minimum number of these liturgical entities became the *icon-trilogy* named *mass, force,* and *energy* (or, as the fundamentalist sect would have it, *mass, length,* and *time*). Few followers have thought of these parameters as simply a set of conditional measures sufficient to the implementation of the self-consistent set of numbers needed to describe the interrelationships that satisfy both the central dictum of universal gravity and the obvious geometries of motion calibrated by the cyclic rotation of the Earth about its axis and its cyclic revolution about the Sun (there being no real measure of time in practice, and no idea of time according to the Holy Principle). Given faith (1) that the properties of euclidean and spherical geometry could be expressed in precise numerical form, (2) that there existed a linear and unidirectional measure of time, and (3) that the proportionalities among such numbers must satisfy not only (a) the universal precept but also (b) the data of Galileo concerning the acceleration of falling objects and (c) the observations of Kepler concerning time (orbital period) and distance, then the rest, so to speak, fell out of the conceptual strategy itself (cf. Gamow 1962; Ovenden 1975, p. 495).

Were such a reenactment true, it should not lessen our admiration for Newton's achievement. It might, however, bring into question the ease with which ideas are adopted, entrenched, and believed in as the ultimate expressions of the scientific mind. In nonlinear dynamics, one often speaks of attractors and trajectories of motion (or phase-space trajectories) within *basins of attraction*. The similarity to the language of gravitational-field theory is not accidental. In the language of nonlinear dynamics and dissipative processes, however, one speaks of *purely numerical structures* (in at least this respect Kepler was more modern than Newton in emphasizing, despite criticism of his astrological propensities, the importance of numerical structure in cosmology; cf. Cohen 1975; Haase 1975a, b). Likewise, in these numerical structures, there are nonattracting sets and repelling sets. The latter are the complementary parts of attraction, while the nonattracting or null sets are analogous to inertial motion.

Experimental nonlinear dynamics, sometimes called *numerical dynamics*, has made it possible to visualize regimes of "orbital-gravitational motion" in almost unlimited detail. This capability has shown that what is neither (1) self-inspired to evolve according to different courses of *self-determined action*, whether attracting or repelling (e.g. critical self-organization), nor (2) determined analogously to achieve a constant or homeostatic course of undeflected action (e.g. fixed-point attractors), ergo (3) must be arbitrary (e.g. an integrable Hamiltonian)—a form of *inertial special creation* (contrast with this third alternative the manner in which numerical structures evolve in the sine-circle map: Shaw 1987a, Cover Illus. and Fig. 2). Such "special creations" have been the fixation of celestial mechanics for centuries, even when Poincaré's questioning of them during the late nineteenth century exposed their ecumenical corruption at its very foundations; and they are the dominant paradigm even today, 30 years after Lorenz (1963, 1964) rediscovered the *principle of sensitive dependence on initial conditions* (cf. Gleick 1987) first enunciated by Poincaré (1952, p. 68) in the form: *small differences in the initial conditions produce very great ones in the final phenomena* (see Endnote 8.1).

Another of the many ironies in the history of science is that none (to my knowledge) of Newton's disciples who have used that discipline to express dissent toward proposals of change have seemed to notice that they were promulgating an inaccurate picture of the thought processes that led to their own system of thought (i.e., Galileo, Kepler, and Newton were not wedded to relatively ancient *scientific* law, though I have taken the liberty of implying, for the sake of illustration, that Newton may have been more wedded than we would like to believe—in our roles as fully conscious and rational modern scientists—to ancient religious law). They would make either the inertial state or the state of special creation dominant by default. In the absence of any attracting or repelling potentials there are none of the *punctuated* fits and starts that characterize trams, trains, airplanes, and elevators—or, more to the point, the punctuated fits and starts that characterize orbital motions, planetary rotations, mass impacts, extinctions, and life. Using Hamiltonian dynamics, one can produce systems full of regular and chaotic motions, but those motions are predicated by the initial conditions and do not evolve, being full of "sound and fury" but signifying nothing other than the ability to stipulate different reference states of independent motions. And a dissenting stance relative to the majority application, couched in either similar or analogous terms, does not offer any other reality, because it does not equate with the repelling part of either the gravitational theory or the numerical theory; and repulsion is an active mode, distinct from the status quo (authority), that implies the existence of attractors beyond the limits of our horizons. Such out-of-sight attractors therefore testify to the role of dissipation as the unifying principle of cosmology.

Few until Einstein noticed that there was something in Newton's *revelation of gravity* that is more fundamental than the notions of mass, length, and time (force and energy being derived quantities). This something is the abstraction of *non-euclidean space-time geometry*. And there is something yet more fundamental than this form of *geometro-dynamics*. That more fundamental something is the abstraction of *number* or *numerical structure*. Lacking the notion of, and properties of, numbers (or their symbolic equivalents), none of the rest of the cosmology could have been described, or even experienced, in a form that we could associate with the scientific method. It might be suggested that *ideographic* representations, as in Chinese characters or molecular biology, would obviate the digital implications of this "Gutenberg Galaxy" (McLuhan). But even in pictures or ideographic models (e.g. protein structures) there must be an ability for systematic discrimination, an ability that depends on some form of accountability, such as the number of characters required to generate a communicable language (just as in the immune system there must be an appropriate number of antigens relative to the number of pathogens that constitute the actual or potential types of communicable diseases to which a given organism is or may become subject and yet survive, in the face of all the possible nonlinear-recursion schemes available to bacterial and viral natural selection).

Yet another of the many ironies of authoritative physical science is that even the truth of the existence of a measure, called *mass*, that can be identified with the firmament rests but on a foundation of abstract number. This is demonstrated not only by the formal device by which mass was originally defined as a Newtonian quantity, but by the fact that the numerical algorithms of nonlinear dynamics can produce identical forms of the trajectories of motion that characterize Newtonian gravitational-field theory *without any statements of material nature whatsoever*.

I have intentionally manipulated Newton's *principia* in order to draw a parallel between manifestos of alternative types: religious, scientific, political, and psychological. In the material realm the Newtonian manifesto concerning *Gravity* gives rise to the Three-Body Problem, which is, paradoxically, intractable within the laws of gravity. Analogously, certain religious manifestos establish the concept of an all-encompassing God, yet they teach the reality of a Holy Trinity that is not (analytically) obvious from the principle. Neither one can be described rigorously, and in that sense neither one is understandable within the context of its own laws. Both exist only in abstraction, or in personal experience. The insolubility of the three-body, or n-body, problems is experienced in the motions of our planet in the Cosmos and in all that attends those motions, including life and the genetic potential. Yet neither the gravitational nor the spiritual potential is explicable in terms of predictable motions in the Cosmos.

The dilemmas of the *impact-extinction hypothesis* (IEH) and of the correlative hypothesis invoked by the concept of *impact-related nonlinearly resonant boundary crises* (INRBC) are not new to the world (Shaw 1987a). They invoke the age-old paradox of clocks and chaos, and the dichotomy between the will-of-man and the will-of-God. Clocks, chaos, and the three-body problem have now all been integrated in a contextual sense by the new/old paradigm of nonlinear dynamics. This occurs at a time in history when mathematicians have shown that such problems cannot be integrated in the literal sense by any calculus that the computer-aided mind of man can or ever could conceive of, by virtue of Gödel's *incompleteness theorem* in mathematics (see Nagel & Newman 1958; Kramer 1970; Hofstadter 1979). In order to be able to integrate numerically (or symbolically) any conceivable function in mathematics, one would have to be able to create an absolutely complete mathematics. But a one-to-one correspondence between an analytical formulation of unlimited complexity and its analytical or numerical integration, for each and every point in time and space, is impossible. What is left? Some say the void, in which all actions represent the symbolisms of the universe operating on, or "seeing," itself (cf. Spencer-Brown 1972, pp. 101 and 105–7). Others say God. Some might say number, or the source of all information that is number (see the biographical sketch of Cantor's work by Dauben 1983; cf. Shaw 1987a, 1988a, b, 1994).

The geological record is in fact something like a poem, or more aptly a system of poetry, in the sense that it embodies the poetic ingredients of number, language, and dream (Sewell 1951, Chap. 10). Although the term *geopoetry* often has been used in the pejorative, it is a perfectly valid expression for the spectral complexity of the study of Earth's history, which includes what we know of the history of life from the fossil record. In the circularity, or *cyclicity*, of this statement lies a clue to the major question that has plagued or blessed, depending on viewpoint, the study of Earth history, or of history in any sense that may lay claim to reflecting qualities of the universal.

Elizabeth Sewell (1951, p. 77) defined the words "number," "dream," and "language"—and implicitly "order," "disorder," and "nightmare"—in the context of poetic structure according to her contextual purpose of revealing the nature of the ephemeral bridge called *communication*, as follows:

Number: An infinite collection of complex mental concepts, forming an independent system of relations based largely on the principles of similarity and succession, which the mind employs to bring the relations of experience into order.

Dream: An infinite collection of complex mental concepts, forming an independent system of relations based largely on the principles of similarity and succession, which the mind employs to bring the relations of experience into disorder.

Language (as it is represented in poetry): a universally recognizable admixture of these ingredients (e.g., Sewell 1951, Fig. 1, pp. 50 and 79ff).

Sewell's definitions reverberate with my own thoughts about the nature of the *Geologic Column*, particularly as it is expressed in the language of biostratigraphy. Or, perhaps, I should say that those discussions in the literature of biostratigraphy that are aimed at *communication* of its essential nature are couched in terms that are recognizable to me in such a context (see McLaren 1970, 1982, 1986; Ager 1980, 1984). The syntax of poetry, and of language in general, is not unlike the "syntax" of biostratigraphy. Number is expressed in meter and in form, in the usage of rhyme (a form of codified redundancy, or reduction of complexity, somewhat analogous to nonlinear resonance), and in the hierarchical relationships of grammar. Sonnets, ballads, and so on are metaphorically analogous in cyclical character to stratigraphic cyclothems (a series of beds deposited in a single sedimentary cycle that typically reflects marine transgression and regression), wherein repetitive superpositions of characteristic strata (stratigraphic bundles) are numerically similar, but not identical, from one cycle to the next. I have performed numerical experiments in which I have transcribed cyclical geological information of other types (cf. Shaw 1987b, 1988a) into English letters, based on abbreviated alphabets, and hence into words, finding that—even though the "vocabulary" was severely limited and the word sequences usually made no sense—the structural complexity of the letter and word frequencies was similar to that of similarly abbreviated samples of the English language (see Shannon & Weaver 1949). In parallel with these explorations, and aided by the far greater abundance of numerical information available from molecular biology, some geneticists have been exploring generative grammars and biological messages inscribed in DNA and RNA sequences, with the hope that they will eventually be able literally to read the language of nucleic acids and proteins (e.g. Searls 1992).

I suggest from all this that a profound conclusion is inescapable. Each of the categories of natural data just mentioned is made up of a rich variety of symbolic entities that I shall informally call *cognons*, by reference to the fundamental units of meaningful linguistic information in a given natural record, or in two or more types of natural record taken together (which would be analogous to dialogues, conversations, etc.). For instance, a given geological record might be made up of volcanologic-paleomagnetic-sedimentologic-paleontologic-biologic aggregations of linguistic data that are analogous to the aggregations and arrangements of letters, words, lines, stanzas, and so on, of a story consisting of several different themes. Both types of story are "put together" or combined according to explicit, though inexact, rules of order to produce numerical structures of diverse types that vary widely in *relative redundancies* (see below) within the guidelines of a hierarchically limiting grammar (e.g. Searls 1992).

This brief outline hopefully is sufficient to the day in making a point concerning what was to have been my intended topic for this volume—the meaning of *linear and nonlinear periodicities*. Language is the general term for the medium of communication—whether poetical, geological, or mathematical—by which the nature of periodicity is conveyed, and language is a combinatorial scheme in which there is an appropriate though not absolutely specifiable mix of the essential elements or components Sewell called "number" and "dream," or "order" and "disorder." Deterministic mixtures of periodic-aperiodic cognons—or, in general, of periodicity and chaos (the latter inherently containing both periodic and aperiodic elements)—are ingredients of (1) written and spoken language, (2) the geologic record, and (3) the dynamical mechanisms that are essential to the universal functioning of *impact-related nonlinear resonant boundary crises*, hence to the putative phenomenologies of the impact-extinction hypothesis, and to extinction phenomenologies of any other kind that are communicative components of the biostratigraphic code.

At about the same time that Sewell (1951) began to develop her theme (at least three years prior to its publication, to judge from her Preface), Claude Shannon had just published his major work on *a mathematical theory of communication* (see Shannon & Weaver 1949;

Pierce 1980). Shannon's theory also has come to be known as *information theory*, or, equivalently, as the *entropic theory of communication*. Without engaging the formalisms of the subject as put forth by Shannon, the essential idea of the principle of codifiable transmissions is essentially identical to the richer codification qualitatively identified by Sewell in her study of the principles of poetic communication. Shannon's great achievement concerned the ability to transmit and receive information purely in the form of *number strings*. These number strings conveyed characteristic information (entropy) contents as the basis for systems of *decipherable codes*. Such codes represent, in effect, the numerical subsets of information abstracted from some language statement within which they can be given meaning. In themselves, the number strings are symbolic sets devoid of context (other than that of the transmitter/receiver system that gives them their identities), in this respect resembling the nature of fossil forms that have been abstracted from their biostratigraphical or other geological settings (Shaw, 1987b, in almost this literal sense, applied Shannon's theory to the frequency structure of the "fossil" earthquake-recurrence history of California). Compared to the more qualitative, but richer, formulation by Sewell (1951), it would be as if a poem had been placed in an acid bath that dissolved away all of the contextual matrix that attended the circumstances of its origin—including whatever residues remained of the DNA-based machinery of its transcription that identified it as a work produced by the genetic code of a living organism—leaving only the numerical parts as a legacy equivalent to the hard parts of fossil organic remains.

What is perhaps most startling is that Shannon came from a background of mathematical science and engineering technology, whereas Sewell came from a background of English literature! Yet *both* viewpoints fit naturally within the context of nonlinear dynamics (*theories of chaos*) as the communications medium that subsumes the disciplines of mechanics, kinematics, and thermodynamics *as well as* the evolutionary principles of genetics, linguistics, and semantics. In a profound way, Sewell—especially when buttressed by the cogency of Shannon's theory—set the stage according to which we can *read*, in the broadest sense (see, feel, touch, taste, and hear in concert), the richness of the process of poetical, hence universal, communication. Shannon provided a system of accounting by which we can transmit and/or receive the symbolic parts of a language in an identifiable way over arbitrary "distances" using actual sets of numbers, given a context. Nonlinear dynamics provides us with the opportunity to glimpse the universal mechanisms by which all languages have evolved, or might evolve.

Shannon's formulae describe admixtures of the numerical contents of language codes in quantitative terms called *uncertainty* and *redundancy*. Thus *relative uncertainty* and *relative redundancy* are normalized values each of which expresses its defined measure as a fraction or percentage of the total range between, on the one hand, "perfect" uncertainty and, on the other, "perfect" redundancy; the former expresses a numerical situation wherein there are no repetitions of the basic units of a code, the most fundamental of which are sometimes called *codons*, while the latter expresses a numerical situation wherein a single codon is repeated endlessly. Neither of these extremes contains intelligible information. They convey, respectively, either an infinite number of unintelligible messages or one unintelligible message repeated infinitely—unless the persons doing the transmitting and receiving have established ahead of time that a given number, set of numbers, or scramble of numbers has some agreed upon meaning, as in the cryptography of clandestine codes (see Endnote 8.2).

Universal language codes, such as English and other written, spoken, or symbolic languages of cultural evolution, fall somewhere in the vicinity of a 50:50 mixture of numerical uncertainty and redundancy, reflecting the same categorical "mix" of order and disorder represented by Sewell's "number" and "dream" components of communicable languages. Shannon's mathematical theory of communication is less comprehensive than Sewell's qualitative theory in the sense that it can compare only degrees of complexity, or particular numerical patterns of complexity; for example, it cannot identify the nuances between two different samples of identical relative redundancy (at this point in a cryptographic analysis, further discrimination is made possible by the grammatical rules of a particular language; cf. Searls 1992). In a sense, Shannon's measures represent a sort of self-similar quantification of Sewell's concept of the proportionality between "dream" and "number" in language. Curiously, it seems doubtful that either Sewell or Shannon knew of the work of the other, or of the remarkable alignments in their thinking, at the time each of them was writing her/his major work.

Nonlinear dynamics can be expressed similarly, by any of an infinite variety of possible demonstration algorithms, in terms of classifiable admixtures of *periodic* and *aperiodic* parts of an evolving spatiotemporal structure. In this case, however, the nonlinear "periodic" part is richer in its structure than is the "redundancy" part of Shannon's construction; it is closer to the nature of "number" in Sewell's construction. That is, redundancy, for Shannon, implies chantlike repetition without variation. In spectral terms, this constitutes a single harmonic signal of exact and constant frequency, and of exact and constant separation in time and/or space. This is what the term *period,* or *periodicity,* seems to mean to most people, not recognizing that in this mode the concept has been stripped of all meaning and is reduced to the information content of a one-note symphony lacking rhythmic, tonal, overtonal, or inflectional variation. Such a conception of periodicity conveys far less meaning than the dial tone we hear when the other party hangs up in the middle of a telephone conversation (admittedly, however, in circumstances loaded with implicit or semantic nonlinear information, even a dial tone can convey a crumbling universe of meaning). On the other hand, an exact nonlinearly *mode-locked periodicity* conveys a *pattern* of variation that is intracyclically aperiodic and intercyclically periodic (cf. Shaw 1987a, Fig. 1). By this I mean that, like an idealized stratigraphic cyclothem, the sequence of iterations or states of the system is irregularly distributed within a certain progression (representing a cycle), but then the same irregular sequence is repeated in exactly the same order in the next cycle. Rather than a monotone, a nonlinear periodic tonal signal of exact numerical character may consist of a repeated sequence of different notes, a sequence, for example, analogous to the repeated multitonal signal emitted by the benevolently alien spaceship in the motion picture "Close Encounters of the Third Kind."

One way of expressing the nature of nonlinear periodicity, in the context of the co-periodic nonlinear-resonance conception of coupled phenomenologies mentioned at the outset of this essay, is to say that it conveys a finite bandwidth of relative redundancy, as contrasted with the usually inferred standard of linear periodic or chantlike forms, *while retaining the same quality of exact cyclic recurrence.* Crudely speaking, this richer context of the nonlinear period is qualitatively analogous to a redundant word, as contrasted with a redundant letter in the Shannon sense, or to a redundant phrase relative to a redundant word.

The equivalent idea can be expressed in terms of the positions and movements of the hands on the face of a linear mechanical clock vis-à-vis those on the face of a nonlinear mechanical clock (Shaw 1987a, Fig. 1). The motion of one hand of each clock stands for the response of one of a pair of coupled oscillators, while the other hand represents the oscillatory motion of the primary mechanism of the clock (see Endnote 8.3).

In the linear (limit-cycle) mode of a mechanical clock, the second, minute, and hour hands tick monotonically around the clock face, each in an endless sequence of proportioned jerks meted out over time by the energy that keeps the pendulum or other fundamental timing oscillator from "winding down" by losing stored mechanical energy to thermal dissipation. Each time the minute hand reaches 12, so does the hour hand. In the overtly nonlinear mode, however, the second hand, say, ticks around the clock according to a set of unequally spaced (timed) disproportionate jerks that is repeated in exact "cyclothemic" bundles in subsequent cycles. In the nonlinear cycle this hand may require several trips around the clock face to define its "period" if the numerical ratio that describes the mode-locking (nonlinear resonance) is not a simple fraction like 1/2, 1/3, 1/5, 1/7, etc., where the numerator tells how many circuits are needed to define the repeat cycle (e.g., 3/5 means that there are five ticks per circuit of the clock face and three circuits per repetition of a cycle, at which point, or "time," the hand is at exactly the same point it was at in the previous cycle).

Another term seen in the literature of nonlinear dynamics is *quasiperiodicity*. This term has been adopted for those components of nonlinear periodicities that cannot be described by the properties of *rational numbers* (i.e., their numerical properties are irrational rather than corresponding to ratios of integers; cf. Shaw 1987a, 1988b). In such a case, a "quasiperiod" is analogous to the fundamental of a linear harmonic function in the sense that a single recurrence mode (*beat*) is activated, but that mode is one that emerges from a more complex interaction in which the operative mechanism responds disproportionately to its stimulus (e.g., a drumbeat that does not fit with the sound or timing of all possible harmonic drumbeats of a given system of drums, and in which the acoustic energy produced is not proportional to that of the linear mode, relative to the amount of mechanical energy supplied to the percussive mechanism).

Theories of chaos invoke the notion that components of *nonlinear aperiodicity*, as well as components of nonlinear periodicity, grace the picture of motions in phase space. Thus, the term *chaos* accords rather well with its more commonplace meaning, as viewed in the context of Sewell's terminology. That is, the components she calls "dream" states are implicated in the norm of communicative languages, providing, in proportion to "number" states, their richness. I venture to identify such communicative structures by the phrase *emergent geometronumerics* (ideographic patterns), meaning spatiotemporal structural possibilities that are consistent with the context of a universal language. Relativity again enters such scenarios, because in the absence of a familiar and *safe* context describable as language, phenomena of closed *logic* can appear on the side of egregious order, and *nightmare* can appear on the side of noncontextual, hence relatively unique and frightening, dream states (see Sewell 1951, Fig. 1, pp. 50–52). Either or both may appear to represent forms of "madness." That they may do so may explain an aversion by some scientists—or, for that matter, by anyone else—toward the nonlinear-dynamical paradigm. But that such disturbing impressions can occur means only that a safe contextual perspective has not yet been developed for the identification of a compatible language of nonlinear perceptions, relative to a given field of familiar observations—whether a "familiar observation" is correctly or incorrectly recognized from the standpoint of more specialized concepts of number (see Endnote 8.4).

Here, I suggest, is where we can return to a sense of the familiar, by retracing our ways through the cognitive minefields wherein we began our training (and/or our "brainwashing") in communication. In physics and mathematics, for example, a critical aspect of the nonlinear-dynamical paradigm is concerned simply with the action or nonaction of dissipation. Dissipation, in turn, is directly related to (entropic) questions of open-system processes far from equilibrium vs. closed-system processes close to equilibrium. In the prior discussion of the historical connection (disconnection) between mechanics and heat, I noted that the tradition in celestial mechanics has been, de facto, to ignore dissipation, except for those aspects of "detached" studies of such things as the internal processes of "stellar evolution," etc. Description of the actual evolving processes of star formation, as noted previously with respect to habitual descriptions of vehicular engines and Earth processes, is treated as something apart from descriptions of the internal thermonuclear reactions that we apply to stellar objects beyond the stage at which they can be called full-fledged stars (cf. Zirin 1988; Maran 1992).

Contemplation of the impact-extinction hypothesis (IEH) and

impact-related nonlinearly resonant boundary crises (INRBC) in the above context persuades me to entertain the possibility that a universal dynamical code or language exists in the record of celestial evolution, just as it has been inferred from the analogy of biostratigraphical constructions with poetical constructions that such a code exists for the geological record of evolution within the Solar System. In yet another example of historical irony, the patterns of such a language are afforded by, or emergent from, apparently violent mechanisms of nonlinear-dynamical evolution scenarios that provide the *Sewellian logic/number and dream/nightmare* (chaotic) components of that language. These "violent mechanisms" constitute the terrestrial-extraterrestrial connection wherein dissipatively coupled dynamical phenomena are exemplified by plasma dynamics, frictional effects of orbits, processes of open-system heat and mass transport involving coupled orbital evolution and collisional aggregation and fragmentation, and mass impacts on the Earth, Moon, and other planets.

The irony is that the IEH-INRBC connection often is couched in terms of nightmarish scenarios characterized by battlefield terminologies that invoke phrases like "killing potential," "kill rates," "mass extinctions," etc. Accompanying these horrific imageries are the more immediately nightmarish implications of terms reminiscent of "nuclear winter" scenarios. They, in turn, remind us of Hiroshima-engendered nightmares with apparitions of "impact winters," "cosmic winters," "cosmic catastrophes," and so on (cf. Clube & Napier 1982, 1990; Silver & Schultz 1982; Chapman & Morrison 1989; Sharpton & Ward 1990; Glen 1990). In actuality, however, an eventual spin-off of the IEH debates may be a newfound serendipity of dream states that will bring new alignment to the separate disciplines of mathematics, physics, geology, and biology. Such an alignment could go far toward mediating the propensities for violence engendered by an atmosphere of intellectual conflict and escalating greed that is only thinly veiled by pretenses of scientific altruism. One need but look at what has been happening to the "science" of genetic engineering to see how any form of scientific specialization will respond to a potential for commercial exploitation. Here we have an example of total disregard for the clear divergence toward the simultaneous extremes of *Sewellian logic* and *Sewellian nightmare*, symbolizing an absolute split between perfect order and perfect disorder, the ultimate form of social schizophrenia and madness (cf. Sewell 1951, Fig. 1; Crichton 1990, pp. ix–xi; Barinaga 1993).

In a way, we are returned to those moments in the history of science when Galileo, Kepler, Newton, Rumford, Joule, Kelvin, Darwin, and others were poised to impose their new disciplines on their scientific descendants. Now, by virtue of a time warp of nightmarish potential (e.g. Crichton 1990), as it were, we may be privileged to see the emergence of a new scientific language that neither reacts nor recants, but acknowledges the context within which the *Sewellian number/dream* states of those scientists can be more fully honored. The states of "perfect order" and "perfect disorder," bordered, respectively, by the insane realms of utter logic and utter nightmare, are at the ends—the global closure—of Elizabeth Sewell's Mercator-like worldview of poetic language (Sewell 1951, Fig. 1). Whereas the balance between *Sewellian number and Sewellian dream* is the prime meridian in her map of poetical language, its antipode, the imbalance between *Sewellian logic and Sewellian nightmare*—kept apart but by the choice of map projection—is closed in identity on a globe. What new sanity or insanity we may make of this in the evolution of science hangs in the balance.

Perhaps Newton's incentive, as one might imagine it in order to preserve one's own individual sanity independently of science at large, was to bridge the chasm between the indescribable realm alleged by the Church to represent spiritual reality and the realm of physical reality describable by the familiar activities of observation and experiment. But such a notion is a non sequitur to the degree that the scientific language either mimics or is mimicked by the evolution of the cultural language. Another irony in the roundabout loops of scientific activities that have led to the emergence, if not explosive breakout, of the nonlinear propensities of scientific thought during the intervening half a millennium may be the possibility that computer-aided experiments and observations made available by the algorithms and interpretations of nonlinear dynamics can lead us safely to that edge of mind where once religious belief seemed to be the only available bridge between utter order and utter disorder.

Heretofore, sanity apparently could be maintained only by either the total embrace or the total avoidance of the dogmas of religious authoritarianism. Now, more diversified choices may be available to us, as individuals, scientists, societies, and cultures, if we can reconcile these dichotomies of language. As in the completion of the *Sewellian map* by the joining of its ends, nonlinear dynamics presents analogous global joinings. Examples are the *bifurcation bubbles* of

Knobloch et al. (1986) and the *bifurcational synapses* of Shaw (1991, Figs. 8.27, 8.34, and 8.35), wherein period-doubling routes to chaos are opposed, coming together in immeasurable, uncomputable, and unknown interfacial folds that pervade all that is cyclical in nature. And these global joinings are distributed holographically throughout the universe, being independent of the scale and nature of its material states. In this there is more than a hope of sanity in the completion of Sewell's map. Nonlinear dynamics potentially makes whole, and of one universal contextual piece, all of science and all of the living history of Earth, notwithstanding the imagined fate along the way of this or that structural form that has returned to the indescribable interface that simultaneously joins the spatiotemporal contexts of logic, number, dream, and nightmare. The transfinite bifurcations of this synaptic join span the void, offering holisms in place of schizophrenic quantum particle-wave dualities (cf. Spencer-Brown 1972, pp. 104–6). This is the realm of selective choice among all of the possibilities offered us by the patterns of the evolutionary potential. How could Kepler and Newton, or Hutton and Darwin, for that matter (see Endnote 8.5), not be pleased with such an outcome?

9

Concepts and the Nature of Selection by Extinction: Is Generalization Possible?

Leigh M. Van Valen

Natural selection
Will lead us to perfection,
As we climb the hill toward our
adaptive peak.
But what if an earthquake
Makes our hill a sad mistake
And drops us to a valley,
Where we dare not dally,
For other species tally
Their gains while we are weak?

—Melody: Adeste fideles

Here's to all endangered species,
The hidden and the grand;
Soon their habitats will vanish
In marshes and on land.
But however loudly we may praise
thee,
The losers of the game,
We know that some would lose
without us,
Though too hidden for a name.

—Melody: college fight song, first at Miami University

Natural selection is the direction-giving process of evolution, and it is manifested in many ways. We are most familiar with its action on individuals, as when a fern plant dies sooner or reproduces more than its neighbor does because the plants themselves differ (rather than dying from some random cause, like being in the way of a wandering moose). But selection occurs within individuals too. Thus there are many initial attachments of nerve cells to a muscle fiber: those attachments that are *used* remain; those that *aren't* used die off.

Similarly, selection operates on *groups* of individuals. As at other levels of selection, this occurs in several ways. For instance, a species whose members disperse a lot is (usually) less likely to give rise to new species than is one composed of stay-at-homes. Extinction, too, is commonly selective, because some groups are more prone to extinction than are others.

In what follows, I will discuss some aspects of natural selection operating by differential extinction. There are a number of conceptual issues that need clarification, and we will see (or at least *I* will) that extinction is not merely loss—and thus a phenomenon to be explained simply for its own intrinsic interest—but one of the fundamental causes of evolution.

Some History

The concept of selective extinction has been known for a long time—not as long as individual selection (of the stabilizing sort), which Aristotle mentioned, and extinction itself wasn't usually recognized until Cuvier. (Thomas Jefferson was one who had argued against the idea of extinction.) The first proponent of selective extinction seems to have been the Italian naturalist Brocchi (1814), whom Lyell (1832, p. 129) criticized for proposing that species were endowed with different amounts of "proliferative virtue" (predetermined species longevity) when they were created.

But Lyell himself, in chapters 9 and 10 of the second volume of his *Principles of Geology*, argued at some length for selective extinction's occurring as a result of both biotic and abiotic causes. He even quoted de Candolle, in this context, on the severity of plant competition. (Lyell's discussion, it seems to me, was a background influence on Darwin's discovery of the importance of individual selection, but that is another story.) Selective extinction permitted an equilibrium number of species in the face of ongoing creation of new species, and these two processes caused the biotal evolution that let geological periods be separated from each other. Phillips (1841) gave a surprisingly modern-looking graph of biotic diversity over geologic time, showing major declines at the ends of the Paleozoic and Mesozoic. Darwin, for his part, took selective extinction for granted, as have many since then, anti-Darwinian (e.g. Morgan 1903) as well as Darwinian (e.g. Simpson 1953; Wright 1945). It has been only since Mac Arthur and Wilson (1967), though, that the phenomenon has been actively studied.

What Is Extinction?

Extinction is the irrevocable death of a group, its elimination rather than its transformation into something else. But what is a group? Some geneticists (e.g. Wade 1978) restrict the term to the groups they are comfortable with, namely local populations or demes. Others (e.g. Wright 1945) take a broader view. The various kinds of biological "groups" differ in important ways, but perhaps no kind of group is entirely discrete from those kinds conceptually near it, and all share some properties.

Any group can go extinct, and almost all species, clades, and such do so eventually. How, then, can extinction be selective? Some extant groups or their descendants do (and will) survive indefinitely, but we don't know accurately which ones they are, and we are rarely able to determine the moment when a particular species (or group) has gone extinct; but more significantly, survival to our particular year or moment of observation has no ontologically privileged status. Most selection by extinction involves, in a basic and inescapable way, the *time scales* of selection. A survivor at one age or year may have vanished by another. A mass extinction considered in isolation comes as close as possible to evading this ongoing process, and on such occasions the apparently pre-extinction biota had itself been partly shaped by earlier extinction, and the survivors are still at risk from later misfortunes.

"If we see, without the smallest surprise, though unable to assign the precise reason, one species abundant and another closely allied species rare in the same district—why should we feel such great astonishment at the rarity being carried a step further to extinction?" (Darwin 1845, p. 176). Extinction is a special case of contraction of a group, and the latter (along with expansion) is how perhaps most selection acts with respect to the economy of nature. For species and other taxonomic and phyletic groups, though, extinction is irreversible, and this finality makes extinction qualitatively separable in an evolutionary context.

The processes acting on organisms determine the nature of the groups that organisms really constitute. Genetical processes give us, for example, families, demes, isolates. Ecological processes form guilds and other adaptive complexes, patches, communities. Geography mostly determines the biota of regions of one size or another. And phylogeny produces a *clade,* or an ancestor plus all of its descendants. Species and other taxa emerge from several processes. Biotas may go extinct in the usual way, especially locally (e.g. by the explosive destruction of the volcanic island Krakatoa, or the killing effects of a dried-up lake), but mostly they unravel, many of their constituent subgroups surviving in new contexts. Should we call such unraveling a form of extinction? It partly is and partly isn't, thus forming one of the fuzzy boundaries of the concept. Other fuzzy boundaries are more familiar to ecologists than to paleontologists, like how small or lacking in potential for persistence a group can be, or the ecologically temporary extinctions following repeated immigrations. Groups can intergrade without losing their reality; reality doesn't demand discreteness.

Natural Selection

Groups and other evolutionary units evolve by change in absolute and relative frequency or size of their components. The causes of such change can be deterministic or random. Deterministic causes of overall relative-frequency change are conventionally grouped as natural selection, whereas random causes, ranging from ordinary mutation or undirected gene conversion to bolides from the sky, have a cumulative directional effect only by chance. Interactions among causes are common. Vrba and Gould (1986) distinguished selection from what they called *sorting,* the latter being a more general concept that they equated (presumably together with production of new variation) with evolution. Darwin and many others after him have made about the same distinction, using other language.

Natural selection can be regarded as having three components: establishment, expansion, and termination, each associated with time. These components apply to every unit, level, and time scale of selection, although they have somewhat different manifestations for different aspects of selection. The three components are correlated in several ways and also partly intergrade. I consider here that sort of termination called *extinction.*

Selective Extinction

For extinction to be an aspect of natural selection, its causes must be selective. In other words, groups with some properties must have a higher probability of extinction than do groups with other properties, and the properties in question must act causally rather than merely being a chance association or being correlated with some other cause. This isn't a severe set of conditions. If we consider the various known causes of extinction, it seems likely that every one, in fact, has a selective component. Take bolides, which I mentioned above as an example of a random cause. The biota vaporized by a bolide, and the biota in the rest of the total-kill area, can reasonably be regarded as having been randomly exterminated with respect to their own properties. (Even this isn't quite certain, on a worldwide basis; it is conceivable that the probability of a bolide's hitting a particular place depends on its latitude, and of course organisms inhabit different latitudes mostly as a result of their own properties.) Outside the total-kill area, though, some kinds of organisms will be more susceptible than others to the various stresses caused directly and indirectly by the bolide. This area of partial extinction is likely to be much larger than that of total death, and the overall extinction should thus have a large selective component.

Whether we think a particular extinction is selective or not may depend on our focus, i.e. on the spatiotemporal scale we are considering, or on the set of comparisons we make. During interglacials such as the present, there have been (and are) trees in Ontario, but at the end of each interglacial it gets too cold for all the snow to melt, and an ice sheet forms, as it will again a while after our brief greenhouse blip. Probably most tree populations there go extinct, rather than surviving by dispersal, for ice sheets form rapidly. Considering the tree vegetation of Ontario by itself, the extinction is more or less nonselective. But considering that of North America as a whole, the extinction is quite selective. Mammoths and mastodons thrived outside the successive ice sheets until a bit after the latest one. Mastodons, unlike mammoths, had gone extinct in Eurasia a million years earlier, but at least in North America the extinction of the two in relation to each other was nonselective, because both succumbed. For North American (or European) mammals as a whole, however, the little mass extinction that occurred at this time was quite selective, the excess above background being restricted to species with large body size (Van Valen 1970).

Correlated Responses

Any selective process is necessarily adaptive at the level and time scale at which it occurs. We often think this is false, because our focus is instead on some other level or time scale. Selective effects do commonly carry over, beyond where they are actually caused. In quantitative genetics this carryover is called *correlated response to selection*, a suitably general term that antedates others, including a narrow use of "sorting" (Darden & Bach 1990). (Descriptively correlated responses can also arise externally, by the same cause or a correlated cause acting separately to produce the effects; Kitchell, Clark, and Gombos [1986] have given a nice example for diatoms.) Heisler and Damuth (1987; Damuth & Heisler 1988) provide equations for interlevel correlated response. Correlated responses at other levels and time scales may retain the adaptive response of the focal selection, but if causation doesn't carry over, then such adaptation is accidental and unlikely. Arnold and Fristrup (1982) have emphasized the interlevel costs. The situation is just like the familiar one where sexual and other reproductive selection is adaptive for itself but is adaptive for survival only by causal carryover or by chance.

Thus extinction at one level, or time scale, need not be decoupled from that at another. How much decoupling there is remains a problem which has rarely been looked at except for the special case of mass extinctions.

Even random extinctions, though, can have consequences for organismal (as distinct from biotal) evolution. Because dispersal out of a population is selectively equivalent to death in that population, such dispersal is selected against within a population (Van Valen 1971). If extinction is frequent enough that long-distance dispersal doesn't disappear, though, each extinction produces (on the average) an opening for a new population to be founded by long-distance dispersal. Such dispersal can also evolve as a byproduct of within-population dispersal, a correlated response. Random extinction can increase or decrease geographic differentiation (McCauley 1991). On a larger scale, even random extinction can leave empty ecospace. The filling of such ecospace selects for different adaptations than does later survival there, the obvious generalization of Schmalhausen's (1946, 1949) concepts now called *r*- and *K*-selection. Random extinction can also produce macroevolutionary drift, but in the context of extinction such drift is undoubtedly most commonly caused by lack of correlation between the character of interest and the adaptations against extinction. For instance, Pojeta and Palmer (1976) found that clams which bored into hard substrate evolved in the Ordovician, but their clade succumbed at the end-Ordovician mass extinction, and the useful trait didn't evolve again until the Triassic.

Groups and Metagroups

Extinction can be, or at least characterize, a way of life when it occurs often enough. Fugitive species can't prevent their local extinctions but survive as good colonizers in a metapopulation, a population of populations. Such a strategy works only for local extinctions, but it is a common observation that some participants in the bloom-phase of recovery after a mass extinction are normally weedy, being invaders after minor crises. Here direct carryover of adaptations is presumably involved.

There is another sense, though, in which metagroups occur at all levels and permit what might whimsically be called fugitive families or communities as well as the opportunity for selection. Some groups are more volatile than others, their subgroups having higher rates of extinction and origination. Most ammonoids (cf. Ward & Signor 1983) had high turnover, with repeated radiation of replacement groups. They differ from mammals, which have had comparably high turnover but no widespread extinctions, and whose replacements are usually competitive (e.g., Van Valen & Sloan 1966; Krause 1986; Maas, Krause & Strait 1988; Stehli & Webb 1985).

Some properties of groups, like density or the probability of speciation, are emergent, i.e. are not reducible to properties of individuals. Other group properties, like the characters of a taxon, or geographic location, are so reducible. The probability of extinction of a group exemplifies properties which have both reducible and irreducible aspects. An extinction is, of course, an aggregation of individual deaths. Ordinarily, these deaths aren't independent of each other, but this lack of independence can still be from the individuals having the same susceptibilities. That this need not be the case is shown by irreducible properties, like population size or geographic range, which also affect the probability of extinction.

Emergent properties are of some interest in their own right, but whether a property is emergent has no influence whatever on the dynamics of group selection, as Damuth and Heisler (1988), Van Valen (1990), and Stanley (1990) have noted. The three components of selection, namely expected establishment, termination, and change in size, operate at any level and time scale, however the causes of these components are determined. The same selection may thereby be at more than one level and time scale simultaneously, which should be confusing only to people who try to impose an artificial discreteness on a continuous and overlapping world.

Time Scales

The recognition of levels of selection has become fashionable, if in a rigid sort of way. Oddly, the intimately related aspect of the time scales of selection has nearly been ignored. An aspect of the relationship can be seen from the geneticist Thoday's (1953) criterion of fitness as persistence for 10^8 years. Almost no allele persists that long, no individual, almost no species. Fitness of alleles is then intimately involved with fitness of supraspecific groups. Cooper (1984) actually argued that fitness is basically the expected time to extinction, but this proposal ignores the time scales and processes of real selection.

The old idea of the evolution of adaptability (the "survival of the unspecialized") involves the cumulative effect of selection by extinction at different time scales. The less adaptable are, by hypothesis, more susceptible to extinction, making the set of survivors more adaptable overall. There is, though, shorter-term selection for less adaptability by the jack-of-all-trades principle. Wilson's (1961) taxon cycle, for example, may work in this way (but cf. Pielou 1979). For adaptability itself to evolve, though, the causes of extinction should be diverse enough that narrowly adapted forms of many sorts are at higher risk.

A remarkable paper by Patterson and Atmar (1986) studied the results of the flooding of low-lying land by rising seas as ice sheets melt. The islands of the resulting archipelagos are too small to support the entire pre-existing fauna, and species actually tend to go extinct in the same ordinal sequence on the different islands. Thus the expected longevity, in such an environmental change, varies predictably among species. Natural selection is occurring, with respect to the composition of the biota, by extinction.

Causes of Extinction

I won't review what we know of the causes of extinction. It is, though, probably necessary to deal with the supposed dichotomy between biotic and physical causes. Sometimes these are easily separable; the ongoing extinction of the American chestnut by a fungus is biotic, and the drowning of an atoll by a too-rapid rise of sea level is physical. A bit fuzzier are, say, episodes of extinction caused by interchange of biotas across a new connection. When we get to changes in competitive ability and the like, as a result of change in the physical environment, a very common situation which even Lyell discussed, in 1832, the choice (if there were a forced dichotomy) would be a matter of taste. I therefore do not accept even in principle methods like the use of correlations between extinction rate and environmental change (Hoffman & Kitchell 1984), or seeing when arms races are interrupted (Vermeij 1987), to estimate the relative importance of the two supposed classes of factors. Each could be necessary in the large majority of cases.

The real point at issue is whether biological influences are important in macroevolution. I know of no qualitative disagreement that physical influences are important at all levels; that biological influences are important to evolution within species shouldn't need discussion in 1991, although there remains some uninformed dispute. Therefore, relevant evidence is evidence which concerns biological influences themselves on macroevolution. I have given some evidence which is difficult to explain without biological interactions (Van Valen 1985b) and discuss some more specific causes below; there are also other arguments which I can't go into here.

I also find too facile the recent tendency to treat most extinctions as occurring independent of the groups which radiated to replace their ecological forebears, as reviewed by Benton (1987). Sometimes this does occur. Too often, however, as with Benton's favorite case, tetrapods in the late Triassic and early Jurassic (Charig 1984), all that the evidence shows is that there wasn't a geologically protracted replacement. There is a period of overlap, which can only be extended by future discoveries; more intense (but still ecologically weak) interactions can produce the same result as more prolonged but weaker ones. The very existence of the spatiotemporal overlap of adaptively similar groups is some evidence for interaction at an evolutionary time scale, not (as is actually sometimes claimed) for independence if the overlap is geologically short.

That major biotas go extinct, or at least unravel so much that they no longer exist, may be surprising to ecologists unfamiliar with the fossil record. It isn't something we would usually expect with biotas composed of diverse sorts of species which coexist by each having its own needs satisfied. Biotas with a keystone species or so do this, however; the best-known case is that of the sea otter (*Enhydra*) on the coast of the northeastern Pacific (Van Blaricom & Estes 1988). It eats (and can regulate) sea urchins, which eat (and can regulate) kelp; the regional extinction of the sea otter by human hunting drastically reduced the nearshore kelp forests in favor of a more open sort of community, and the process is now reversing as the sea otter expands. There are also other ways for a biota to unravel.

Biotas

Local communities are not unique. They rather represent fuzzily bounded kinds that recur at ecologically appropriate places. Communities may grade insensibly into geographically adjacent communities, so that not only where to draw a boundary, but also where to put a center, may be arbitrary. The species in communities of the same kind often differ in different geographic regions without this much affecting the structure of the communities.

Similarly, communities and community kinds extend through time. Community kinds are, of course, more stable than their local avatars, and their detailed composition changes over time as well as across space. They too can unravel, though. If Guthrie's (1983) analysis is correct, as I suspect it is overall, a productive community kind usually called steppe-tundra has existed near the northern ice sheets for most of the Pleistocene. It isn't with us now, nor was it in the other brief interglacials, but it will presumably reappear in a now depauperate form with the next glaciation. Community kinds, like small populations on islands, can thus have repeated and temporary extinctions. It is the structure of the community or other biota which goes extinct in cases like the steppe-tundra, not necessarily the component species themselves, which may remain elsewhere. That doesn't affect the reality (as distinct from the effects) of the extinction; biotas have their own evolution (Van Valen 1991).

Community kinds commonly persist for millions of years. Even longer-lasting are what Boucot (1983) called ecologic-evolutionary units, one unit containing all the contemporaneous community kinds in the ocean or on land. Ecologic-evolutionary units (which are real, and partly represent stasis in biotal evolution) last until disrupted by a mass extinction. Community kinds do go extinct, but how this happens hasn't been studied, and the selectivity of these extinctions is unknown. Despite there being only one contemporaneous unit (except, probably, briefly at a few boundaries), even the extinction of ecologic-evolutionary units may be selective. This is true because a mass extinction may need to be stronger to disrupt some such units than others, so the former units would have a greater expected persistence. Such selection would have no later evolutionary effect, though.

Mass Extinctions

There is now some evidence (Van Valen 1984; Raup 1986; Stigler 1987; McKinney 1987; Raup & Boyajian 1988), contradicting earlier work, that mass extinctions represent a sample from the extreme of a continuum of intensities. (Gilinsky and Hubbard [1990] claim to find a difference using rates for specific taxa, but just how many extinctions will be called "mass" by any method like this depends on the statistical power of the test. Bambach [1990] finds that only by removing mass extinctions is the probability of extinction correlated with length of geologic stages, but there are various possible confounding variables not mentioned in this abstract.) Even if there is a continuity, though, that doesn't mean that mass extinctions don't differ qualitatively from others (or that different causes may not predominate), because there may be threshold effects, as Jablonski (1989) has noted. Such thresholds need not be at the same level for different phenomena, nor even give the same ranking for them. We will see that there are indeed such qualitative differences, and Kitchell (1990) has used their existence as a means to distinguish mass extinctions from others. Thresholds may operate at any level, however, including extinctions of local populations, although this hasn't been explicitly studied.

Evolution by Extinction

One surprising pattern (Van Valen 1984, 1985a; Kitchell & Pena 1984; Gilinsky & Bambach 1987; Thackeray 1990; Gilinsky & Good 1991) is an exponential decline in the probability of extinction of marine families during the Paleozoic, and another such decline later, separated by the late-Permian extinction. These declines turn out to be caused by selection for extinction-resistant higher taxa; for single taxa, there is no directional change over time, on the average. The Permian extinction is the only one above a threshold here; we don't

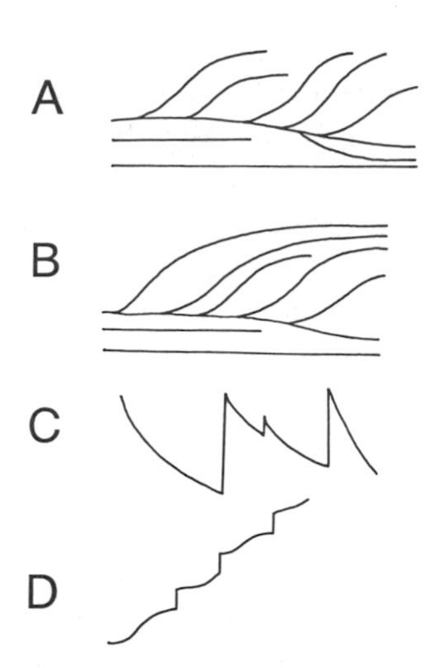

Figure 9.1. Simple diagrams of some patterns of evolutionary response to extinction. **A.** Cul-de-sac effect: what is predominantly advantageous (up) over a shorter time interval is disadvantageous over a longer interval, giving long-term group selection for adaptability. **B.** Guillotine effect: like A, but the extinctions occur together in a mass extinction. **C.** Decay, during normal evolution in one lineage, of the adaptations selected by mass extinctions: the obverse of B. **D.** Reinforcement: what is adaptive during normal evolution in one lineage is also adaptive in mass extinctions.

know just how it reset the declines. Thus the world marine biota has been selected for extinction resistance and has responded to this selection strongly. As Sepkoski (1990) notes, such resistance is selected for even by local extinctions, and resistant taxa in normal times also tend to be resistant to mass extinctions (McKinney 1987).

There are several reasonable ways in which selective extinction can interact with other components of the evolution of larger groups. Figure 9.1 diagrams some of these, and I have already discussed drift. In case A, relatively short-term evolution of some sort sends lineages in predominantly one direction. On a longer time scale, though, this change usually proves deleterious and so is selected against by extinction. An equilibrium distribution results, perhaps varying through time. I have given theory and equations for case A (Van Valen 1975), which can be called the *cul-de-sac* effect.

Case B is similar in structure to A except that the extinctions must be concentrated in a mass (or other concentrated) extinction. In case B what is adaptive, perhaps even from shorter-term extinction, between mass extinctions is irrelevant or inadaptive in the mass extinctions. This can be called the *guillotine* effect. The usual way of looking at the guillotine effect is that fitness of the inclusive group increases between mass extinctions and is reduced when they occur. Diagram C shows another perspective. Here we consider first what the mass extinction does in the context of the character under consideration, namely increase the mean fitness of the inclusive group with respect to a repetition of the mass extinction. (Other things, like the number of subgroups, may of course be adversely affected.) Between mass extinctions this adaptation *decays* and may well be lost. Both the decay and guillotine perspectives are appropriate; adaptation is always relative to some set of environments, which differ between the two perspectives.

In case D, *reinforcement*, the concentrated extinctions select in the same direction as do processes between them, and evolution is thus not reversed.

Group Properties and Selective Extinction

I will briefly review some group properties, emergent and not, which have been involved in selective extinction, and will comment where appropriate on their relation to the cases I have discussed. Most extinctions, even of species and higher groups, don't occur in mass extinctions. Raup (1990) estimates that more than 95 percent don't, but of course this is a ballpark figure, and it also depends on where one's cutoff for mass extinctions is.

Extinction can select in opposite directions at different times. People from Lyell (1832) to Wolfe (1990) have noted the extinction of heat-susceptible and cold-susceptible plants at different geological times, in relation to change in the area of land favorable for them. The classic example of the cul-de-sac effect is loss of biparental sexual reproduction. It appears, from the phyletic distribution of obligate asexuality and parthenogenesis, that this loss eventually increases the probability of extinction, perhaps from the reduction of adaptability that some theory associates with loss of sex. Under hard selection, which is independent of the composition of a population, the expected selective extinction against clones with new deleterious mutations can incapacitate Muller's ratchet, which can otherwise lead to irreversible deterioration (Melzer & Koeslag 1991). The evolution of adaptability, which I discussed above, is another general case of the cul-de-sac.

A similarly general example of the guillotine effect is what Vermeij (1987) calls *escalation*, the arms race in various forms among competitors and between phagers and their prey. This increases (escalates) during normal evolution, but is selected against by at least some mass extinctions. One visible correlate of escalation, which has been strikingly selected against in several groups in several mass extinctions, is morphological ornateness (e.g. Anstey 1978, 1986; Flessa 1986; Sheehan 1988).

As one might expect, groups with a narrow geographic distribution have a higher probability of extinction than those more broadly distributed. Lyell (1832) first noted this, and mass extinction to some extent reinforces normal extinction (including that of local populations: Mac Arthur & Wilson 1967) in it. However, there seems to have been a threshold area like the Atlantic Coastal Plain of the United States for mollusks at the end of the Cretaceous, below which no effect of area is detectable (Jablonski 1989). Similarly, taxa with few species are usually more at risk than the more speciose, but this is not the case for terminal Cretaceous mollusk genera. (Thus species of these genera didn't go extinct independently, but rather in relation to properties of their genera. This phenomenon isn't unusual [e.g. Van Valen 1973], but it is unusually strong here.)

Mass extinctions, except again for that of the Permian, just produce blips in graphs of diversity through time. Because the patterns of the evolution of diversity are similar at both local and worldwide scales, ecological factors are probably more important here than biogeographic ones (Sepkoski 1990).

Darwin (1859) noted that rarity should, and does, enhance the probability of extinction. There are various forms of rarity, Rabinowitz (1981) distinguishing seven combinations of low density, endemicity, and habitat specificity. A low population size is the overwhelmingly predominant risk factor for local populations (Pimm, Jones & Diamond 1988). The latter authors found an interesting interaction with body size: for populations with fewer than about seven pairs, species of large size have a lower risk of extinction than smaller forms, because of their longer lifespan, while for larger populations the reverse is true, because large size is correlated with a low rate of increase. For whatever reason (and others operate also on longer time scales), large size gives an increased probability of extinction during both normal evolution (Van Valen 1975) and mass extinction (Van Valen & Sloan 1977). This seems reinforcing, but, as in some other cases, shorter-term evolution often predominantly increases body size within lineages, and the overall effect is thus the cul-de-sac.

Not surprisingly, small population size is a major risk factor beyond the level of the local population (Diamond 1984). Boucot (1975) extended this to genera in geologic time, attributing the effect to drift. The numbers of individuals involved are much too large for this, but the effect (with a faster evolutionary rate for rarer taxa) may still be real, despite Darwin's opposite prediction and Williamson's (1989) agreement with him. There seem to be no quantitative studies yet. Similarly, as Small (1946, 1948) may first have shown, taxa with few species have a greater probability of extinction than those with more.

Some other population attributes selected for by extinction are low variability of density, low isolation, and low death rate (e.g. Karr 1990). These are difficult to study in the fossil record except by their correlates, so we have no data on their long-term effect. Other attributes selected by local extinction, such as a broad niche, high dispersal ability, and susceptibility to stress (Diamond 1984; Patterson 1984), do seem to carry over to a longer scale (Van Valen & Sloan 1977; Hansen 1980; Signor & Erwin 1990; Edinger & Risk 1990; Erwin 1990). Vrba (1980) has emphasized the cul-de-sac aspect of the common evolution toward stenotopy.

Tropical biotas, most notably those of reefs, are disproportionately affected by mass extinctions, probably because of their greater emphasis on adaptations to the biotic environment (e.g. Jablonski 1986; McGhee 1989; Erwin 1990). Reefs have disappeared entirely a number of times during the Phanerozoic, as Newell (1971) probably first pointed out, just how often depending on the criteria one uses (Fagerstrom 1987; Talent 1988). Raup and Boyajian (1988) nevertheless concluded that reef genera have the same extinction pattern as nonreef genera. Perhaps this unusual result comes from their treating pseudoextinctions, where one taxon evolves into another (Van Valen 1973), as real extinctions; genera are more susceptible to pseudoextinctions than are higher taxa (Van Valen 1973).

Obligate mutualisms, or those with a strong effect on fitness, may increase the risk of extinction for their components (Fowler & MacMahon 1982), although cases like lichens and mitochondria suggest caution here. Symbiotic corals did seem to be more susceptible than nonsymbiotic ones in the Cretaceous-Paleogene extinction (Rosen 1990), although they were mostly in the (rudist-dominated) reefs. Similarly, a narrow diet, pollination system, seed dispersal, and the like should increase risk. Janzen and Martin (1981; but see Howe 1985 for a critique) proposed that mastodons dispersed some large Central American fruits, and that their trees are among the living dead, in an irreversible population decline.

What Johnson (1974) called perched faunas are especially at risk during lowerings of sea level. These faunas occupy broad epicontinental seas and commonly evolve much endemicity, having species with narrow geographic ranges, before their habitat disappears. This endemicity is important for organismic, as distinct from the biotal, extinctions (Sheehan 1988). Very-near-shore species aren't affected (McGhee 1989), and the end-Ordovician extinction, although with a major regression of the sea, was more severe for open-ocean groups (Fortey 1989).

During at least some mass extinctions, but apparently not at other times (but see Levinton 1974, for an apparent counterexample), the plankton and plankton-based food chains, including most sessile filter-feeding benthos, are more susceptible than the mobile, predaceous and detritivorous, benthos (Van Valen & Sloan 1977; McKinney 1987; McGhee 1989). The pelagic conodonts managed to escape the Permian extinction unscathed (Erwin 1990), although they suffered severely in others (Ziegler & Lane 1987).

One plausible factor, age of a taxon, isn't actually selective, despite a claim for it by Boyajian (1986) which some others have accepted uncritically. Previous and later work (Van Valen 1973, 1985a, 1987; Gilinsky & Bambach 1987; Gilinsky & Good 1991) has shown that this is an artifact, caused entirely by more susceptible subtaxa predominating early and being selected against, and by combining for analysis groups with different probabilities of extinction, younger taxa being more frequent in groups with greater risk. Oddly, Boyajian (1987, 1991) has repeated the fallacy, adding one by Raup (1978) that age-independence at one taxonomic level requires age-dependence at others. The latter is required only if the levels are independent in extinction, which (as discussed above) is empirically false.

Fresh-water biotas have been remarkably resistant to at least some mass extinctions (Van Valen & Sloan 1977; McGhee 1989), for unknown reasons. Sheehan and Hansen (1986) suggested that predominant reliance on predation and detritivory was responsible. This may be true, although today plankton are important. Possibly the escalation is lower overall than that of marine biotas (Vermeij 1987) because of greater resistance to physical stresses, a matter which can be investigated even if it sounds as implausible as the known difference in escalation.

So, is generalization possible? Yes, with care. Groups come in diverse forms—populations, clades, taxa, guilds and other adaptive complexes, and biotas of various kinds. Extinction is common and is usually, perhaps always, selective in one way or another. Extinction occurs at all levels, and time scales are especially important in such group selection. Correlated responses to selective extinction, which often occur, let selection at different levels and time scales be coupled together. Even possibly random extinctions are evolutionarily significant, and metagroups sometimes occur. Whether a property of a group is emergent is irrelevant to selection on that group. Our focus can be important in what we see, but it doesn't determine what really happens. Biotic and physical causes of extinction broadly overlap. All kinds of biotas go extinct. Mass extinctions can differ from others by threshold effects, noncoincident for different properties. Adaptability, extinction resistance, and other group properties have evolved on various scales. Drift, cul-de-sacs, guillotines, decay, and reinforcement exemplify some modes of evolution by extinction.

Cauda

We are currently in the early part of a mass extinction, one with different causes from those of the past. There are nevertheless some similarities in the selective aspects of the extinction. It would be interesting and not altogether ghoulish, perhaps even productive, to have a better idea of how they all compare.

Acknowledgments

I thank William Glen for inviting me to participate, and him, Virginia Maiorana, and a referee for comments.

10

Uniformitarianism vs. Catastrophism in the Extinction Debate

Kenneth J. Hsü

When I came to the United States in 1948 to start my graduate studies, I was asked by my mentor, Edmund Spieker, if I had read G. K. Gilbert's monograph on the Henry Mountains. No, I hadn't. I did not know where the Henry Mountains were. I did not know who G. K. Gilbert was. Furthermore, I did not understand why that should be the first question asked of an incoming graduate student. Now I understand. G. K. Gilbert was the hero of many geologists of Spieker's generation in North America, and he is still the hero of many of us. G. K. Gilbert's work is the "ultimate ideal" of the uniformitarian approach in geology.

T. C. Chamberlin, a contemporary of Gilbert's, was considered a "villain" by Spieker. "He has done more harm to geology," in the words of Spieker, "than any person when he published his paper in 1907 on "Diastrophism as the Ultimate Basis of Correlation." Chamberlin was a catastrophist. Ironically, both Spieker's "hero" and his "villain" speculated on the role of meteorite impacts in geology. "If Gilbert, with his towering prestige as Chief Geologist of the United States Geological Survey, had favored an impact origin for the feature we now call 'Meteor Crater,' impact processes might well have been integrated into the mainstream of geology by the early 1900's. Unfortunately, Gilbert's finding for volcanism reinforced the opposition to impact and stopped virtually all research by American geologists for nearly 60 years" (Marvin 1989, p. 381). In the anti-impact atmosphere of the time, Chamberlin's planetesimal theory for the origin of Earth passed into oblivion with his demise, until the theory was exhumed by students of lunar exploration.

Spieker was a representative of the geological establishment of North America during the first half of the century. Many of us inherited the establishment *Leitbild* of geology. We tended to forget that substantive uniformitarianism is but an Occam's razor of the uninformed. Blind faith in the dogma lies at the root of numerous controversies in modern geology. After years of tutoring by Spieker, I too became a dyed-in-the-wool uniformitarian. I also tended to confuse ignorance with objectivity. I thought I was being scientifically objective when I indulged in simplistic speculations on the calcite dissolution of ocean bottoms by assuming that the chemistry of oceans has never changed (Hsü and Andrews 1969). If I had been less ignorant, I would not have lightly dismissed the more valid alternative of postulating ever-changing calcite-compensation depth (see Hay 1969). I was jolted out of my complacency, however, when I had to write a cruise report on the 1970 deep-sea drilling expedition to the Mediterranean Sea. That this inland sea had dried up during Late Miocene time was the only conclusion that is consistent with the implications of a wealth of data on Mediterranean geology (Ryan, Hsü, et al. 1973). The reactions of many colleagues were, however, negative. They said either: "I don't believe your story," or "You are way off base, because your conclusion contradicts Lyell's uniformitarianism, the fundamental principle of geology."

To believe or not to believe is irrelevant; faith is a question of religion, not of science. The second critique did, however, trouble me. Either I was wrong, or Lyell was wrong in his substantive uniformitarianism. This paper is a summary of the reasons that led me to the latter conclusion when I joined the K/T extinction debate.

The Reality of Extinction

Carl Linnaeus established the foundation of taxonomic biology. Linnaeus lived in the Age of Enlightenment, but he still retained a conservative weltanschauung. His paradigm was Christian theology: God creates, and He does not take away what He has created. Linnaeus wrote in 1751 in his *Philosophia Botanica* (cited in Hsü 1986, p. 88), "There are as many species as the infinite being created diverse forms in the beginning, which, following the laws of generation, produced as many others but always similar to them. Therefore, there are as many species as we have different structures before us today." Biologists in the decades after Linnaeus thus believed in the immutability of organic species, and the fixity of the number of species. William Smith's brilliant observation in the late eighteenth century that ancient strata are distinguishable by their fossil faunas is the first seed of all modern theories of biologic evolution. The leaders of French biology, Georges Cuvier and Jean-Baptiste Lamarck, were especially quick to appreciate the implication of Smith's discovery for the history of life on Earth.

Cuvier questioned the Linnaean paradigm of a fixed number of species. His studies of the faunas of the Paris Basin gave undisputed evidence of faunal extinctions. Their apparent abruptness led Cuvier to postulate catastrophic, or revolutionary, changes in biota. Catastrophists like Cuvier were considered creationists, but I have not found any reference to creationism in Cuvier's writings. Like all of his contemporaries, he probably accepted the dictum that God has created life. He probably also believed, like most of his contemporaries, in the immutability of species. But he did not accept the second Linnaean dogma. There have been extinctions, consequent upon catastrophic changes.

Lamarck, the first evolutionist, wrote (cited in Hsü 1986, pp. 25–26), "In this globe which we inhabit everything is subject to continual and inevitable changes. These arise from the essential order of things and are effected with more or less rapidity or slowness, according to the varying nature or position of objects implicated in them. Nevertheless they are accomplished within a certain period of time. *For Nature, time is nothing, and is never a difficulty: she always has it at her disposal, and it is for her a means without bounds, wherewith she accomplishes the greatest as well as the least of her tasks.*" I emphasize the last sentence because it constitutes an ideology. Time was nothing, and was never a difficulty to Lamarck or to other uniformitarians, because they held that particular belief. To many others, time was not nothing, and there was the geologic record of apparently catastrophic changes. No physical measurement was available during the nineteenth century to date the past; the Lamarckian/Lyellian uniformitarian assumption could thus not be verified or falsified. Whether to postulate evolutionary or revolutionary extinction was an article of faith, the choice independent of scientific judgment.

Lamarck became an evolutionist, because he was a creationist. He had no reason to oppose the Church's teaching that the Lord does not take away what He has given, nor did he disagree with Linnaean dogma on the fixity of numbers of species. As a confirmed uniformitarian, however, Lamarck could not agree with contemporary pleas for extinction through a diluvian disaster or by any other catastrophic means. After abandoning hope that extinct species could still be encountered by geographical explorations, Lamarck took the unusual step of questioning the Linnaean dogma of immutability. He had then to propose a theory on the transmutability of species. Synthesizing creationism and evolutionism, Lamarck suggested that changes in species crept in gradually after their creation, resulting ultimately in discrete differences between progenitors and their offspring. By invoking lineages, Lamarck could thus sidestep the issue of extinction. There was no extinction, he claimed, only "pseudoextinction." None of the older species had died out; all had been converted into new species by slow and gradual changes during the immensity of time since the Creation.

One or the other Linnaean dogmas had to go when Charles Lyell came on to the scene as a young rebel to question the French masters. Like Lamarck, Lyell was a radical uniformitarian. He thought, as a young man, that large areas remained unexplored, and that the descendants of apparently extinct fossils could be found in faraway places. With advancing age and better knowledge, this excuse of ignorance was no longer credible, and Lyell accepted the reality of extinction. Still, he saw no reason to postulate catastrophes. Extinction of groups could be a slow and gradual process, as natural as the death of individuals.

Lyell proposed a geochronology in terms of percentage of species still extant in an assemblage of fossil cohorts. He proposed the Epochs: *Eocene* (dawn of recent time), *Miocene* (less recent), and *Pliocene* (more recent) (see Endnote 10.1). There must have been gradual extinction, so that less and less of the Eocene cohorts, or conversely, more and more of the extant species, are present in younger fossil assemblages (Figure 10.1). Lyell had no idea of the rate of various extinctions. Using Occam's razor, he implied linearity in the half-life of faunal decay.

Falsification of Substantive Uniformitarianism

It was known to Lyell and Darwin that the end of the Mesozoic is marked by a mass-extinction datum-horizon, when 75 percent of the organic species then living were wiped out. What is the meaning of this *when*? The *when* is represented by the Cretaceous/Tertiary, or

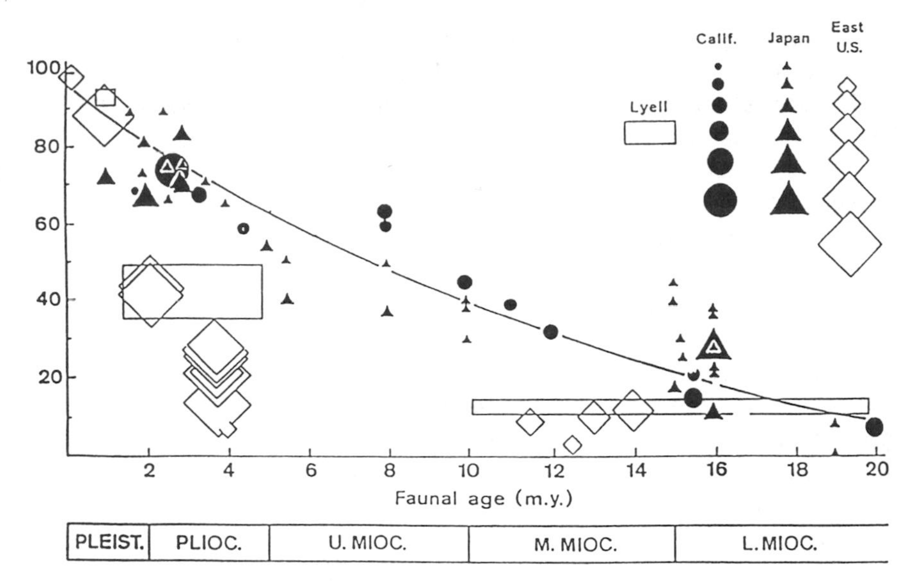

Figure 10.1. The continuous decay of faunal cohorts during the Cenozoic. The vertical axis denotes the percentage of species that are still living.

K/T, boundary. What is the duration of the *when*? This has been a key question in the history of life on Earth. This has been the key debate between the Cuvierian revolutionists and the Darwinian evolutionists.

Charles Darwin chose the paradigm of evolution, because he was influenced by Charles Lyell's interpretation of the K/T boundary. Remarking on the faunas in the chalk at Maastricht, Lyell wrote (cited in Hsü 1986, p. 48), "M. Deshayes, after a careful comparison, and after making drawings of more than 200 species of Maastricht shells, has been unable to identify any one of them with the numerous Tertiary shells in his collection. The belemnite, one of the cephalopods not found in any Tertiary formation, occurs in the Maastricht beds: an ammonite has also been discovered in this group."

The investigation of fossil shells by Lyell and Deshayes thus confirmed the great dying at the end of the Mesozoic era, verifying an observation by Cuvier a few decades earlier. Yet Lyell could not give up his abhorrence of violence and revolution. In order to hold onto his passionate belief in evolutionary processes, despite overwhelming evidence to the contrary, Lyell used circular reasoning and came up with the postulate of a big gap in the geologic record of the K/T transition. Writing in his *Principles*, he stated (cited in Hsü, 1986, p. 48):

"There appears, then, to be a greater chasm between the organic remains of the Eocene and Maastricht beds than between the Eocene and Recent strata: for there are some living shells in the Eocene formations, while no Eocene fossils are in the newest Secondary Group (=Maastrichtian). *It is not improbable that a greater interval of time may be indicated by this dissimilarity in fossil remains.*"

I emphasize the last sentence because the statement implies a uniform rate of faunal evolution. Radioactive decay had not yet been discovered; Lyell had to define geologic time on the basis of another decay phenomenon—the decay of faunal cohorts. Assuming a constant half-life of decay, and a fixed number of species in a cohort, the extinct species being replaced by extant species, the theoretical percent of still-extant species in fossil faunal assemblages is that shown by Figure 1. Lyell's prediction for the Cenozoic was more or less verified by the data on Pacific molluscan faunas compiled by Stanley and others. Studies by David Raup and his associates at the University of Chicago have also illustrated the logarithmic decay of cohorts with time (Raup 1991).

If we accept, on the basis of the data in Figure 1, an assumption of a constant half-life of 8 m.y. for the faunal decay of Lyell's Eocene cohort, we could explain why his Eocene fauna of about 48-m.y. age should have only 3 percent extant species. Assuming the same half-life for the decay of the Maastrichtian cohort, the unconformity marking the K/T boundary should represent a time-interval of more than 72 m.y. This is the logic behind Lyell's interpretation that the hiatus marked by the K/T unconformity had a longer duration than the time-span of the whole Tertiary.

The rise of neocatastrophism, or actualistic catastrophism, is a consequence of precision stratigraphy. Lyell's assumption that there have been no catastrophic extinctions has been proven wrong by radiometric dating. Yet Lyell's *Leitbild* of substantive uniformitarianism remains the prejudice of numerous biologists and paleontologists; they continue to insist on the imperfection of the fossil record when the record clearly shows mass extinction at the K/T boundary, and at other era boundaries.

Uniformitarianism and Natural Selection

There might be various proximal causes for extinction, but the ultimate cause is habitat destruction. Habitat destruction can be a consequence of catastrophes, and that was Cuvier's postulate. In his fanatic distaste of revolutionaries, Lyell denied that there could be catastrophes in history. His friend Darwin had to postulate habitat destruction in evolution through biotic interactions, "natural selection or the preservation of favored race in the struggle for life," because Darwin was misled by Lyell that there could be no natural catastrophes in Earth history.

Intraspecific and interspecific competition can certainly result in habitat destruction and natural selection. But in a seminar at the University of Chicago two years ago, I asked for fossil evidence of such selection in the history of life on Earth. The distinguished group present on that occasion came to the conclusion that natural selection is too complex a process to be verified by paleontology. The premise of speciation or extinction by natural selection is thus a matter of belief or faith, rather than a conclusion dictated by historical evidence.

Was there catastrophic destruction of habitats at the end of the Cretaceous? There have been two different approaches to the study of the problem. At the First and Second Snowbird Conferences, physicists and chemists used the deductive approach, making predictions on the basis of computer models to predict what would have happened to environments if a billion-ton (10^{18}-gm) meteorite hit the Earth. We geologists have applied the inductive approach, searching for geologic evidence of catastrophic changes in the sedimentary record.

I do not have to cite the literature that appeared during the last decade in support of the bolide-impact theory of K/T extinctions. Numerous predictions by the modelers of the First Snowbird Conference have been verified. The most impressive, in my opinion, was the discovery of tsunami deposits at the K/T boundary, first in the Gulf Coast and then in Haiti (see Glen, this volume). Many of us can recall the discussions at that conference, at a time when the apparent absence of such deposits at the boundary was considered the strongest evidence against the impact hypothesis.

My own work on that problem has been centered on the evidence of carbon-isotope anomalies at the boundary. The geochemical data so far can only be interpreted on the basis of a Strangelove ocean, when habitat destruction was so severe as to render nutrient-rich waters no longer fertile for plankton production (Hsü & McKenzie 1987, 1991). The first observation, made on the basis of studying a few localities, has been verified by all subsequent studies of K/T and other era-boundary sections. I see no need to get involved in debating with anyone who continues to insist that there was no habitat destruction by catastrophic events at times of biotic crisis, until they can come up with a satisfactory alternative explanation of the carbon-isotope data.

Death or Population Decay

When we speak of catastrophes, we have an image of mass mortality. Killing is, however, not a very effective mechanism in causing extinction. Dave Raup, speaking at the First Snowbird Conference, told us that only a few percent of extant species would become extinct if a catastrophe should hit the Earth and wipe out all living organisms on one-quarter of the surface of the Earth. We see that in lakes: every Alpine lake is a Strangelove lake in winter seasons. Lake planktons go into dormancy in winter, only to bloom again next Spring. Scandinavian lakes lost their aquatic faunas, not because they were all killed by one acid rain, but because the continuous acidization of lake water finally made the habitat no longer viable. I prefer, therefore, Dave Raup's "quieter extinction scenario," postulating that the birth rate fails to keep up with the death rate. After all, the last remnants of any group of living organisms become endangered species before they die out, even if the decimation of a booming population took place at a catastrophic rate. Most endangered species soon become extinct. On the other hand, some may continue to survive in local refugia, like the "Miocene" *Metasequoia* in the primeval forests of southwest China and the "Devonian fish" in the deep Indian Ocean.

Thanks to precision stratigraphy, which gives us relative chronology down to millennium range across the K/T boundary, the sedimentary record can be examined to falsify theoretical predictions of alternative scenarios. If the mass extinction was the direct result of mass-killing by catastrophe, be that a bolide or volcanic explosion, the extinction event should be considerably shorter than the lifespan of living organisms, and thus much less than a century. If, however, the mass-killing catastrophe served only to reduce the population, while the species thus "endangered" gradually died out in an environmental crisis, then the extinction event should span a time interval of several or many generations, in hundreds, thousands, or even millions of years.

K. Perch-Nielsen and others (1982) found successive *last appearance horizons* of Cretaceous nannoplankton species in the earliest Tertiary sediments. The apparent last stragglers may have been dead skeletons reworked from the Cretaceous, rather than "survivors of endangered species" in the earliest Tertiary. To resolve this question, my student Matthias Lindiger studied the isotope geochemistry of foraminifers from the classic El Kef K/T boundary section of North Africa (Ph.D. dissertation, ETH Zurich, 1989). He found that the stragglers lived in the Tertiary; they had skeletons with isotope signatures indicative of their existence in the earliest Tertiary ocean. Lindiger's findings reinforced my conviction that mass-killing by meteorite impact was not the direct cause of plankton extinction; Cretaceous plankton species became extinct successively, but within a very short geologic time interval of thousands, or tens of thousands of years; the proximal cause of their extinctions was habitat destruction brought on when the prevailing ocean chemistry curtailed the reproduction of calcareous planktons (Hsü 1986).

Meteorite Impact or Volcanism

At the first K/T debate after the Alvarez discovery, I presented evidence of a carbon-cycling anomaly at the K/T boundary as evidence of a bolide impact. Dewey McLean advocated a scenario of catastrophic volcanism, and he correctly pointed out that my isotope data cannot discriminate between the impact and the volcanism scenarios (McLean 1982). Since I have been more concerned with the tempo and mode of evolution, I have also been more concerned with the proximal cause of habitat destruction, and have not actively engaged in the debate whether the ultimate cause was a bolide impact or explosive volcanism. I was invited, however, to give my opinion at the Second Snowbird Conference. After consideration of the temporal aspect, or, more precisely, the frequency of mass extinctions, I reiterated my preference for the impact hypothesis.

Uniformitarianism and catastrophism are positioned at the two ends of a spectrum, but the pattern of natural temporal changes has a fractal geometry.

Anything that can go wrong will go wrong, and the inevitable will come sooner or later, however improbable the odds. Whether you call it "Murphy's Law" or "the Peter Principle," this is the common wisdom of the fractal geometry of time. Statistics have shown that the frequency of occurrence of natural processes is inversely related to their magnitude (see Figure 10.2). Processes occurring daily are harmless, but when on rare occasions they operate at great intensity, they

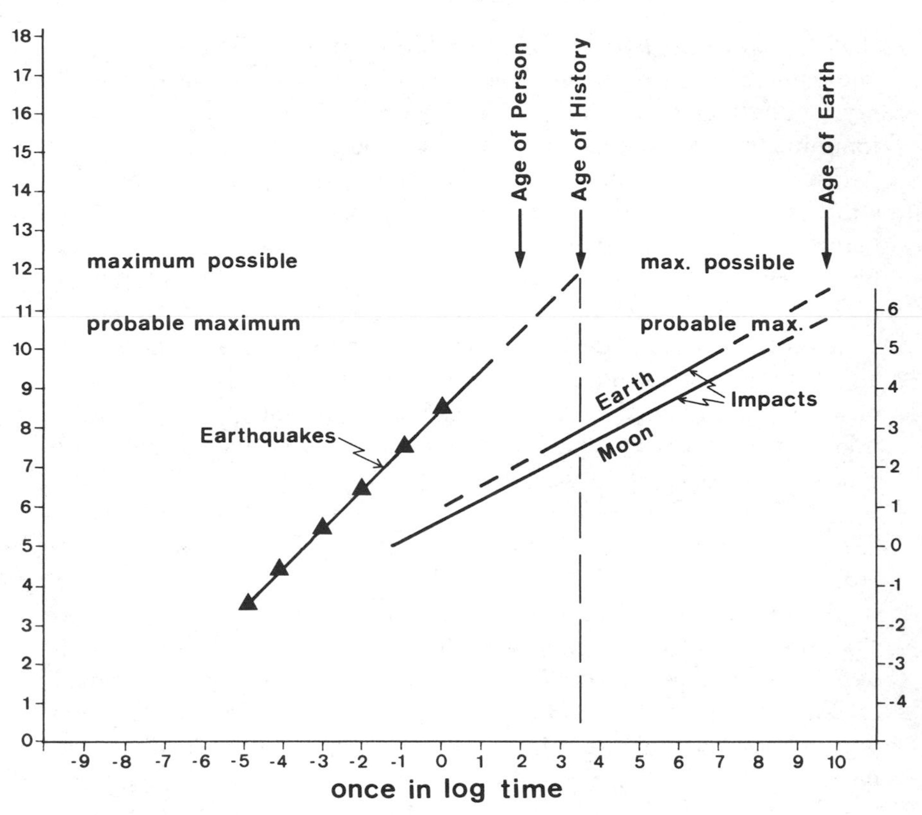

Figure 10.2. The waiting time of catastrophic events. The vertical axis on the left side indicates magnitude of earthquakes; the vertical axis on the right side indicates crater diameter in meters expressed by a logarithmic scale. An earthquake event with maximum magnitude is likely to take place within historical time, if not in a person's lifetime. Lyell was correct in assuming that mountain-building processes during the past had about the same energy as those of the present; substantive uniformitarianism is applicable to tectonic interpretations in geology. But a meteorite-impact event with a maximum energy release has occurred only a few times in Earth history; those were the rare events postulated by catastrophists to explain mass extinctions.

become natural catastrophes. Small stones fall, but when a mountain falls, we have a landslide. Small streams of water trickle across a lawn, but an occasional flood devastates a community. Small volcanic eruptions damage meadows or forests, but only rarely have cities like Pompeii and Herculaneum been destroyed. Meteors decorate a summer sky, but when a trillion-ton bolide hit the Earth, three-quarters of the living species became extinct. The statistics relating the magnitude and frequency of earthquakes, river discharges, volcanic eruptions, and meteorite impacts have been given in numerous publications (see Hsü 1989).

Mandelbrot's (1977) fractal concept is an alternative to determinism and chaos, and the fractal geometry is defined as

$$N = c/r^D \qquad (1)$$

where N is a dimensionless number, r is a linear dimension, c is a proportionality constant, and D, the fractal dimension, is a fraction, and not a small integer.

My work on the temporal frequency of natural and man-made catastrophes has shown that the frequency occurrence of natural events is inversely related to their magnitude (Hsü 1982)

$$\log f = c/\log E^D \qquad (2)$$

where f is the frequency and E the energy of the event. In the fractal geometry of natural hazards, then, N is an expression of the logarithm of occurrence frequency and r is the logarithm of the energy of a natural event.

The time interval T, called waiting time by Raup (1991), between successive events is the inverse of the frequency, and equation (2) can thus be expressed by

$$\log T = c \log E^D. \qquad (3)$$

Habitat destruction is related to energy flux in environments. Energetic events cause more havoc and thus more species extinctions. Raup's killing curve (Figure 10.3) is thus an expression of the fractal geometry of natural catastrophes. The waiting time of era-boundary events such as the K/T catastrophe, which caused the extinction of some 75 percent of marine species is of the order of 10^8–10^9 years. The most energetic meteorite impact on Earth was probably a head-on collision with a long-period comet with a mass of 10^{18} gm (Weisman 1982). The waiting time of such events is indeed the same order of magnitude as the duration of Phanerozoic eras. The frequency of large-bolide impacts is thus indeed what could be predicted on the basis of the fossil record, if mass extinctions are related to impact events. The same cannot be predicted by the "volcanists." The most energetic volcanic eruptions are far more frequent than era-boundary extinction events. The waiting time between eruptions of maximum energy is in the order of 10^7 years (Hsü & McKenzie 1991). If volca-

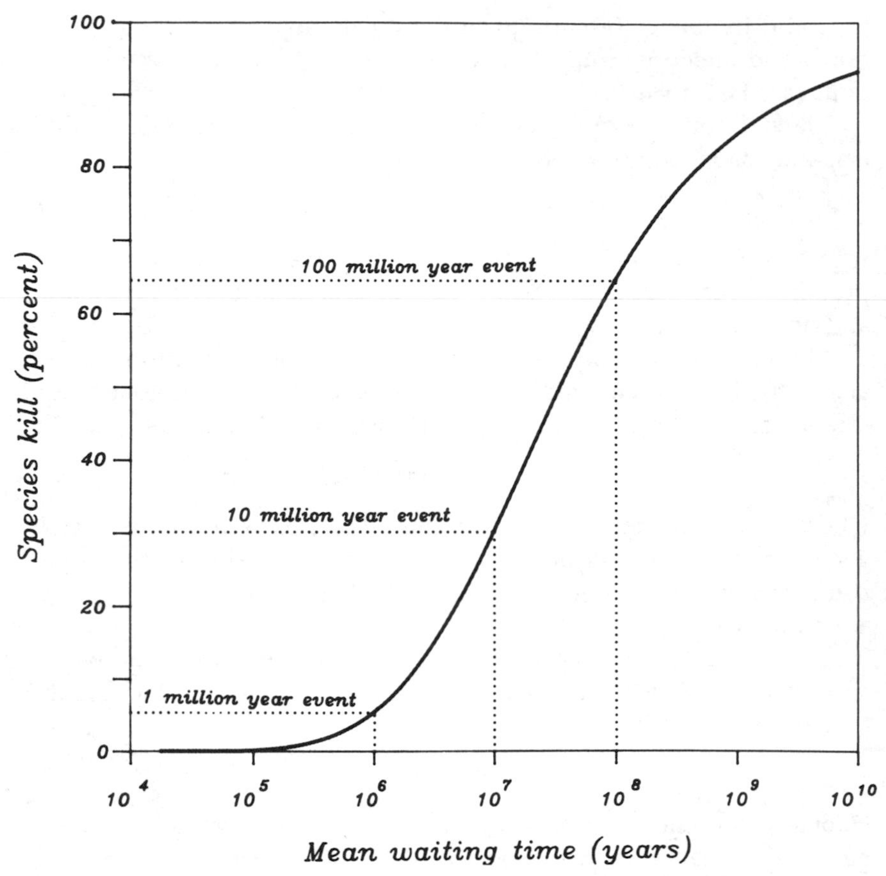

Figure 10.3. Kill curve summarizing the history of Phanerozoic species extinction (after Raup 1991). The curve shows the mean waiting time for events of varying extinction intensity. For example, a species "kill" of about 5 percent occurs about every 1 million years. The major Phanerozoic extinction events are approximately 100-million-year events.

nism had been responsible for mass extinctions, the duration of geologic eras should have been an order of magnitude less. This consideration on the fractal geometry of time further hinders the postulate of relating the K/T extinctions to volcanism.

A Historical Perspective

Substantive uniformitarianism is the epitome of the ignorance and arrogance of nineteenth-century scientists. They had no clock to measure geologic time, yet they were sufficiently arrogant to believe that what is not observable cannot have happened, forgetting the brevity of human life compared to the course of Earth history.

The K/T debate is likely to continue until the present generation of antagonists has met its fate, when the new *Leitbild* of the fractal geometry of time can prevail.

11

Mass Extinctions: Fact or Fallacy?

John C. Briggs

The terms "mass extinction," "mass killing," "mass murder," and "catastrophic event" have all been used in the recent literature to describe historical extinction episodes. The use of such terms implies evidence of sudden disasters that wiped out significant portions of the Earth's animal and plant species. They give the impression of a devastated globe littered with the remains of dead organisms. For example, Allaby and Lovelock, in their book *The Great Extinction* (1983), said that 65 million years ago (m.y.a.) the Earth had collided with a small planet. They described the impact as a conflagration when "volcanoes erupted, tidal waves swept the oceans, and earthquakes shook the continents. For years, a dust cloud shrouded the earth blocking out the sun. Three quarters of all species died, including the mighty dinosaurs." In their *Scientific American* article, Alvarez and Asaro (1990) referred to a "sensational crime" that killed off half of all the life on Earth and provided evidence that supposedly showed that a giant asteroid or comet had committed this "mass murder."

Are we, in the light of our present knowledge, justified in using such lurid descriptions and terms? The following questions are important: (1) How did the mass extinction idea get started? (2) Did the historical extinctions occur so suddenly that they need to be called mass extinctions? (3) Were they so widespread that they deserve to be termed global events? And (4) did they really have drastic effects on the world's species diversity?

Historical Development of Extinction Hypotheses

From a historical standpoint, it may be said that we are living in a time of neocatastrophism, a repetition of an era that began in the first part of the nineteenth century and died out about 30 to 40 years later. The original idea of catastrophism began with the famous French anatomist and paleontologist Georges Cuvier. He became convinced that the Earth had undergone a series of great catastrophes, the most recent being the biblical Deluge. After each catastrophe, the Earth was repopulated by remnants that had somehow survived the crisis. The new species that subsequently appeared were supposed to have come from parts of the world previously unknown. Others, such as William Buckley, argued that each catastrophe was worldwide and was followed by an entirely new creation. The early theory of catastrophism finally succumbed to the concept of uniformitarianism introduced by James Hutton and effectively championed by Charles Lyell and Charles Darwin.

Charles Lyell viewed life as a continuous fluctuation of living populations which expanded or contracted their boundaries as geological agents altered local topography and climates (Browne 1983). Wallace (1855) observed that new species gradually arose to take the place of those that had become extinct. Darwin (1859) emphasized the imperfection of the geological record and noted that the sudden appearance or disappearance of fossil species was probably due to the fragmentary nature of that record. By the twentieth century, paleontologists had recognized two important times of change in the history of life, one at the end of the Permian and one at the end of the Cretaceous. But these changes were not, for many years, considered to be sudden catastrophes. For example, Dunbar (1960) described the Permian/Triassic change as "orderly and gradual, not cataclysmic."

More recently, changes in viewpoint about the rapidity of extinction began to be expressed, particularly in regard to the Cretaceous/Tertiary (K/T) boundary. Schindewolf (1962) announced his concept of "neocatastrophism" in reference to an essentially synchronous annihilation of major groups of Mesozoic organisms. Bramlette (1965) described "massive extinctions" in biota at the end of Mesozoic time; Newell (1967) referred to the boundary as marking a "mass extinction"; and Percival and Fischer (1977) discussed a "Cretaceous-Tertiary biotic crisis." These more dramatic terms, as well as continued speculation in popular-science articles about the fate of the dinosaurs, apparently set the stage for acceptance of theories involving catastrophic events.

A new era of speculation about historic extinctions began in 1979, when Luis Alvarez and co-workers discovered a level of enriched iridium in the clay of the K/T boundary. At first, they thought a supernova had been responsible, but in 1980 they changed their minds and decided that an asteroid had struck the earth about 65 m.y.a. The impact was supposedly followed by a cosmic winter lasting several years, which caused a biotic catastrophe. This startling news received extensive coverage in the scientific and popular press.

Several paleontologists, who had worked with fossils from the K/T boundary era, protested the widespread assumption that the extinctions came about with catastrophic suddenness (Kauffman 1979; Clemens et al. 1981; Hickey 1981; Archibald & Clemens 1982). But they were paid little heed, and Alvarez (1983) expressed his frustration at his inability to convince them that an asteroid had done the job. General acceptance of the impact theory was aided by the fact that, at first, there seemed to be no other reasonable explanation for the iridium enrichment in the boundary clay.

The next important event was the publication of the periodicity theory of Raup and Sepkoski (1984). They analyzed the entire marine fossil record and found evidence of a 26-m.y. cycle, which suggested that most historical extinctions were caused by extraterrestrial impacts. This, plus the advent of volcanism as an alternative high-energy theory (Officer & Drake 1985) put us firmly into an era of neocatastrophism.

The Tempo of the Extinctions

The evidence, which has been recently reviewed (Briggs 1990), now indicates that all seven of the major extinction events took place over extended periods of time ranging from about 1 to 10 m.y. or more. Furthermore, it seems likely that all may have occurred in the form of a sequential series of minor episodes that, only collectively, constitute a significant extinction. The Ashgillian extinction extended over a period of 1 to 2 m.y. near the end of the Ordovician; many taxa, belonging to complex benthic and pelagic communities, became extinct (Brenchley 1989, 1990). The general marine-invertebrate data pertaining to the Frasnian extinction toward the end of the Devonian were reviewed by McGhee (1988, 1989, 1990). He found multiple periods of high extinction spanning an interval of 2 to 4 m.y.

The Permian/Triassic boundary marks the time of the greatest of all extinctions. High extinction rates took place during the final three stages of the Permian (Vermeij 1987), a period of about 10 m.y. The late Triassic is often noted as a time of significant marine extinctions. In his review, Benton (1990) concluded that there were three separate extinctions, none clearly larger than the other two. The first occurred during the Scythian Epoch over a period of 5 to 6 m.y., the second from the Scythian to the end of the Carnian Stage (15–19 m.y.), and the third during the Norian Stage (12–17 m.y.). Thus there may have been three protracted extinctions which, together, occupied most of the Triassic.

The Cenomanian Stage of the mid-Cretaceous has recently become recognized as the time of an extinction event. Shoemaker and Wolfe (1986) referred to this episode as a stepwise extinction that probably spanned about 2.5 m.y. In the K/T event, most of the extinctions among the marine benthic animals took place over a 2.5- to 2.75-m.y. interval (Kauffman 1986). The decline in the calcareous plankton apparently occurred during an interval of widespread volcanism that lasted several hundred thousand years (Herman 1990). This plankton extinction probably began some 350,000 years prior to the K/T boundary (Hansen 1990). Earlier, Keller (1988) had reported that the diversity decrease in planktonic foraminiferans took place over a period of 0.8 to 1.0 m.y. As these works indicate, the plankton extinctions took place relatively rapidly but certainly cannot be considered sudden or catastrophic events.

On land, it has been observed that the dinosaur extinction was probably a gradual process that began some 7 m.y. before the K/T boundary (Sloan et al. 1986). Possibly to the contrary, Sheehan (1990) maintained that dinosaur-family diversity did not decline during the last 3 m.y. of the Cretaceous. However, Dodson (1991) found that, within the final stage of the Cretaceous (a period of about 6 m.y.), trends toward a reduction in generic diversity were evident. Some dinosaurs quite possibly remained alive as much as 1 to 3 m.y. into the Paleocene (Van Valen 1988). The flying reptiles (pterosaurs) had been gradually diminishing in diversity for about 70 m.y. (Carroll 1987), and by the end of the Cretaceous, only a few members remained. In regard to all nonmarine vertebrates, Archibald and Bryant (1990) concluded that a gradual, noncatastrophic change had taken place.

For the terrestrial plant life, palynological data have been interpreted to indicate either five sequential floral changes that predate and postdate the boundary (Sweet et al. 1990) or only a single abrupt change (Nichols & Fleming 1990). In general, the published data on megafossils appear to indicate a reduction in plant biomass, at least in North America, but no decline in species diversity (Briggs 1991a). Charig (1989) concluded that plants, brachiopods, insects, and bony fishes passed through the K/T boundary virtually unchanged. The one major extinction of the Tertiary took place in the Priabonian Stage toward the close of the Eocene. In a comprehensive review of both the marine and terrestrial events, Prothero (1989) provided evidence that the extinctions took place in five steps over a period of 10 m.y.

Impacts by giant extraterrestrial bodies are still being proposed as the primary cause for almost all of the extinction episodes (McLaren & Goodfellow 1990). Although impacts and volcanic eruptions may have occurred during or near times of biotic decrease, the tempo of the extinctions is inconsistent with hypotheses involving single high-energy events. These were gradual deteriorations in organic diversity that developed in response to environmental changes. Terms such as "mass extinction," "mass killing," or "catastrophic event" are misleading when applied to extinction episodes in the geologic record.

The Scope of the Extinctions

Of the seven episodes in Phanerozoic history that have been identified as times of great extinctions, only the Permian/Triassic event deserves to be considered a truly global phenomenon. This is the only extinction in which there was a drastic diversity decrease in *both* marine and terrestrial environments. For the K/T boundary, which has often been described as a global catastrophe, the overall effect on the terrestrial biota was small compared to the reduction of the tropical marine life (Briggs 1990). Certainly, the relatively few remaining dinosaurs eventually died off, but the other vertebrate and invertebrate groups survived quite well. The angiosperm plants received only a temporary setback in biomass, one that had no discernible effect on their long-term expanding diversity (Niklas 1986).

Effects on Global Species Diversity

Estimates of the severity of so-called global extinctions have often been biased because of their dependence on fossils from the shallow waters of the marine tropics. Samples from high latitudes or the deep sea almost always show a better survival during times of environmental crisis. Another problem is that the fossil record as a whole may not be a sufficiently dependable basis for an accurate assessment of the magnitude of extinction events (Smith & Patterson 1988). For some years still, information about the magnitude of widespread extinctions probably should come from detailed studies of individual groups where the histories of discrete phyletic lines have been determined.

There are indications that the effects of widespread extinctions on the global species diversity have been greatly exaggerated. For example, estimates of species extinguished at the K/T boundary range from 50 to 80 percent of all species (Hsü 1986; Raup 1988; Courtillot 1990; Alvarez & Asaro 1990). However, the number of terrestrial, tropical arthropod species in the world today has been estimated at between 6 and 9 million (Thomas 1990). An additional 2 to 3 million arthropod species probably occur in the temperate zones. There are possibly a million species of nematodes, a predominantly terrestrial phylum (May 1988). These figures provide a total of about 9 to 13 million species of metazoan land animals. Raven (1990) referred to a minimal world total of 10 million. There are about 300,000 species of vascular plants (Burger 1981). In contrast, the total number of marine metazoan species is about 160,000 (Briggs 1991b). This means that these huge groups of terrestrial organisms comprise at least 98 percent of the world's species. Yet, there is no evidence that any of them suffered much extinction at the K/T boundary.

It appears that the marine metazoans may have about doubled their species diversity in the Cenozoic (Signor 1990). It also appears that the terrestrial arthropods and vascular plants may have about doubled their diversity (Niklas 1986). If we assume that the metazoan species diversity of the latest Cretaceous, on land and in the sea, was about half that which exists today, this provides about a 5 or 6 million to 80,000 ratio, with the marine species still comprising less than 2 percent of the total. Even if the K/T extinction destroyed half of all the marine species, which is doubtful, the decline in the global species diversity would have been less than 1 percent (Briggs 1991a).

Although the fossil evidence does not support the concept of historical mass extinctions or mass killings, there is a catastrophic extinction event occurring in contemporary time. Raven (1990) has estimated that, by the first quarter of the next century, the world will have lost 2 million out of a minimal world total of 10 million animal species and about 65,000 out of 300,000 species of vascular plants. These losses, due to habitat destruction by humans, are occurring with a rapidity that is unprecedented in Phanerozoic time. Historic extinction episodes were so gradual that many lineages were able to accommodate in an evolutionary and ecological sense. The tempo of the current extinctions precludes any such adjustments.

Conclusions

Our present era of neocatastrophism is a recapitulation of the original theory of catastrophism that was popular in the early nineteenth century. Neocatastrophism is predicated on the assumption that there occurred a historic series of global catastrophes called mass extinctions or mass killings. However, study of the tempo of the extinction episodes reveals that they took place over extended periods of time ranging from about 1 to 10 m.y. or more. The deliberate pace of the extinctions was, in reality, the antithesis of catastrophic. They took place over evolutionary, not contemporary, time.

The scope of most of the extinctions was far less than has usually been described. Only the Permian/Triassic event deserves to be considered a global phenomenon. This is the only extinction in which there was a drastic diversity decrease in both marine and terrestrial environments. There are indications that the effects of widespread extinctions on the global species diversity have been greatly exaggerated. At the K/T boundary, for example, estimates of species extinguished range from 50 to 80 percent of the world total. Yet, when the evidence is examined, it appears that the actual demise could have been less than 1 percent. Certainly, many important groups of marine animals disappeared, and some terrestrial groups did so also, but their relative diversity was very small. The only event that deserves the title of mass extinction is that which is going on right now. It appears likely that, within the next 35 years, the world will lose about 20 percent of its total species diversity.

There is an ongoing debate about what caused historical mass extinctions—asteroid impacts or volcanic eruptions? But this debate is based on the false assumption that mass extinctions or mass killings have taken place. Aside from the present human destruction of the Earth's tropical ecosystems, there is no evidence that global mass extinctions (defined as short-term, catastrophic events) ever took place.

12

On the Mass-Extinction Debates: An Interview with William A. Clemens

Conducted and compiled by William Glen

William Glen: What was your own mindset or gestalt regarding the Alvarez-group impact hypothesis when you first heard of it in 1979?

Wm. Clemens: The question of what were the causal factors of dinosaurian extinction was one that had not really commanded the rapt attention of the vertebrate paleontological community. During the three decades preceding publication of the Alvarez group's first paper, the end of what Benton [1990] has dubbed the "dilettante phase" and the beginning of the "professional phase" of the study of dinosaurian extinction, a number of articles on this topic appeared. Some were quite competent, others most fanciful. De Laubenfels [1956] probably was not the first to suggest the impact of an extraterrestrial body as the cause of the extinction of dinosaurs. I think Jepsen's [1963] tongue-in-cheek summary of speculations on the causes of the extinction of the "terrible lizards," caught the spirit of these times.

Because hypotheses testable with the data at hand had not been framed, investigation of the possible causal factors of dinosaurian extinction often was regarded as a second-class issue. There were notable exceptions, of course. Early on, Norman Newell [1982 and references] addressed the problems of identifying the processes involved in mass extinctions. For many years Harold Urey "was fascinated by the 'sudden' disappearance of dinosaurs" and "hoped to show that this mass extinction event . . . was related to some catastrophic event involving changes in temperature" [comments made by Urey in 1949, quoted in Lowenstam & Weiner, 1989, p. 221]. Later, in a paper that did not attract wide attention, he attributed their extinction to the impact of a comet (Urey 1973).

The Alvarez group's [1980] paper in *Science*, which primarily reported the intricacies of their studies of a rare platinum-group element, iridium, provided yet another explanation of the extinction of these ancient reptiles. In a brief section of the lengthy report, an awe-inspiring, impact-produced catastrophe was advanced as "the killer."

Well before publication of their paper in *Science*, journalists had got wind of the story. Unlike the introduction of major changes in paradigms of the earth sciences, such as recognition of the role of plate tectonics in shaping the Earth's history, the assertion by my colleagues at Berkeley that the impact of an asteroid caused the extinction of dinosaurs provoked widespread scientific interest outside the earth and biological sciences. Soon, with varying degrees of embellishment, the popular press reported the story. During the early 1980's, it appeared as a cover story in *Time* magazine, was related by radio and television reporters, and filled many column inches of newspapers and popular tabloids. Quickly a bandwagon was rolling.

Glen: Digby McLaren's presidential address to the Paleontological Society in 1970 is a piece of curious architecture. He more or less beats around the bush and ruminates on what might be called innocuous issues before presenting—in the most soft-peddled way at the very end of the paper—his hypothesis of meteorite impact to explain the Frasnian-Famennian extinction about 365 million years ago. McLaren invoked a bolide impact, on the basis of his analyses of the selectivity of extinction of the faunal elements. He strongly anticipated resistance to such an idea, and his fears can easily be read in his presentation (he told me in conversation in July 1992 about his fears of being openly explicit about his hypothesis at the time of presentation). What was the nature of the resistance that McLaren anticipated? How much of such resistance was engendered by the untestability of the hypothesis? How much because impacting raised the specter of catastrophism? How much simply because novel ideas impinge on long-established prevailing "truths"—other than catastrophism—in which the community has a large stake, whether intellectual, personal, and otherwise?

Clemens: Obviously, I cannot put myself in McLaren's shoes and comment on his motives. As to the reception of his presidential address, several factors probably are involved. Unlike the Alvarez et al. paper [1980], which presented their hypotheses in an exceptionally long, lead article in *Science* that attracted the attention of many scientists, McLaren tucked his ideas in at the end of his address almost as an afterthought, and without rigorous justification. Presented in that fashion, his speculations were taken by me, and apparently many others, as trial balloons to provoke colleagues to think about new interpretations. They were presented at a time when most earth scientists accepted the view that impacts of extraterrestrial bodies were extremely rare events during the last 600 million years of Earth history. Also, he lacked geochemical evidence of the kind used so effectively by the Alvarez group.

You raise the issue of a "specter of catastrophism" as impeding acceptance of the Alvarez group's hypotheses. I would take issue with the view that vertebrate paleontologists or other scientists routinely reject any hypothesis that invokes catastrophic events of kinds that have not been experienced during historic times.

Surveys of paleontologists taken during the 1980's revealed that the group, particularly the vertebrate paleontologists, had doubts about the Alvarez's hypotheses. During these years I talked with many colleagues about the bases of their skepticism or support. Rarely did I hear someone dismiss the possibility of impact-produced extinction out of hand. By far the most typical remarks were couched in terms of their research experience; the hypothesized impact-produced catastrophe did not or did explain the evolutionary history of the group they studied. I encountered the greatest skepticism among vertebrate paleontologists.

Interestingly, the greatest level of support came from paleobotanists, particularly palynologists, who were specialists on the floras of the interior of western North America, where a major extinction of many lineages of plants can be used to define a K/T boundary. My reading of the subsequent literature and discussions with paleobotanists indicate that the floral change marking the K/T boundary in the western interior currently is recognized as much more extreme than contemporaneous floral changes now documented in other parts of the world.

Glen: What was your initial assessment of the probability that the (Alvarez) hypothesis might prove true?

Clemens: In their article in *Science*, the Alvarez group presented not one but two major hypotheses. One was that an asteroid hit the earth some 65 million years ago. Presentation of geochemical evidence, development of this hypothesis, and falsification of a competing hypothesis, took up the vast majority of the article. This was a new area of research for me, and I spent many hours searching the literature for information concerning iridium and its geochemistry. I became concerned by the clear message that at that time little was known about the distribution of iridium in the Earth's crust.

The second hypothesis suggested that an impact was the cause of a mass extinction, extinguishing the dinosaurs and many other lineages of organisms. Of the almost 14 pages covered by their article, except for some introductory remarks the discussion of the biological effects of this hypothesized impact occupied less than two-thirds of a page.

Circumstantial evidence, apparent synchrony of impact and extinctions, was the only link between the two hypotheses. The end of the Cretaceous is defined by the extinction of such groups as terrestrial dinosaurs or marine foraminiferans or ammonites. The geochemical signals of the hypothesized impact came from rocks that either contained the last representatives of these lineages or from stratigraphically higher, therefore younger, strata.

Both hypotheses were well constructed and open to falsification. Likewise, the circumstantial evidence cited to link them was open to testing. A number of us immediately accepted the challenge. During the following five or six years my research and that of some of my students focused on the question of contemporaneity of the extinction of dinosaurs and the time of formation of the iridium enrichment in Montana, where we had been carrying on paleontological field work throughout the 1970's. This question required a new, more refined scale in our geological fieldwork. No longer was it sufficient to measure sections in units rounded to the nearest foot; critical boundary sections had to be restudied on a centimeter scale. We debated and tried to determine the relationship of the stratigraphically highest (youngest) dinosaurian remains to the actual time of the extinction of this group. The emphasis was to attempt to wring the most precise information on the sequence of events possible from the rock and fossil record.

Our shift in the focus of our research to analysis of patterns of survival and extinction came later. In part, this resulted from the realization that we could not determine the sequence of geological and biological events on a biologically significant time scale. Also, the extensive samples of latest Cretaceous and earliest Paleocene faunas that we had accumulated in these early years had grown into an obvious basis for studies of patterns of survival and extinction of lineages.

The early 1980's at Berkeley were times of many special seminars and lectures by guest speakers discussing aspects of the Alvarez group's hypotheses. For several months my graduate students and I met with Luis and Walter Alvarez, Frank Asaro, Helen Michel, and Dale Russell, who was on leave at Berkeley, for weekly discussions. These meetings ranged from interesting exchanges of data and wide-ranging discussions to full-fledged fiery debates.

Unfortunately, at times, personal attacks, some finding their way into the press, clouded the scene. Looking back, it appears many of these flare-ups were sparked by differences of opinion on the relative merits of apparently conflicting physical and biological data. Jastrow [1983] correctly highlighted this in his comments on "pulling rank." On the other hand, this also was a time for learning and exchange of ideas. For example, I remember with great pleasure the hours Frank Asaro spent introducing me to some of the mysteries of isotope chemistry, and discussions of asteroids and the formation of craters during a field trip with Gene Shoemaker to Meteor Crater.

Glen: Do I hear—implicit in your concerns at that time about contemporaneity—the question of geologic instantaneity? The Alvarez-group hypothesis really called for a geologically instantaneous event, which flew in the face of the party line at that time regarding the time frame for mass extinctions. I've examined textbooks and monographs on that subject, and the time frame for extinctions, since the last century, seems always approximated to that for orogeny.

Clemens: Yes, the hypothesis of global contemporaneity caught my attention, but not for the reasons you suggest. I received my early geological education in California, an area of profound tectonic activity. I can see a bit of the trace of the San Andreas Fault from my dining-room window, and memories of the Loma Prieta earthquake still are fresh. The assertion that short-term geological events can have catastrophic environmental effects was and continues to be an obvious assumption. The claim of global contemporaneity caught my attention because it appeared to be open to testing and potential falsification.

Extinction is a biological phenomenon. If purported global contemporaneity of a mass extinction is to be used as evidence supporting a hypothesis of a particular set of causal factors, then the evidence must be collected and studied at biologically significant time scales calibrated in units of days, years, or centuries. The paleontological evidence of the history of biological changes before, during, and after the K/T extinctions is organized on a geological time scale calibrated in increments of a million years, or rarely, hundreds of thousands of years. Applications of magnetostratigraphic and radiometric techniques have provided a wealth of data that shows iridium enrichments, and last records of many lineages whose extinction marks the end of the Cretaceous are found in rocks of essentially the same age, about 65 million years, on a geological time scale. It must be stressed that the western interior of the United States and Canada is the only area in which the time of extinction of dinosaurs has been determined radiometrically.

Even application of the very recently developed $^{40}Ar/^{39}Ar$ method, which can provide age determinations with uncertainties measured in a few tens of thousands of years, cannot test the hypothesis of global contemporaneity on a biologically significant time scale. Demonstration that the extinctions and iridium enrichment events occurred within the same, for example, 50,000-year interval is not sufficient to support the hypothesis of global synchrony on a biologically significant level. If one, for example, were able to show that dinosaurs became extinct in the Northern Hemisphere a century after their demise in the Southern Hemisphere, this asynchrony would falsify the extreme hypothesis of an instantaneous global catastrophe, the so-called "Dante's Inferno" scenario [Alvarez 1986].

By the mid-1980's we were coming to realize that, with the tools available, we could not conduct a biologically significant test to determine whether extinctions had occurred on land and in the sea within a time frame measured in days to hundreds or thousands of years. The very valid question of global contemporaneity of events on a biologically significant time scale had to be abandoned.

Shifting from the question of global contemporaneity, in the early 1980's we also attempted to test the assertion that the extinction of dinosaurs and the formation of an iridium-enriched "impact" layer in eastern Montana were "precisely" contemporaneous events in eastern Montana. Our fieldwork had demonstrated that the stratigraphically highest dinosaurian fossils were about 3 meters below the iridium-enriched layer. Subsequently, we have been able to find a few bones of these reptiles at higher levels, but the uppermost meter still has not yielded bones of dinosaurs or any other vertebrates.

The iridium-enriched layer is preserved in the basal lignites of the Tullock Formation. Sedimentological studies by David Fastovsky (1987) and others demonstrated that the depositional environment of the Hell Creek Formation was a relatively dry coastal plain. The inception of deposition of the Tullock reflects a shift to a much wetter environment, one characterized by the development of high water tables and extensive swamps. The "ghastly blank," the unfossiliferous meter or so separating the stratigraphically highest dinosaurian bones and the iridium-enriched layer, might well be the product of leaching of fossils from the uppermost Hell Creek by acidic ground waters derived from the widespread Tullock swamps. Whatever the cause, the absence of vertebrate fossils in the uppermost strata of the Hell Creek Formation has robbed us of a record of changes in the fauna during the last days of the dinosaurs.

The possibility that gaps in the fossil record of other origins were distorting the fossil record was investigated by Lowell Dingus [e.g. Dingus & Sadler 1982]. Deposition of sediments, particularly in terrestrial environments, is episodic. The remains of organisms are preserved from destruction only when they are buried rapidly. In a terrestrial environment, where major accumulations of sediments occur rarely (during a 100-year or 500-year flood, for example), and preservation of vertebrate remains is an infrequent event, the fossil record might document the composition of the fauna at only thousand-year, ten-thousand-year, or longer intervals. Such incompleteness of the fossil record would distort the history of gradual evolutionary change or gradual elimination of various members of a biota into one suggesting rapid, episodic change.

Lowell demonstrated that the deposition of sediments entombing fossils in the Hell Creek and Tullock formations probably was so intermittent that we could not hope to distinguish between patterns of gradual or abrupt change in composition of the biota just before or after the extinction event. Later work in other areas, particularly in marine sections [e.g. MacLeod & Keller 1991], has shown that hiatuses in the stratigraphic record are also common in many of the marine sections that have figured in studies of the K/T boundary. I think Bob Dott [1983] accurately summed up the limitations of this type of study when he noted that stratigraphic sections, and the fossil records they preserve, are akin to a poor grade of Swiss cheese—more air than reality.

Glen: An early paleontologic rationale for rejecting the Alvarez-group impact theory held that the K/T boundary was not marked by a sudden mass extinction. Philip Signor and Jere Lipps [1982] countered by defining what is now called the Signor-Lipps artificial range truncation effect. What effect in the mass-extinction debates did the Signor-Lipps effect have?

Clemens: In the early phases of testing the Alvarez group's hypotheses, many paleontologists working in both terrestrial and marine sections placed particular emphasis on the stratigraphic levels of the last records of particular lineages. The realization of the potential incompleteness of the fossil record cast doubt on the significance of the last record of a lineage. Signor and Lipps' contribution showed that given the incompleteness of the rock and fossil records and collecting biases, high-resolution studies of last occurrences would tend to show a pattern of gradual extinction of different lineages. Again, what initially had appeared to be a potentially rewarding line of inquiry proved to have major limitations when addressing questions on a biologically significant time scale.

A different approach to analyzing the fossil record had to be developed, one that would not be affected greatly by gaps in the rock and fossil records and did not require age determinations on a biologically significant scale. David Archibald and Laurie Bryant [1990] took what has proved to be a fruitful tack in their research on the patterns of change of the terrestrial fauna of Montana that circumvents many of these problems. They built on and increased our stratigraphically controlled sequence of well-sampled fossil localities in the Cretaceous Hell Creek Formation and the Paleocene Tullock Formation. Systematic analyses of each lineage of vertebrates in these faunas were undertaken. Different kinds of fishes, amphibians, nondinosaurian reptiles, and mammals were considered to establish an overall pattern of survival and extinction. This pattern was subjected to various kinds of analyses. The records of relatively rare lineages and those that were abundantly represented were discriminated, and greater weight was given to the latter; the absence of members of a rare group from some samples has a higher probability of being an artifact of preservation or collecting. They determined which lineages crossed the K/T boundary and survived in Montana, in contrast to those that became locally extinct but survived into the Tertiary somewhere else, usually at lower latitudes.

Because their fossil record is so meager, Archibald and Bryant omitted one group of terrestrial vertebrates, the birds, from their detailed analysis, but on two counts birds deserve special recognition. Throughout this interview I have spoken and shall speak of the extinction of the dinosaurs, for we can all envision the demise of *Tyran-*

nosaurus, Triceratops, and their lumbering or fleet-footed contemporaries. Recent research has established that birds evolved from small, bipedal dinosaurs, and has revealed that they are a group of dinosaurs characterized by the evolution of feathers and the ability to fly. Thus, dinosaurs are not extinct. One dinosaurian group, the birds, survived into the Tertiary and prospered.

There is more to this than just a play on names and the intricacies of biological classification. Consider the genealogy of a family in which, five generations ago, the "founders" of the clan had seven children. Today only the descendants of one of them, great-great-aunt Susan, are among us; all the other branches of the family died out earlier. An obvious question is, what did Susan and her progeny have going in their favor? Similarly, why among the diversity of different kinds of dinosaurs were the birds the only group to survive? It was not just their ability to fly. The pterosaurs, distant relatives of the dinosaurs, were flying reptiles, but were extinct by the end of the Cretaceous.

Early versions of the Alvarez group's second hypothesis argued that darkness and cold temperatures produced by the impact's dust cloud were the causal factors of dinosaurian extinction. Later, other hypothetical causal factors were added. Walter Alvarez [1986] styled these as "a whole Dante's Inferno of appalling environmental disturbances." In addition to darkness and cold—an "impact winter"—acid rain, global wildfires, and greenhouse warming after the dust cloud settled were added to the horrors. The resulting hypothesis is one of a brief period of short-term, catastrophic changes of the global environment.

If such drastic, short-term changes in the environment took place, how did birds survive? All too frequently the answer to this question by those who favor the "Dante's Inferno" scenario is that birds huddled in burrows or fluttered about in caves for weeks or months until the conflagration had subsided. Given what we know about the environmental tolerances of modern birds, this kind of response is less than convincing.

Without going into further details, I think the results of studies of patterns of survival and extinction of terrestrial vertebrates fully falsify the hypothesis that an impact caused the terminal Cretaceous extinctions of terrestrial vertebrates through the series of environmental catastrophes embodied in the "Dante's Inferno" scenario. Ancestors of groups that today are known to be unable to tolerate major climatic change, such as frogs, salamanders, lizards, turtles, and birds, survived whatever caused the extinction of the other dinosaurs. Some critics have argued that it was just a matter of chance that during the "Inferno" the dinosaurs died out and many other lineages of vertebrates survived. David Raup [1991, p. 102] analyzed the data on survival and extinction from Montana and concluded that this was not the case, remarking that the dinosaurs "had some extra susceptibility to whatever caused the Cretaceous extinctions."

Again, remembering that extinction is a biological phenomenon, patterns of survival and extinction provide data that can be used to test any hypothesis of causal factors of the terminal Cretaceous extinction. The current challenge is to expand the data base from the western interior of North America to include information on patterns of survival and extinction from other parts of the world.

Glen: At the time that the extinction debates were triggered anew by the Alvarez-group hypothesis in 1979, the idea of punctuated equilibrium had been out for about seven years. How did the idea of punctuated equilibrium—which was quite controversial and still is to some extent—impinge on the notion of paleontologic discontinuities, especially the major breaks in the faunal record that were used to define the geologic time scale?

Clemens: Mass extinctions have produced many of the major breaks or changes in the fossil record that are used to characterize the boundaries separating units in the geological time scale. Gould and Eldredge asserted that a dominant macroevolutionary pattern involved brief periods of rapid change separated by long periods of essentially evolutionary stasis.

For those of us brought up on George Gaylord Simpson's works, the concept that rates of evolution had varied was nothing new. What was intriguing to me was Gould and Eldredge's emphasis on evolutionary stasis. This certainly provoked thinking about ranges of tolerance to environmental changes and developmental constraints.

The macroevolutionary pattern of punctuated equilibrium supported the view that mass extinctions played a major role in directing the course of evolution. Concurrent geological research led to the discovery of many hitherto unrecognized impact craters, and negation of the view that, in contrast to Precambrian times, the Phanerozoic [the last 500 to 600 million years] was characterized by relatively few impacts of extraterrestrial bodies. The next step was to hypothesize that all mass extinctions—or major punctuations—were triggered by the impacts of extraterrestrial bodies.

Currently a great deal of effort is being focused on studies of sections containing records of major boundaries in the geological time scale. In some instances, craters of appropriate age or other geological data signal the possible occurrence of an impact at some of these boundaries; for others, the implicating evidence has not been found. It is intriguing that well-documented impact events, involving the formation of a crater such as the Nordlinger Ries in southern Germany, are not contemporaneous with times of major or even minor extinctions [Heissig 1986].

Glen: How did David Jablonski's conclusion—to which he came as early as 1983 in large part—that mass extinctions are qualitatively different from background extinctions accord with your own views? Did his conclusion impinge in some way on punctuated-equilibrium arguments in your own thinking?

Clemens: Thanks to a university fellowship, Dave spent two years at Berkeley during the early 1980's and was deeply involved in the discussions concerning the possibility of impact-caused extinction. His differentiation of the nature of background and mass extinctions definitely sparked my interest in studying the contributions of different biogeographic patterns of distribution and species richness.

We now have evidence of the presence of dinosaurs and some other groups of terrestrial vertebrates on all continents, including Antarctica, at some time during the Late Cretaceous. In a few areas the latest Cretaceous terrestrial fauna is relatively well sampled, but only in the western interior of North America is there a detailed record of faunal change during the transition from the Cretaceous to the Tertiary. Currently, our knowledge of the distribution of Late Cretaceous vertebrates is not sufficient to undertake analyses similar to those of Jablonski's on marine invertebrates.

The available records of latest Cretaceous and Paleocene floras are much more abundant than those of terrestrial vertebrates. A global survey of their histories during the Cretaceous/Tertiary transition reveals an intriguing pattern. The severity of extinctions of plants at the end of the Cretaceous appears to have been greatest in western North America and Asia, of lesser magnitude in eastern North America and Europe, and least in the Southern Hemisphere. This is in part the basis for my earlier suggestion that the patterns of survival and extinction of vertebrates now well documented in western North America might not be a sample of a globally uniform pattern.

Glen: Did Jablonski's work put a dent in your own thinking about the likelihood of an extinction mechanism that was of a magnitude or severity and geographic extent different from what you'd been disposed to thinking about before, in terms of handling the comings and goings of your own taxonomic groups?

Clemens: Certainly working with the extensive data base provided by groups of marine invertebrates, he was able to show that differences in biogeographic range influenced the probability of survival or extinction of groups, and that these probabilities differed between background and mass extinctions. This was something that we could not determine from the relatively meager record of Late Cretaceous terrestrial vertebrates. His work raises the question, did the causal factors of the Late Cretaceous extinctions vary geographically, or were they globally uniform and the tolerances of the organisms varied?

Glen: How did the community react to Jablonski's surmise about the qualitative difference between mass and background extinction? Did they tie it to punctuated equilibrium or to other questions in paleontology?

Clemens: Yes, Dave Jablonski's contributions did and continue to play a part in discussions of macroevolutionary patterns in general, and of extinctions of various levels of intensity. Remember, the community you cite is characterized by a remarkable breadth of scientific interests, with the scientific and popular press in a significant role providing messengers communicating between disciplinary camps. I doubt that Jablonski's findings had as much influence on the thinking of astrophysicists or other physical scientists who were concerned primarily with the first Alvarez-group hypothesis, that an impact occurred. For those interested in the second Alvarez-group hypothesis, yes, his work pointed out that extinctions could be the result of different kinds of processes.

Glen: How did you receive the Dave Raup and Jack Sepkoski claim in 1984 that their pioneering statistical analysis of the marine-fossil record showed that mass extinctions are periodic about every 26 million years?

Clemens: In 1983 I participated in the conference at Flagstaff on the dynamics of extinction. Here Jack Sepkoski presented their paper on periodicity of extinctions, based on an analysis of the fossil records of families of marine invertebrates. Not too long thereafter, at a seminar in Berkeley, Gene Shoemaker expressed his doubts about the significance of the apparent periodicity. He was the first of many to challenge the statistical techniques applied to the data.

I was not in a position to evaluate the statistical techniques, but I was concerned by the limitations of the data base. The families recognized had been established with differing philosophies of classification. Actual records of stratigraphic ranges had been modified to fit into an internationally recognized geological time scale. I am glad to see that Jack continues to refine the data base, and that Dave Raup and he continue with their analyses.

Glen: Of late, there is evidence accruing that is bringing an increasing number of people to view impacting of the Earth as a much more significant factor in shaping the history of Earth's biota through time than it was, say, up until a decade or two ago. More and more people are paying attention to impacting, including paleontologists. How do you view this influence as regards the course of research in paleontology?

Clemens: Soon after Raup and Sepkoski advanced their case for periodicity in occurrences of mass-extinction events, others pulled together the then-known records of impacts of extraterrestrial objects. An apparent coincidence of mass extinctions and the impact of very large bolides led to the "Nemesis" hypothesis.

Given the research funding and effort being put into the search for evidence of prehistoric impacts, it is not surprising that the number of documented events is increasing rapidly. The old view that the rate of impacting during the Phanerozoic was significantly lower than during the Precambrian has been challenged effectively. I suspect that it is too early to make any definitive statements about rates and patterns of impacting, but it is probable that during the Phanerozoic the impacts were one kind of physical cause of change in the environment.

Because the impact of a really large extraterrestrial body has not occurred in historic times, we lack direct observational data on its biological consequences. The observation that some mass-extinction events appear to be contemporaneous with impacts warrants testing the hypothesis that all mass-extinction events were the consequence of impacts. In spite of considerable recent interest in studies of sections containing the Permian/Triassic boundary, for example, the time of the greatest mass extinction in the marine environment, clear evidence of an impact has yet to emerge. I think this hypothesis of constant linkage of major impacts and mass extinctions is going to prove to be an overstatement.

One area where I would like to see more research is in the physical consequences of impacts and their biologically significant effects on the environment. Over the years, interpretations have changed markedly. In 1980 the impact of an asteroid at the K/T boundary initially was purported to have produced a dust cloud that lasted for a matter of several years. Soon thereafter the estimates of its duration were reduced to a matter of weeks or a few months. The models for other elements of the "Dante's Inferno" scenario have been advanced with emphasis on what they would kill off, and little concern for the many lineages that survived. Paleontological data on patterns of survival and extinction must be employed to test the validity of these and other, similarly modeled, causal factors of extinction.

Regardless of their causes, mass-extinction events and background extinctions have played a significant role in shaping the course of evolution of life. Analysis of their differing contributions continues to be a fruitful line of inquiry.

Glen: Do you, at this time, see sea-level changes or climatic change of a non-instantaneous-catastrophic kind, that is, a climatic change other than that possibly produced by massive volcanic episodes or impacts, as being able to cause what has been interpreted as a mass extinction of the sort that we see at the Cretaceous/Tertiary, or Permian/Triassic boundaries?

Clemens: You have covered a large number of topics in your question. All play a significant part in our attempts to identify the processes that resulted in the terminal Cretaceous extinctions.

Tony Hallam and others have explored the effects of changes in sea level and their effects on both marine and nonmarine environments. The end of the Cretaceous was a time of a major marine regression that significantly altered the configurations and connections of continents. It no doubt had an influence on patterns of circulation of oceanic currents, which would have influenced climatic patterns. Invertebrate paleontologists point out that during the late Cretaceous the oceans appear to become more homogeneous. The loss of environmental diversity easily could have contributed to the extinction of some groups of marine organisms.

Charles Officer and others have called our attention to the intense volcanic events that occurred during the latest Cretaceous and the Cretaceous/Tertiary transition. The great outpouring of the Deccan Traps over what is now peninsular India, and volcanic activity in other parts of the world, had the potential to significantly alter global climates. In historic times the eruption in 1815 of Tambora in Indonesia has been associated with severe climatic changes in New England and western Europe. What were the environmental effects of terminal Cretaceous volcanic activity?

It needs to be noted that the environmental effects of volcanism and impact show some significant similarities. Both result from clouds of dust and various aerosols. Also, both volcanism and, probably, impact can produce acid rain. Now evidence drawn from a crater in Yucatan indicates that a large extraterrestrial body impacted the Earth at the end of the Cretaceous. This impact occurred during a period of volcanic activity. We cannot avoid the challenge of determining the relative contributions of the two to modifications of the environment.

Without trying to designate a causal factor, the evidence available documents a period of climatic cooling during the last 5 to 10 million years of the Cretaceous. This is reflected in modifications of both the terrestrial fauna and flora. The dinosaurian fauna of western North America that lived about 7 million years before the end of the Cretaceous was significantly more diverse than the last dinosaurian fauna of the region. Loss of diversity is a predictable consequence of long-term cooling of the climate.

In the debate about impacts of extraterrestrial bodies as the causal factor of mass extinctions, other biologically significant, physical changes all too often are overlooked. It can be demonstrated that global climates cooled during the last 10 million years or so of the Cretaceous. Probably, changes in size and patterns of circulation of the oceans contributed to global climatic change; certainly they resulted in increased areas characterized by continental climatic regimes. Regression of the oceans also would have resulted in the connection of long-isolated land masses, and would have provided the opportunity for dispersal of many kinds of organisms. The geologic record documents an increase in volcanic activity in the latest Cretaceous that could have influenced global climate.

I feel that any hypothesis of the causal factors of the extinctions used to mark the end of the Cretaceous must take all of these physical factors and their potential environmental consequences into account. Were they sufficient to cause the extinctions by gradually changing the environment until conditions exceeded the ranges of tolerance of many groups of organisms? Now we have evidence of the impact of one or more extraterrestrial objects at the end of the Cretaceous. Did the physical effects of the impact(s) jolt the environment and cause the extinctions of some groups?

We can draw an instructive parallel from current studies that suggest the global climate is warming. A wide variety of causal factors including everything from human activities to the eruption of Mount Pinatubo and El Niño events appear to be implicated to some degree. I suspect the resolution of the question of what caused the extinction of the dinosaurs and some other organisms is going to be equally complex and require testing of a hypothesis that involves a number of environmental variables.

13

On the Mass-Extinction Debates: An Interview with Stephen Jay Gould

Conducted and compiled by William Glen

William Glen: What was your own mindset or gestalt on receiving the Alvarez-group impact hypothesis in 1979? I'm particularly interested in how your punctuated-equilibrium ideas impinged on the notion of paleontologic discontinuities—especially the major breaks in the faunal record that served to define the geologic time scale? From conversations with the Alvarez group, and especially Luis, and a number of letters between you, it is clear that you were an early, outspoken supporter of their hypothesis.

S. J. Gould: Of course there was much in my background beyond punctuated equilibrium. For one thing, my very first paper was on uniformitarianism. I thought a lot about the logical analysis of the concept and realized that the mixing of the methodological postulates with the substantive claim for gradualism was wrong, and that therefore there was no a priori reason whatsoever to accept gradual change. Lyell had pulled the wool over the whole profession's eyes in [applying] uniformity. If there is anything coming out of my own past work, it is that on uniformity, which made me realize the potential validity of catastrophic alternatives to the methodologically tainted gradualist credo. After all, I had been Norman Newell's student [at Columbia University]. I don't think I got that much from him about uniformity (for I had done this work as an undergraduate), but I'd like to think that [he provided] an environment in which you could talk about mass extinctions as a reality. For most of the paleontologists, [mass extinctions] hardly existed. Most of paleontology regarded them as longish periods of accelerated extinction. Punctuated equilibrium is, of course, a phenomenon of a different scale; it's about the extinction of individual species. Punctuated equilibrium has a broader theme that I like to call the punctuational style of change. In working on punctuated equilibrium I just became much more friendly to punctuational styles of change at all scales. Punctuated equilibrium is about the extinction and origination of species; mass extinction is a different scale of event within the punctuational model, but I had already developed a personally coherent and broadly ranging critique of gradualism, so not only did I not have in my head the false notion that gradualism is preferable a priori as part of the definition of science, but I was even empirically predisposed to rapidity of change. So when I heard about Alvarez, that part was terrific.

Glen: How did David Jablonski's conclusion, as early as 1983 in part, that mass extinctions are qualitatively different from background extinctions accord with your own views—especially with respect to punctuated equilibrium?

Gould: I think it is almost a logical consequence of Alvarez. I don't think it relates a whole lot to punctuated equilibrium per se, which is a different-scale phenomenon. If you read my article on "The Paradox of the First Tier," it goes through that more explicitly. But 1983 was already well into the debate. I, having been favorable to Alvarez—at least for the Cretaceous—had to expect that different rules would prevail. Because again, the one thing that Alvarez [theory] does, with respect to tradition, it makes it impossible to see mass extinction as a mere intensification and therefore in continuity [of background extinction], and therefore there would have to be different rules than just turning up the gain of ordinary processes. When things are truly catastrophic, the rules change—you get a different pattern. So I was delighted to get Jablonski's information, but I was expecting it.

Glen: How did you view the community reaction to Jablonski's surmise that mass extinctions are qualitatively different from background extinctions? Did they tie it to punctuated equilibrium?

Gould: It's so secondary—it's not an enormously important issue. It was so secondary to whether you were for or against Alvarez to begin with. Once you're for the catastrophic theory, it follows as a logical necessity. I think it got subsumed into basic Alvarez theory.

Glen: An early paleontologic rationale for rejecting Alvarez-group impact theory held that the K/T boundary was not marked by a sudden mass extinction. Philip Signor and Jere Lipps countered by defining what is now called the Signor-Lipps artificial range truncation effect (which you discussed in your *Natural History* article of March '92). What effect in the mass-extinction debates did the Signor-Lipps effect have?

Gould: In retrospect, the Signor-Lipps effect is a powerful argument. I certainly saw people were using it. It's one of those things that are rather obvious once you accept Alvarez. You just didn't think about it before. As an abstract issue, anyone could have appreciated it. But unless [they had] some predisposition to accept—or at least want to test—the catastrophic mass-extinction hypothesis, most paleontologists would have said, well, that's true, but we pretty much know that the extinctions weren't sudden. So since gradual petering out is interpretable two ways, either as Signor-Lipps superimposed upon a truly catastrophic termination, or read more literally, as petering out—since we know petering out is true anyway—what's the big deal? I think that's what most people probably would have said, but I don't really know. So again, with Signor-Lipps you almost have to accept Alvarez first.

Glen: I take it that you took Signor-Lipps at face value on its advent?

Gould: It was a good argument. It was a good way to quantify.

Glen: Did you see the Signor-Lipps effect relating to punctuated equilibrium?

Gould: Look Bill, I'd like to sell punctuated equilibrium as widely as I can. You put it in every question, so you obviously want me to draw parallels. Punctuated equilibrium is a different-scale phenomenon—it's just not about simultaneity in extinctions and originations, which is what the issue of mass extinctions is about. It's a punctuational model for the origin of individual species. What they have in common is a general philosophical approach to change. So again, anything that looked favorable to Alvarez was happy for me, because of my philosophically favorable inclination toward punctuational change. But I don't think there's a direct relationship with punctuated equilibrium.

Glen: Do you recall how you were disposed to the paper by Al Fischer and Michael Arthur, in 1977, that suggested that mass extinctions occurred at 32-million-year intervals during the Phanerozoic?

Gould: I remember that, because I was quite friendly with Ida Thompson, who worked with Al on that. If I'd not known Ida I'd probably have been disposed to dismiss it, because I have a prejudice against cycle making, because I know how tempting it is, how easy it is, to see cycles in messy data. I have enormous respect for Al Fischer and for Ida as well, so when it came out I was very skeptical, but I didn't dismiss it for those two reasons—mainly because Al was right so often.

Glen: How did you receive the Dave Raup and Jack Sepkoski claim in 1984 that their pioneering statistical analysis of the marine-fossil record showed that mass extinctions are periodic about every 26 million years?

Gould: I loved it. I've written a number of essays on it. To this day I feel it's just a brilliant use of statistical arguments. However, there just isn't enough evidence to tell one way or the other. That's where it still sits. But once you have plausibility of extraterrestrial impact established for the Cretaceous evidence, then long-period cycles suddenly become much, much more plausible than something internal to Earth-based mechanisms—at least there are plausible mechanisms.

Glen: By this date we sort of know how the question of periodicity has come to a virtual standstill. The bulk of the papers debating the question came out between 1984 and 1987. We have about half for and half against, with equally good, statistically credentialed mooters represented on both sides of the question. But I wonder if you recall what the paleontologic community's response was to the Raup and Sepkoski paper when it appeared, and has the community realigned itself on that question since it appeared?

Gould: I'm not a good person to ask about these things. I'm very much of a loner. I don't talk a lot to my colleagues. I'm not much into the daily life of the profession. I do go to meetings [but] I've never been politically involved. I've sat on the council and been president of the society, but I'm not actively involved. I don't really have a sense of it. I think what's easier to answer is the paleontologic community's reaction to Alvarez [theory] itself, because there was so much hostility to the notion when it first came out. So many paleontologists thought that they knew for a fact that the Cretaceous extinctions had been long-petering before the final blip.

I only remember Dave Raup and myself as being strongly favorable [to the Alvarez theory]. I remember Dave calling me up when Alvarez was circulating his preprint and we first learned of it and his saying to me, "You know this is different from all the other extraterrestrial hypotheses—from Schindewolf to Digby McLaren," which were just guesses and didn't mean a damn thing. Dave said, "This is different" and I said, "Yes, Dave, I understand that. You've got the iridium, you've got something you can test for." What's more, most of the others had been overt speculations. [Alvarez theory] had been forced by evidence originally generated to test the opposite hypothesis. I mean, if you are testing sedimentation rates at the end of the Cretaceous, the assumption initially made was that the iridium influx was a constant, cosmic gentle rain from heaven that could test differences in earthly rates. A totally different phenomenon arose from an empirical surprise, not from an attempt to speculate a priori that extraterrestrial causes [were involved]. In retrospect we were right, but I'm not sure it was for the right reasons. Dave liked it because he was interested in random processes; I liked it because I'm interested in punctuational processes. In other words, we had predisposing biases that made us look upon Alvarez favorably. By the time Raup and Sepkoski came along, it was enough years later that at least the plausibility of Alvarez [theory] was established. (By the way, the 26-million-year cycle period is not totally, happily, within the original version of Alvarez theory, because he wanted the effects to be produced by [the impact] of Apollo objects, which by their mechanics can't be cyclical. So you have to figure out some other way to get periodicity.)

Tony Hoffman [was] apoplectically negative [to the Alvarez theory]—but then he was negative toward almost any idea.

Glen: Digby McLaren's presidential address to the Paleontological Society in 1970 is a piece of curious architecture. He more or less beats around the bush and ruminates on what might be called innocuous issues before presenting—in the most soft-pedaled way at the very end of the paper—his hypothesis of meteorite impact to explain the Frasnian-Famennian extinction about 365 million years ago. McLaren invoked a bolide impact based on his analyses of the selectivity of extinction of the faunal elements. He strongly anticipated resistance to such an idea, which can easily be read in his presentation (he told me in conversation in July 1992 about his fears of being openly explicit about his hypothesis at the time of presentation).

Gould: That's exactly what he did at the meeting.

Glen: What was the nature of the resistance that McLaren anticipated?

Gould: I love Digby dearly, but I think that he lucked out with that [hypothesis]. The resistance was entirely appropriate—he didn't have a shred of evidence [for impact]. It was pure conjecture, the old pre-Alvarez style of guesswork. What was interesting about it was his argument for the rapidity of the intra-Devonian extinction, but the meteorite speculation was just guesswork.

Glen: What proportion of the resistance that he met was fueled by the untestability of the hypothesis?

Gould: I think most of it. It just wasn't a scientific suggestion.

Glen: How much of the resistance came from his raising the specter of catastrophism?

Gould: That's always there, but even if it had been a uniformitarian, perfectly orthodox hypothesis, it wouldn't have got people convinced at that point. What can you do with something you can't test?

Glen: How much resistance comes simply because novel ideas impinge on long-established prevailing "truths," other than catastrophism, in which the community has a large stake—intellectual, personal, and otherwise?

Gould: That's a key issue. It's even worse for paleontologists because of a priori gradualism. Again, most paleontologists didn't even realize that gradualism was an empirical viewpoint with plausible and methodologically sound alternatives. They thought [gradualism] was part of the definition of what it meant to be a scientist. Because Lyell had virtually gotten away with a fast one in arguing that because uniformity has methodological meanings that all scientists must accept, then therefore he could sweep under the rug of acceptance the empirical meanings that have such a totally different status.

Glen: A comparative framework in which to examine the role of institutions in the process of theory reception, and also in the application of standards in assessing evidence, might be based on a comparison of closely related subdisciplines—which can be considered institutions—that share much of their information, techniques, methods, and theories. Comparisons among such subdisciplines, in which the fund of commonalities is great and the differences few, might allow differences in cognition between individuals from different subdisciplines to be correlated with specific subdisciplinary (institutional) causal factors. The search for the institutional bases of cognitive differences is an old one.

As early as 1739 David Hume spoke to the social origins of the categories of time, space, and causality, and showed how it would be impossible for life and society to exist if all men "did not have the same conceptions of time, space, cause, number, etc."—he felt that there was "a minimum of logical conformity beyond which [society] cannot go [without destroying itself]. For this reason, [society] uses all its authority upon its members to forestall such dissidences." Emile Durkheim, in 1893, wrote that "Cognition is the most socially conditioned activity of man, and knowledge is the paramount social creation." Ludwig Fleck (1935), an eminent scientist who took his examples from science, enlarged upon Durkheim and defined the "thought collective" as "a social group with a thought style that shapes perception through training and produces a body of knowledge." Fleck's definition seems to fit a modern community of subspecialists in science. The "thought collective" appears also close to the term "institution" as defined by Mary Douglas as "any legitimized social grouping." Douglas noted in her book of 1986, *How Institutions Think*, that the old view of human cognition as a process divorced from the social milieu is obsolete but still prevalent. Fleck believed that the thought style of the collective truly conditioned perception to the extent that even the dialectic framework treating objective reality was constrained for its members. The "style," he emphasized, decides which questions are admissible and whether answers are true (the style becomes the institution or a most integral part of it). Fleck held (p. 41)—most importantly, I believe—that "The individual within the collective is never, or hardly ever, conscious of the prevailing thought style which almost always exerts an absolutely compulsive force upon his thinking, and with which it is not possible to be at variance."

My own experiences strongly support Fleck's ideas—even his surmise that the control of individual cognition by thought style may preclude or encumber communication with those inculcated with different thought styles. Groups do have distinctive thought styles and strongly influence the thinking of their members. Although there exists no satisfactory theory of behavior that fully explains how groups come to have and maintain such power, we know that they do. But still I find a dilemma in the purported role of the thought collective. It is an old problem that concerns the basic question of how two individuals, similarly conditioned within the same community of scientific practitioners, and imbued with the same cognitive framework, including standards of appraisal—or loosely, in Kuhn's view, with the same paradigmatic imprinting—can come to assess new ideas or new evidence bearing on old ideas in totally opposing ways.

Do you think that the thought styles of subspecialties within the same discipline—such as micropaleontology and vertebrate paleontology—are sufficiently different to engender conflicting receptions of theory or clearly different assessments of evidence bearing on debated issues?

Gould: It is interesting that even the more iconoclastic of dinosaur paleontologists, Bakker and Horner, until very recently were very much opposed to Alvarez, whereas most iconoclasts within invertebrate paleontology were pretty favorable much quicker, I think if anything because the geological time scale is based on these extinction events and invertebrate paleontologists knew more about them and knew them better. The tradition of gradual disappearance is even stronger in vertebrate paleontology.

Glen: How do you think institutions figure in generating resistance to novel ideas, that is, if you regard an institution as a "thought collective"? The institution's participating members simultaneously take up its thought style, surrender freedoms, and give it their allegiance. The institution and its thought style become inseparable, as some have presented it. The institution can almost be taken as a way of thinking. So considered, how might scientific subspecialties such as micropaleontology or vertebrate paleontology influence the reception of novel ideas?

Gould: I just don't know how to parse paleontology and its institutions. If I think of some other debates like continental drift, then there's geographic variation. There are Southern Hemisphere geologists, like Du Toit and the South Africans and the Australians, who were not only feeling marginalized with respect to the centers of power on northern continents, but also living on old Gondwanaland, where the evidence [for drift] is better. They see a geographic division. Nearly everybody [in paleontology] was opposed to Alvarez in the beginning. I suspect the invertebrate people came around quicker for the most part. I'm not sure that the reason is the different material we study, or the different traditions. It could be that in such a small profession there were a few key people around, like Raup and me, who liked it and happened to be invertebrate paleontolgists, whereas the main thought makers of vertebrate paleontology didn't like it.

Glen: To what extent do the institutions of science support orthodoxy or prevailing views by such devices as discouraging published dissemination of radically new ideas, especially by control of the editorial and refereeing processes, or by control of research through refereeing grant proposals such as to exclude research that seeks to undermine, rather than further articulate, the discipline's paradigm(s)?

Gould: I think orthodoxy is enormously supported. In fact, I would make an argument—and I think that anyone who argues against this is not being quite honest—that institutions, universities in particular, are very conservative places. Their function is not—despite lip service—to generate radically new ideas. There's just too much operating in tenure systems and granting systems, in judgmental systems—usually older upon younger people [with] the pretenure needs to conform. I speak as someone who never had a whole lot of trouble with editors, because I could make my cases for writing things the way I wanted to and usually get away with it. In a funny sense, at least within our funky, little paleontology journals, there's probably a little more flexibility in publishing possibilities. I think more of the conformity comes from what you get grant money for, and from what universities press you to do. But think about how many really great thinkers from Darwin back and forward—Einstein for that matter—did their work outside of universities.

Glen: Do you think it possible for a scientist to achieve a successful career without adhering to the rules and strictures of conduct as defined by his community's power-holding institutions?

Gould: Oh yes! I think it's possible, but it's hard, damn hard. You have to have a whole lot of luck, and if you're independently wealthy it doesn't hurt—that was Darwin's way of doing it. If you can forge a popular career the way Jack Horner and Bob Bakker have done, for example, and don't care that much about the inevitable opprobrium that will be heaped on you by some members of the official professional community, yeah, you can do it, but most people won't. Because there are more failures than successes [among those who try it]. People have their families to feed, and most people are conservative anyway.

After all, it's another myth that institutions of science somehow encourage a free range of iconoclastic thought. The problem is that we have a tremendous bias to celebrate, and we focus upon the very few successful iconoclastic papers of the past. Pick up a seventeenth-century volume of the *Philosophical Transactions*. Most items are not world-shaking reports. They're little descriptions of this and that from the bulk of the community of gentlemen. It's always been a conservative institution, and I think even more so now. Let's face it, for many people science is a good job that is interesting, pays pretty well, has reasonable prestige, good hours, good vacations—you name it.

Glen: How and to what extent do magisters or doyens in the community exercise authority in defense of the prevailing community view when that view is their own?

Gould: Bill, I don't think they can do it by direct fiat, because there is so much mythology about not doing it by fiat. I think it's different in bigger professions. Paleontology is such a little profession—there are only a few movers and shakers, and some of them are very forceful people. Think of the way in which Ernst Mayr, by force of personality, has influenced evolutionary thought. Now I'm not saying this to criticize him, I actually agree with most of his views. Think of the influence that [George Gaylord] Simpson had in paleontology. Simpson wasn't even a nice man, and he didn't do it by force of personality. He did it by wonderful writing, by clarity of prose. They have enormous influence with people. Ed Wilson once said to me, slightly cynically, "Real originality in science is a titration, it's being just original enough that it's interestingly different from orthodoxy. But not so original as to be incomprehensible." I think that in a small profession like paleontology the phenomenon of the magister or doyen is extremely important. I think that Osborn dominated in vertebrate paleontology. I think that Walcott and Schuchert dominated invertebrate paleontology. Simpson really dominated vertebrate paleo, and Mayr dominated evolutionary theory with Dobzhansky.

Glen: How about the opposite case, when the magister's own view is novel and threatens community paradigms?

Gould: [They] marshal forces. Look at the response of Mayr to cladism. Look at Clemens and others vs. Alvarez.

Glen: Do institutions of science, by their nature, tend toward maintenance of the status quo, resist change other than that of self-enlargement and thus engender attitudes, policies, and an intellectual climate that reduces Kuhn's essential tension? That is, does the life of the institution—which is deeply vested in its paradigms—contribute to a less-than-conscious defense of the paradigms? Is the institution any better able than the individual to maintain a balance while living under the tension between the scientist's need to illustrate his professional competence and cleverness by articulating and strengthening the paradigm while simultaneously meeting the charge to seek anomalies and illuminate paradigmatic weakness? Does the institution, as the formulator and keeper of standards of appraisal, which I view as largely inseparable from the paradigm, help to sustain the essential tension, or does the institution contribute to the reign of the paradigm, even in the face of a growing anomaly load?

Gould: It's such a funny question, because lip service is always given to the role of the tradition breaker. Everyone always says, "We want bright young people, we want to give grants to beginning scientists—and then it turns out that they don't really want it for the most part. The institutions of science are quite conservative, maybe not as much as some other institutions. At least [scientific institutions have] methodologies of data collection and affirmations that present ways of challenging long-held truths. But no, I think the institution is conservative beyond the individual; primarily it controls the three key things: mainly jobs, money, and access to publication. Jobs through tenure, money through granting systems, and access through publishing systems.

Glen: In March '92 in *Natural History* you wrote that "usually, theories act as straitjackets to channel observations toward their support and to forestall data that might refute them. Such theories cannot be rejected from within, for we will not conceptualize the potentially refuting observations." What is the relationship between ruling theories and institutions, and could you expand on what you meant when you said, "We will not conceptualize"?

Gould: That's the most powerful source of bias of all. Because it's part of the ethos of science that you are approaching things with an open mind, with absolute objectivity. Because everyone is controlled by the theories they accept—you may be objective with your observations, but there's an infinite number of observations you could be making—the ones you choose to make are theory-bound. They may be so theory-bound that there is no conceivable way that they can refute the theory. That's why I wrote that piece on Signor-Lipps [that you refer to], because it's such a simple example and the obvious test. If you believe in catastrophic mass extinction, especially if you understand the Signor-Lipps effect, then search the last few meters of sediment with extraordinary care to see if those things that are going to be hard to find by Signor-Lipps might turn up. This is what I call the needle-in-the-haystack method. But that's not usually what you do in science. In science you usually sample—of course you do that, and it's good scientific procedure. But each theory engenders questions within the theory, modes of test and refutation within the theory, and you needed something like Alvarez to jumpstart people into doing the other kind of work. Alvarez's prediction is that most things should persist up to the [K/T] boundary. Maybe if you really look hard enough you'll find them.

Glen: You have remarked to me during the past year that "institutional stupidity" fueled theory rejection. Your remark implies that you may subscribe, as have a number of sociologists, to the idea of a supra-personal intelligence that resides in or *is* the institution. The individual mind/institution metaphor has long been criticized on various grounds. Some sociologists of science (e.g. Latour 1987, 1988; Gerson in press) find untenable the assumption that there is a ruling consensus within disciplines; they instead hold that the normal condition of any discipline is that it is riddled with debate even about its paradigms. My own experience disposes me to the view that the nature of the stuff the science treats—say ecology, as contrasted with physics—limits the probability of consensus. Consensus within a discipline seems a function, in part, of the ease with which its data at least appear to lend themselves to quantification and interpretation. The science that treats the apparently simplest, most empirically accessible stuff and systems is likely most often to approximate consensus. However, in any case, I see much sense in Nelson Goodman's idea (1978), which seems similar to Fleck's idea, of reality deriving from the social experiences of the thought collective. Goodman believes that it is the organization that decides the standards of appraisal for rightness. The standards need not be reliable, says Goodman, but they must be authoritative (p. 139). Is Goodman's view subsumed by your surmise about "institutional stupidity"?

Gould: Yes, that fits pretty closely to my view. I was at a meeting once with a Jesuit, and he said, "Look, the whole institution of the Church is a debating society. Ultimately, we have to conform to official decisions. But, at best, it works like democratic centralism was supposed to operate under communism." Of course there's some debate in science. After all, lip service is given to iconoclasm, and there's a certain amount of encouragement given to young workers who have different ideas. But I would never say that the institutions of science are debating societies. I wish they were more so. I don't think we operate an orthodoxy in the sense of the Catholic Church. But we do have an orthodoxy that debates within limits. Momentary catastrophism of extinction is way outside those limits—that's the essence of all of this. That's why Signor-Lipps is such a good example. Again, it's a good argument that anyone can understand, but if you're convinced a priori that there was no suddenness of mass extinction then you have another explanation for the Signor-Lipps petering out. The most you would do if you were an exceptionally honorable character would be to say, O.K., they're right, all this petering-out evidence that I thought was the proof of gradualism isn't, it has an alternative explanation, but gradualism is still true. I think that would have been very honorable, but I'm not sure that most people even got that far.

Glen: Although we don't have a satisfactory theory of behavior that explains how institutions acquire and hold power, and influence the thinking of their members, we know that they do. Goodman is simply describing such authority within the institution. He believes that it is empowered to select standards, decide what constitutes valid induction, and rule on correctness in erecting categories. What do you think such power to impose—what appear to be—subjectivities on the practice of science derives from?

Gould: I frankly don't know how you'd answer that. I suppose it derives from the hierarchical nature of any human institution. Some people have more authority than others, some universities have more authority than others. Most of these subdisciplines are small, and the authorities know each other very well.

Glen: It's a curious situation. My own view is that the institution grows up to serve in part as a device or set of mechanisms that is emplaced as a guarantor of objectivity, whereas, in fact—what some sociologists are making a case for and I have witnessed—is that institutions of science impose subjectivities, and their power to do so seems to fly in the face of their philosophic purpose and stated goals.

Gould: Yeah, but you see, there are intermediary positions, which would be mine. I don't think they impose subjectivities—that's too strong. What they impose are limited objectivities, that is, they impose biased subsets of alternatives—they don't explore the full range of potential explanations. Lyell would never say you must believe this because I said so, he would say you must believe this because it is empirically true, and yet the range of what might be empirically true could be far broader.

Glen: You wrote a letter in November 1983 to Luis and Walter Alvarez, David Jablonski, David Raup, Adolph Seilacher, and Jack Sepkoski. The letter is fairly explicit in showing that, by that date, you had already come to believe that mass extinctions have been "more frequent, more unusual, more intense, and more different [from background extinction] than we had ever suspected." You were outspoken about mass extinctions as "special controlling events in their own right that will not be fully explained by the evolutionary theory we have constructed for interaction among . . . populations of normal times." You were also, by that 1983 date, favorably disposed to Raup and Sepkoski's conclusion of a 26-m.y.-cyclicity of mass extinction for the past 225 m.y. since the great Permian kill. And you embraced Jablonski's conclusion that there is a profound qualitative difference between background extinction and "catastrophic zaps." You explicitly derided the notion of a Darwinian "vector of progress" in life history and expressed the very same view about the "faulty expectation . . . of a progressivist bias in Western thought about evolution." You held that "punctuated equilibrium dominates the extinction pattern of normal times," but that there are "different effects produced by separate processes of mass extinction that can be broken up, dismantled, reset, and dispersed by mass extinction." In concluding, you virtually came out and said that mass extinction disposed of groups as a result of "bad luck," rather than bad genes.

How was this set of views of yours—written immediately after your return from the Geological Society of America meeting in 1983—received by the paleontologic community in general at that time?

Gould: Things moved fast. Alvarez's first paper was 1979—and 1983 is a long time later, actually. By that time, at least a lot more people were favorable—a lot of the younger people, especially. I don't remember that well, and I don't think you'll get the answer by asking all the prominent people now, because everyone will say that they supported it earlier than they actually did—and that will be a true memory. It's a false fact, but it will be an honest memory!

Glen: How do you think the paleontologic community has shifted since 1983 on the handful of issues that I just raised from your letter of that year?

Gould: The evidence has come in so strongly for impact that people—no matter how strong their visceral disgust about it—now at least must take it seriously. The evidence came out in stages. Things can change gradually! There are still many dinosaur people who say that, "Yes, there was an impact, but there were hardly any dinosaurs left by then." It's possible, but my first-class prejudice is not to accept coincidence on that scale. More people will say, "Well" (and this may be true) "if the biosphere hadn't been weakened for other reasons [the extinction] wouldn't have been as big. Others of course will say that there wasn't one impact but several impacts, but to me that's a variation of the Alvarez scenerio, just as several floods in the channeled scablands is a variation on Bretz's original single flood, not an opposition to Bretz. And others will even take refuge in the hypothesis: "O.K., they are right about the Cretaceous, but they are wrong for other mass extinctions—it's not a general theory." Luis [Alvarez], as we well know, did not have an enormous or abiding interest in natural history as a descriptive science. If someone had told him, "Well, Luie, you are absolutely right about the Cretaceous, but it's the only one you are right about, and therefore you have not developed a general theory of mass extinction, you've just developed a historically contingent explanation for an event," he would have been very disappointed and disturbed, because he was looking for a general theory of mass extinction.

14

Epilogue: A Panel Discussion on the Mass-Extinction Debates

Convened and compiled by William Glen

The following is an edited transcript of tape recordings of the panel discussion of the session on Lessons from the Mass-Extinction Debates held at the biannual meeting of the International Society for the History, Philosophy, and Social Studies of Biology, which convened at Northwestern University, July 12, 1991. The panel members were John Briggs, Elisabeth Clemens, David Hull, Digby McLaren, David Raup, John Sepkoski, Leigh Van Valen, and William Glen, moderator. A number of attendees, some indicated here simply as "Unidentified," commented from the floor.

William Glen: This panel discussion has been planned as a one-hour presentation following our all-day session on the mass-extinction debates. We have a number of questions to deal with, some that I have drawn up, others that were prepared by the panel members. We will also entertain some questions from the floor.

I'd like to open with what I think is an appropriate question: Have the impact and volcanist hypotheses undermined the orthodoxy of a uniformitarian earthly context in which neo-Darwinian processes are supposed to have shaped our biota? Some argue that the specter of a newly informed neocatastrophism is rising.

Leigh Van Valen: Well, for me it has no bearing on the validity of neo-Darwinism. Natural selection is the core of neo-Darwinism and it occurs at many levels and time scales, and this is a part of it.

Glen: Is this pretty much in keeping with what you said earlier today about mass extinctions and whatever forcing agent is responsible for them as being simply end members of a spectrum or a continuum?

Van Valen: I do think that is correct, but I don't think it's necessary. Even if there were discontinuity I still would not think there would be any remote undermining of neo-Darwinism; natural selection would still occur. There are discontinuous mass extinctions that occur because of thresholds as well as mass extinctions that are part of a process continuous with other extinctions.

Glen: A larger idea of your own that has been called Van Valen's Law suggests that the physical and biotic environment of an organism deteriorates approximately at a constant rate, and that each adaptive type has a distinct extinction rate. That is Van Valen's Law [Van Valen 1973, 1974]. How does that jibe with David Jablonski's recent view that what was effective for survival in normal times does not apply in times of mass extinction [Jablonski 1986]?

Van Valen: What Raup called "Van Valen's Law" (which I would prefer to see applied to a result in evolutionary energetics, just as "Van Valen's Test" in statistics isn't what I would prefer to be known for there) is an empirical generalization, that the probability of extinction is normally independent of the age of the group. This is an odd result, for which I proposed the law as you stated it. The Red Queen's Hypothesis was meant to explain this law by introducing a zero-sum constraint on the absolute fitnesses of interacting evolutionary units. The age-independence, the law, and the Red Queen don't apply in the largest mass extinctions. I explicitly, at the time, said that they apply at normal times, by which I meant not during mass extinctions, and gave some evidence for this.

Glen: In one sense, then, you place a mass extinction and whatever forcing agent you choose to embrace as being responsible for it on a continuum of cause, but somehow its *effects* are not on that continuum.

Van Valen: Yes, how much of an effect on that mass extinctions would have would of course depend on how strong the mass extinctions are, and there are degrees of mass extinction, as everybody knows. Some may be detectable and some may not. There are also threshold effects. There are some effects that are sufficiently strong that we know they don't fit.

Digby McLaren: I strongly agree with Leigh [Van Valen]—he's talking about reality. You [Glen] have started forcing things back into models again. I find argument about uniformitarianism very dreary, and I think we ought to get rid of it. If Darwin had known about mass extinctions he would have accepted them as part of natural selection, because, if you kill animals and plants, you are obviously encouraging evolution, or at least you're opening up new niches and so forth. But don't let's force ourselves into defending the semantics of old-fashioned ideas that have outlived their usefulness.

Glen: I was simply trying to take the whole business of how things work and fit them into a philosophic framework in a larger sense. That philosophic framework has been developed in latter-day terms in rather sophisticated ways, and people continue to invoke it in—as you say—old-fashioned ways. I was simply trying to move us away from the science per se.

McLaren: By philosophic framework, you mean construct another model.

Glen: Oh yes!

McLaren: I don't find that necessary.

John Briggs: I think mass extinctions, or extinction episodes, as I'd rather call them, are certainly an evolutionary force; there's no doubt about it. They are an evolutionary force in a different sense than background evolution, both timewise and in their effects upon the evolutionary process. But they are an evolutionary force. And because they are an evolutionary force, then they are an exception to uniformitarianism.

McLaren: So what?!

Briggs: There's no use sticking very strongly to the uniformitarian concept if extinction episodes have important evolutionary effects.

James H. Shea (Editor of *Journal of Geological Education*): All this talk about uniformitarianism is horribly outdated. It's not modern uniformitarianism that talks about constancy of rate and catastrophes, that's Lyellian [Lyell 1830, 1832, 1833]. That's been abandoned for god knows how long. People who have written about uniformitarianism have pointed this out. All this talk about overthrowing uniformitarianism is nonsense. There is no empirical observation that could be made that could overthrow what really is uniformitarianism, which is simply Occam's razor.

McLaren: Of course Whewell [1832] introduced the word as a derogatory term and was poking fun at geologists.

Clifford N. Matthews (University of Illinois, Chicago): How about "impactology" instead of catastrophe? It's a term that was used last week at the meeting at Flagstaff. New subject.

Glen: It's curious that you bring that up, because the volcanists found some of the same difficulties when they moved to the explosive-volcanism hypothesis, as did the impactors early in their history in the 1950's and 1960's when Bob Dietz [1961] was trying to get interest in impacting under way. The volcanists simply found that people were not familiar with that idea. They faced some of the same kind of prejudice as the impactors. People were simply closed down to thinking about massive, cataclysmic volcanic episodes in the past. And yet it turned out that it looked as if the volcanists outwardly were on the side of uniformitarianism in the strict sense.

I'd like now to take the time to read a paragraph or two here on another question that will take us to what looked like the community view in paleontology regarding the cause of mass extinction at the advent of the Alvarez-group impact hypothesis in 1979. What I'll read was published in 1979, so we happily know that the question I put is appropriate for that date. It is from the 1979 *Encyclopedia of Paleontology*, which appeared just prior to the iridium-prompted impact hypothesis. Karl W. Flessa wrote the section on "Extinction." He noted that the term "mass extinction" is an unfortunate one, and that mass extinctions "are not instantaneous events but can take place during intervals of five million years or more." Flessa grouped the causes that had been proposed for extinction up to that time into "three general categories:" biological, physical/environmental, and extraterrestrial. It seems noteworthy that within the extraterrestrial he included radiation from solar and stellar sources—including radiation from a nearby supernova—and magnetic reversals, but no single mention is made anywhere of impact theory. That is particularly curious since less than a decade earlier, Digby McLaren, in his Presidential Address of 1970 to the Paleontological Society, which was published as a 15-page paper in the *Journal of Paleontology*, had postulated a bolide cause for the Frasnian/Famennian (Upper Devonian, 370 m.a.) mass extinction on the basis of his analysis of the global pattern of extinctions at that horizon. Furthermore, the impact hypothesis of extinction had been published as early as 1742, by Pierre de Maupertuis, a French scientist who proposed that "comets have occasionally struck the Earth, causing extinction by altering the atmosphere and oceans." And in 1797 the French astronomer Laplace wrote that a meteorite of great size striking the Earth would produce a cataclysm that would wipe out entire species and destroy "all the monuments of human history." M. W. De Laubenfels, in 1956, did calculations of physical effects of an impact and invoked the Tunguska bolide event as an analogue to explain the dinosaur extinction. René Gallant's book of 1964, *Bombarded Earth*, is subtitled: *An Essay on the Geological and Biological Effects of Huge Meteorite Impacts*. Furthermore, Flessa never mentions volcanism as a possible extinction cause. Recall also Peter Vogt's 1972 paper, which contained the first volcanist hypothesis and invoked massive Deccan Trap volcanism to dispatch the dinosaurs. In short, none of the pre-Alvarez impact hypotheses, earthly or celestial, was testable. My question is—on the heels of this little exposition—what does Flessa's omission of both impact and volcanism as possible extinction causes tell us?

David Raup: [Flessa] may have omitted them because they were unthinkable. I'm reminded of Harold Urey's paper in *Nature* in 1973, which was at the time a remarkably good statistical analysis for comet impacts as the cause of the major smaller extinctions in the Tertiary. At the end of the paper he predicted that if we looked a little farther we'd find evidence at the K/T, although there was none at that time. That paper, in a major journal by a Nobel laureate who ought to have attracted attention no matter what he wrote, was to the best of my knowledge ignored, until it was picked up in the Alvarez paper. Some ideas are so outrageous that they are not even worth thinking about, let alone debating, and that may have been why Flessa left them out.

Glen: I'd like to ask a question of Digby McLaren, since it was his 15-page paper in 1970 in the *Journal of Paleontology* that presented the first paleontologically sophisticated impact theory. How was the paper received? What do you recall from your own memory about how people responded to you when the paper come out?

McLaren: I would say with silence. My paper was on boundaries in general in the Devonian System, and considered two kinds of boundaries: quiet boundaries and boundaries across which something happened. I suggested that boundaries across which something happened might be rather useful in stratigraphy, and finally produced Dietz's meteorite in the paleo-Pacific as the extinction cause at the Frasnian-Famennian boundary. During the next ten years, when I was doing other things, I was told by student field assistants that they had been recommended by their professors to read the paper, but not to take any notice of that ridiculous idea about impacts.

Unidentified: Karl Flessa may have overlooked some papers. In hindsight he shouldn't have, but that easily happens—you can't be aware of everything—especially with something that is so rare in the thinking of the community.

Glen: Flessa's article on extinctions in the *Encyclopedia of Paleontology* is a well-balanced and well-written piece in which he comes to the conclusion that none of the causes that he's looked at for mass extinction seems capable of effecting the great changes he sees at things like the K/T boundary. He covered almost all the bases in the game. That's why it was all the more surprising to me that the two extinction causes that Flessa omitted turn out to be the two that we are left with sitting around here discussing a decade later.

Unidentified: You've structured the argument that way. It's not that way within the biological community; it's more climate change of some sort that is Earth-based rather than volcanology per se as the driving force of that climate change. They've structured the panel here and the dichotomy a little bit.

Briggs: Yes, I agree with that. I think there is too much emphasis on impactors on one hand and volcanologists on the other hand, without considering other scenarios that have been suggested or hypothesized.

Glen: But that's what the mass-extinction debates have really come down to. The other causes have been more or less moved to the side. They're not dead, but they're certainly not at the center of the K/T debates.

Elisabeth Clemens: Can I just push the statement you [Raup] made that these two causes [impact or volcanism] were not mentioned because they were unthinkable, and get some of the panel to explain to me why extinctions by the explosion of a supernova and radiation were thinkable when the others were not, because it seems to me that those other causes are vulnerable to all of the same objections as impact and volcanism. They are sudden and exogenous.

Glen: That struck me the same way.

McLaren: Schindewolf [in 1954] suggested the supernova.

John Sepkoski: And Schindewolf was the preeminent German paleontologist of his day.

Raup: I suggested this morning that most of us had learned that there were no impacts. That was out, whereas with supernovae we always knew—although our knowledge of astronomy was vanishingly small—that there could be a nearby exploding star.

Van Valen: Not all of us thought about supernovae in this context until [Dale] Russel's advocacy [Russel & Tucker 1971]. I was not familiar with the astronomical evidence, but I had kept up a little with the controversy on what were then called cryptoexplosion structures. It seemed to me that Dietz and other impact advocates had the better argument. Such impacts, though, seemed too small for mass extinctions, as we now know empirically to be the case for the Ries crater. Major, worldwide effects from impacts and the likelihood of larger unknown impacts were less than obvious. Remember that wild speculation was the rule for discussion of mass extinctions until Sloan's and my work. Impacts seemed just part of this speculation and indeed were so, because there was no evidence linking them with extinctions, or evidence that they would be strong enough to produce some major extinctions.

McLaren: One should remember that Opik [in 1958] had already calculated the rate at which the Earth could collect asteroids in Earth-crossing orbits.

Raup: Opik wrote about 15 papers in one way or another arguing for comet impact as the cause of extinction. They were entirely contained in the literature of astronomy, and so the rest of us didn't see them.

Matthews: There was one other line of thought, that of John Oro [Oro 1963]. He is very active in the origin-of-life studies, and held that comets could have introduced organic matter to the Earth. So he wasn't thinking of impacts in the sense of extinctions; on the contrary, he was thinking of benign impacts, but nonetheless he did bring that up and now that's the basic paper that's going to be discussed at a meeting in Wisconsin in September called "Comets and the Origin of Life."

McLaren: And Fred Hoyle has suggested a cometary cause for flu epidemics.

Unidentified: You [McLaren] made the comment that your students were told to read your paper but told not to listen to what it said [about impacts]. In the early 1960's someone was coming down from the University of Utah to talk about continental drift—and all the professors in the department said listen to these guys, but don't pay attention to what they have to say because it's all untrue. I just wonder if you have that kind of mind-set in geology.

Shea: That's what they said to Darwin, wasn't it: read Lyell but don't pay any attention to him.

McLaren: I had the same experience at the University of Michigan in 1949 when I gave an evening's talk on continental drift and debated the geophysicist J. G. Wilson (not Tuzo Wilson). It was quite amusing—it really embarrassed people, and next day they wouldn't meet my eye and crossed the street rather than face me!

Joel Cracraft (ornithologist, American Museum of Natural History, New York): Now that impact theory is coming along, to what extent is it easier for people to accept related ideas such as periodicity of extinctions and other subhypotheses?

Raup: In other words, is there a coattail effect?

Glen: It's an obvious phenomenon that has happened in other cases of debate. When a debated idea comes into ascendancy, all sorts of stuff that had earlier been considered marginal or even flaky gets attached to the ascendant idea. Suddenly a new kind of respectability can accrue to something merely by its resemblance to or connection with what is gaining ascendancy. You could see that in my talk earlier today that touched on the success of the mantle-plume modelers. They are a very esoteric community; very few people can do the math modeling that's required for plumes, so they developed a tight, closed circle of referees. A few papers came out, some got into *Science* or *Nature*, and within two or three months an industry of citation of very tenuous equational constructs arose. Their coefficients—from what I've surmised—are not well-constrained. Suddenly three months later people are citing a model, but you'd think that they were citing a piece of empirical evidence that had been derived by naked eyeball observation of the effects of a hammer on a rock; they're really talking about a model of the mantle and its processes that is little understood and tenuous. It's often quite amazing how the level of certainty of evidence offered up in theoretical debate is not discussed directly. The error bars for theory wouldn't fit on the page! Evidence is frequently presented in a context and mode that suggests greater power for it than can be allowed by even the most generous impartial assessor. That appears to be a common style of combat in debates. It certainly has been in these.

Unidentified: I happen to know one guy who created the three-dimensional model. Those things are as real ontologically as what any physicist does in modeling nuclear phenomena.

Glen: I interviewed some very good math modelers right after the three-dimensional model came out. I said to one of them, "What does this mean? Should I believe in it? Should I follow up and do interviews?" The math modeler said, "It's a lot of junk." He quickly went through much explanation that I didn't understand, but it was his surmise that I shouldn't follow up on this, that it had gotten too much attention and didn't deserve to be published in the prestigious journal in which it had appeared. I am a child of faith when in distant fields, as most of us are, so I didn't really follow it very closely and waited for the mantle-plume modeling to take another tack before going around to ask other people about the whole thing. Further interviews would likely have yielded different opinions about the virtues of the three-dimensional model—after all, it had to get through the refereeing process.

Briggs: I think citation citing or dividing people into groups and saying you've got so many people in this group and for this theory, and then a group on the other side, is like trying to prove a hypothesis on the basis of its popularity or by vote. This of course is a very nonscientific procedure that should be avoided, because certainly hypotheses should be either weakened or strengthened on the basis of new discoveries by individuals or groups. I'm afraid that this is the sort of mode that we've gotten into in this meeting. This business of popularity of concepts is not the scientific process.

Unidentified: I've always thought that the analysis of citation patterns was one of the most useful things.

Glen: What Jack [Briggs] said is profoundly true. There were people at the highest levels of editorship in the major journals who were whipsawed by these influential, vocal partisans in the debates. I watched one theoretical camp seemingly gain control of one journal's editor, and he believed in their side for two years. The editor was pushing papers through on that side. He changed his stance when evidence came out in favor of the other theoretical camp. You could see these journals perturbing the whole research front of science. Science was being decided by editors to a certain extent. Because these interdisciplinary debates were so broad, because scientists of every stripe had become so utterly dependent on authorities whom they did not understand, more and more did people appear to rely on little information and superficial sources of information to make judgments about what constituted valid science. This was reflected all the way down to the level of refereeing for NSF proposals. Something ought be done in science to better control editorializing on debated issues in scientific journals. An editor's first duty should be to know when to abstain.

Unidentified: Caveats get pared away just due to lack of space.

When you cite someone, you can't repeat everything they said in their paper. Then the premise of the paper sort of gets carried along to the point where something that was quite tenuous becomes the received wisdom. [Such a phenomenon] is being viewed as a negative thing to be excised, but I wonder if sometimes what happens isn't something more like this: the tenuous, unquestioned idea gets carried along as an assumed truth until some dead end inevitably gets hit or someone finds that the assumed truth has no basis. But by then the community investigating the problem has moved to a new perspective, because lots of stuff has happened in the intervening period. People go back with a fresh perspective and look at the assumed truth, and new insights come out of that. It's unavoidable not to assume things that haven't been tested, because lots of that is on the fly. It might be part of the scientific creative process.

Glen: Undoubtedly what you say is very true, but I think we're really talking about trying to make the process more economical. There's a real lack of economy in the stuff we've been talking about.

David Hull: Discovery in science is inherently inefficient. If you try to make it too efficient, you might end up killing it off. My favorite example is that of date palms. Early on people discovered that half the date palms didn't produce dates so they started killing off the ones that didn't produce dates until they killed off the last male—then all the others stopped producing dates. You may have to watch out; maybe discovery is inherently a very inefficient process. Compare science to art. All these artists are producing so many lousy pictures. We've got to make this process of producing works of art more efficient and economical. Now that sounds crazy—but maybe it should sound crazy for research in science as well.

Glen: I still never have had anyone give me a terribly adequate answer to the question I keep asking about just what purpose philosophers and sociologists and historians of science are supposed to be serving. For whom are they working and toward what end? I've been asking that one for 15 or 20 years.

Hull: I am one, and I keep asking what is it we're doing. We're trying to do something worthwhile, but what? The people at this meeting study biology, but from a host of different perspectives, and the purpose of this meeting is to allow people to use different methodologies and talk to each other and learn from each other.

Sepkoski: Beware of Jeff Levinton's [1981] uncertainty principle.

Hull: Which is?

Sepkoski: That scientists never behave the same when they have philosophers watching them.

Unidentified: I'd like to ask a question—that is sort of perverse in some ways—predicting the outcome of what this debate might be. There is some pretty strongly supportive evidence on both sides, or even on three or four sides. Is there any way in which each side might be able to come out half right? Does anybody see a way to reconcile the different views?

Glen: I found it remarkable from early on in these debates that Tom Ahrens and John O'Keefe had shown in 1983 that a 10-km bolide impact on Earth would produce a 13 Richter-scale-magnitude quake—an impact that would have enormous crustal relaxation consequences. There'd be all sorts of signatures of such an event. Naturally, the first thing that people started batting about was volcanism; if you really did have a massive impact then you should expect volcanism. I went out and tried assiduously to get the volcanists to talk to me about the possible connection between an impact and volcanoes—both camps could seemingly have their cake and eat it. The volcanists would change the topic whenever I brought it up—it seemed as if they didn't hear me. Then the impactors did the same thing when I worked in the other direction with them. It was as if each of them would have the force and power of their argument or hypothesis detracted from by association with the other, and that went on for years.

McLaren: I'd like to add something to that. Our field is much too narrow at the moment to come to the conclusion you're looking for—and I'm sorry to say this, Bill [Glen], but we keep coming together pretending to talk about impacts and all we've talked about is the K/T. If there's anything to be learned, then we've got to broaden our horizons. Few if any are asking questions about other impact events. In our recent paper, Goodfellow and I [McLaren & Goodfellow 1990] suggested that there were six or seven good candidates waiting to be demonstrated or tested, and currently we're setting about trying to do this. For example, why has no one discussed the Acraman Crater in Australia? This is very important to us and to these theories. We'll probably learn more from the Acraman impact than we ever will from the K/T about the natural history of iridium, because we know that it's definitely fallout from an identified impact. There's no possibility of a volcanic cause.

Glen: The answer for me is that my present work in history, although not limited to the K/T, is mainly focused on it. I mostly ask questions about things which people are likely to know. Few, other than paleontologists and stratigraphers, are informed about other mass-extinction boundaries to the extent that they invoke them during interview responses.

McLaren: Yes, that's quite true, but it's exactly like the story of continental drift in North America when people asked questions of each other about continental drift without realizing that there was a large body of empirical evidence elsewhere in the world that was being ignored. How much of that is happening now? I feel strongly that we've got to look at other impacts and try to model, if you like, the question of biomass killing. I would furthermore state that when you said modeling for impactors, you left out the most important aspect. You modeled for shocked quartz, for iridium, for stishovite, and other immediate things, but you left out biomass killing completely, although this is the way you find impact horizons. You can't find all these other things unless you find the biomass killing. This is how the iridium anomaly was found at the K/T boundary at Gubbio. It had been defined long before by Italian geologists at the obvious biomass-killing horizon. If you want to find an iridium horizon in the prairies or in the Raton basin [U.S.], you have to find the pollen kick before you can find the iridium. So don't let's pretend that modeling evidence for extinctions is only geochemistry—it's not.

Glen: I think your announcement this morning that you have a siderophile anomaly and new platinum-group evidence at the Frasnian-Famennian boundary is going to go a long way toward directing people's attention to that. As soon as that paper comes out, research teams will be directed to the Devonian.

McLaren: Well, not just the Devonian, which is already defined at a global biomass-killing horizon. There is the Ordovician/Silurian, the basal Cambrian, as well as the Devonian/Carboniferous, which is a beautiful example in our opinion. I'm being brash, but we expect the Ordovician/Silurian to come in any time now—my colleagues are collecting in great detail in the Arctic. And of course Dave Raup in his modeling shows clearly that one would expect to find impactors down to stage level, and he now tells me down to zone level. I think that's very important. We have to use every conceivable piece of evidence we've got, and that includes biological evidence—fossil evidence.

Raup: Digby, you raise an interesting point in terms of the development of the field. You're quite correct that for some reason everybody has avoided the non-K/T events, even though the Devonian (Frasnian-Famennian) anomaly was published five years ago.

McLaren: Yes, but we were wrong.

Raup: That doesn't make any difference; there is a good iridium anomaly right where you said it ought to be. The same is true for the Eocene-Oligocene boundary, the same is true for the Cenomanian, [and] also for the Callovian. There are about seven good iridium anomalies, some with tektites, some with shocked quartz—but mostly not—associated with known extinctions. For some reason we've all avoided them. That might be quite an accident; it may be a fixation on the K/T, or it may be something in the gestalt problem keeping us away. I've no idea what's doing it, but it's a real phenomenon. Perhaps we can escape from the nineteenth century, but I have to submit that I think that Lyell is alive and well. Maybe we can ask for one [impact horizon], but nobody thinks we can ask for twenty.

McLaren: I think that this is an American phenomenon.

Unidentified: If you have an immense amount of lava associated with the K/T, and you're assuming that a 13 on the Richter scale loosens up the crust, do you get the same thing at the end of the Devonian?

Glen: Mike Rampino [1987] and also the Alt group [1988] in Montana have written papers in which they say that there are enormous basalt flows that seem to coincide with a number of mass extinctions. They believe that impacting caused the volcanism that matches the mass-extinction boundaries.

McLaren: I believe very strongly that the immediate causes identified at an impact event in the world—and it's not just looking for iridium—will be legion. The ultimate cause must be sought in a coeval horizon that is detectable on all continents, and I'm talking about the continents because for the Paleozoic, at least, there is no remaining seafloor. But the derivative from an impact will give you many local causes which are detectable. The trouble has been that individual workers at individual sites in the world are getting different answers all over the place, and therefore they doubt the reality of a sudden event. Whereas in fact, the only way you can demonstrate a sudden event is by applying Occam's razor to a large number of independent sections where the events are at least approximately synchronous, and it's a reasonable postulate that they are indeed synchronous and that the hypothesis demands a common ultimate cause.

Glen: For the benefit of the nongeologists in the audience, the oldest seafloor crust we have on Earth is around 180 million years. That is, on cycles of about 60 to 100 million years, typically, an ocean floor is recycled as on a conveyor belt and goes back down into the mantle—the oceanic lithosphere dives into deep-sea trenches and takes its craters with it. So the only old rocks we have on the planet are those of the continents. Only the continents have old impact craters. All the ocean floor was virtually created yesterday in geologic terms.

Unidentified: What about the Deccan Traps? A number of years ago they were explaining the Deccan Traps by the impact of India into Asia or the separation of Madagascar from Africa—the separation of the Seychelles Platform.

Glen: It's now the passage of the Indian plate over the Réunion hotspot in the Indian Ocean that's invoked to make the Deccan Traps.

Unidentified: What I'm saying now, is there any way of relating the Deccan Traps back to impacts?

McLaren: If it was related back to an impact initiating the Deccan Traps, it could be caused by a megashock anywhere on Earth, and not necessarily an impact at that spot. Today, two known magma domes rising in the western U.S. would blow as a consequence of a relatively small impact anywhere on Earth.

Glen: The bulk of the people whom I've interviewed who do mathematical mantle modeling and try to bring up superplumes from the D double-prime layer—where they postulate bumps of heated material on the surface between the lower mantle and the outer core—are dead set against the possibility of initiating a mantle plume by impact. They mean that it's not possible to initiate a long-lived, stationary plume that will be a source over which you can pass lithospheric plates and pump out nemataths or thread ridges, which are age-graded volcano chains. But if you go to the impact/flood basalt people instead, there is a considerable number of them who think that the energy of massive impacts can excavate the lithosphere and locally disturb the asthenosphere just below. Some hold that deep lithospheric excavation by impacting will produce a chaos effect, and a plume can thus be built downward rather than upward.

Unidentified: I'd like some clarification on stepped extinctions vs. mass extinctions. The mass extinction at the K/T boundary is global—and then some people talk about stepped extinctions that are not global, that are more local and not synchronous. So how do you reconcile those two kinds of extinctions?

Sepkoski: The local observation of stepped extinctions, in many cases, at least in my opinion, is due to overinterpreting data. It's a problem that arises when one has some tens of species and only some hundreds of specimens in any collection. Not all of the species will be represented in every collection, and if there is some razor-sharp extinction event, the last appearance of many of the species will be well below that. And if there is variation in the fossiliferousness of the beds, as there is in virtually all sections, it's going to give the appearance of steps in any local area. Some of the cases of stepped extinctions have been demonstrated on modeling to be artifacts of data, some have been demonstrated by further collection not to exist, and some have been put in question where people have different local areas and similar steps with different species going out on different steps. The real question about stepped extinctions is: do they really exist or are they an effect of sampling?

Van Valen: There's also a problem with some of these because there was a lowering of sea level going on at the same time, so species which have different degrees of approach to the shore would go out in a single vertical section at different times.

Clemens: I'd like to follow that with a question. Given these observations on stepped extinctions, I suggest that the really telling fossil data should be patterns of survivorship across the boundary, because that's the information that would tell us most about the biological mechanisms. What is the state of research on survival following the major K/T extinction horizon?

Sepkoski: It's really been ignored, for the most part. There is some work slowly coming out on the Danian marine sections, but for the most part people have just been looking at what comes up to the event, and not what comes after the event. There are some detailed studies, centimeter by centimeter, of marine foraminifera, and those data often look confused because there has been bioturbation. Worms and other burrowing organisms in the sediment behave like a conveyor belt; they move older stuff upward in any section. Thus some of the data used to argue for the survival of certain species can be attributed to that effect.

Van Valen: Patterns of survival have been a major focus of work on terrestrial biotas around the Cretaceous-Paleogene [K/T] extinction.

McLaren: Well, here again there is good evidence from other horizons, and the best I know of is Pete Palmer's work on biomeres in the Cambrian [Palmer 1982, 1984]. Many years ago he published a meticulous description—literally inch by inch over certain stratigraphic thicknesses—and found extinction horizons that corresponded with major zonal boundaries. These horizons had no physical manifestation; some of them were in the middle of beds. He would slice the beds vertically and you could see the break. He disintegrated these beds and counted the number of species of trilobites. (The fauna was essentially trilobites. The rocks themselves were made up of trilobite shells and echinoderm remains that were always fragmented.) He got a very interesting story from all of this. Most forms come up to the biomere boundary and disappear; four kinds go through: individual species known in the bed below go through the boundary a meter or two, and die out; others will go through the boundary, live longer, but not proliferate or radiate; a third will go through and begin radiation, which initiates a new fauna; and finally a morphologically more primitive form will appear that resembles a form that began the previous zone. The last are presumably from the deep sea, and have come up slope when they had the chance. Above the boundary there are also forms that have no obvious ancestors [Palmer 1982]. This happens several times. I don't know what the causes are. This is on a considerably larger scale. As far as I know, Palmer's story is the most meticulous that's yet been done on what happens after an event.

Briggs: The difficulty I have with that biomere business is that, first, a number of people have tried to identify biomere boundaries as mass-extinction periods.

McLaren: Well, they *are* mass extinctions.

Briggs: Well, I don't think so, because all it's based on is trilobites, nothing else.

McLaren: Well, then, a mass extinction of trilobites.

Briggs: All right, if you call it a mass extinction of trilobites, that's fine, but it's not a global mass extinction.

Sepkoski: Trilobites constitute 80 to 90 percent of the local preserved species diversity in the fossil assemblages of the Cambrian. In the post-event pattern there seem to be some generalizations coming out. There appears to be an interval of time in which there is a mixture of some survivors and also some new species that are becoming very abundant. Usually this seems to be within a small amount of time. And then this is followed by intervals of very rapid evolutionary turnover which no one understands, during which a few species are booming and crashing, booming and crashing. It takes anywhere from a fraction of a million years to maybe five million years after the Permo-Triassic event for the whole system to settle down again.

McLaren: The Permo-Triassic event is particularly interesting because, in fact, in spite of all you read about it, when you look at it there are forms that go through. Some brachiopods go through and last for a while; some are derivative, and some become extinct immediately. The Permian-Triassic event is unfortunately not very commonly recorded in the stratigraphic column because it happened at a time of minimum sea-level and maximum continentality. The Lower Triassic is highly depauperate, a time with a small amount of life remaining.

Van Valen: The conodonts go through the Permian-Triassic extinction with virtually no extinctions. It would be interesting for someone to compare their absolute abundance in rock before and after the extinction.

McLaren: The conodonts are not affected by the Frasnian-Famennian event either, fortunately, because they allow us to find out where we are.

Unidentified: In the work that Dave [Hull 1988] has done, he studied the systematists, and during that whole history there was what non-systematists perceived as vitriolic nastiness among all the opponents involved. I'm interested in knowing to what extent, throughout this whole history, that's the norm for people trying to get their views across.

Glen: I think I can talk about only a little of what I know in many areas. Much of it borders on the unspeakable. Most of the sociology and social psychology of science is not different from that of the butcher's or baker's shop, except for one exception. Scientists have a much greater personal investment. Much more of their dignity and esteem are tied up in what is at stake here, and their reactivities are so much stronger than those of the less deeply invested.

Unidentified: How does that extend in print? Is that reflected in their argumentation there?

Glen: Of course the whole thing is mainly civil in print, almost gracious at points. When you go around talking to people involved, you can see how deeply concerned they are. It's not so much a matter of attacking people; what you see is almost the phenomenon of a person having an idea who ceases to be a person—he becomes that idea. Very often the adversaries are not badly disposed to the person as such; it's the *idea* that's so terribly threatening.

Clemens: Let me contradict that.

Glen: I said what I did as a general truth.

Clemens: It seems to me that the ideal for the gracious response in academic debate is: Yes, you have a good point there. But! As you read articles and letters in the published literature of the impact debate, much less attention is paid to the first component, "You have a good point there," than to an elaboration of the "But!" I really was struck by the reluctance to give an inch. The opening rhetorical move, the acknowledgment of the opponent's contribution, was strikingly absent.

Glen: I would think that the latter mode that you just cited was prevalent half or more of the time. I think that less than fully civil responses are alarming, and come to occupy much more of one's memory than the expected.

Clemens: I guess what I want to say is that it is not necessarily the attack ad hominem. It's that the civil move wasn't made in advance, so that there wasn't give and take, at least in a form that I'm more used to in my own field.

Hull: People have been becoming less polite. You might also look at various stages in the process. The first draft that is submitted to a journal is likely to be less civil than the one an editor will let you publish.

Glen: There was a surprising find that I was unable to discuss in the last talk I gave for lack of time [the power of gestalt, standards of appraisal, and the polemic mode]. It was that Thomas Kuhn and others hold that people in different disciplines don't speak each other's language and thus can't communicate. They don't have commensurable views for a portion of reality. What I found was, that when such people—from widely separated disciplines, having nothing to do with each other in practice, with different vocabularies and much else—embraced the same hypothesis and started a research problem toward a common end, it was suddenly as if they had been raised in the same household by the same parents. That included their eating together, hanging out together, and learning each other's language with a passion, a fervor. It was as if they had been imbued with the same subdisciplinary gestalt. It was an amazing thing that resulted from the embrace of the same hypothesis.

Unidentified: How would you teach uniformitarianism to a geology class? I always thought that the present was the key to the past. How would you present the whole hypothesis?

Glen: The present can instruct us about the past, and I think the past can instruct us about the present, but we should be alert to what may be exclusive to one or the other.

Unidentified: Is a gradualistic approach locking us into something?

Glen: I think that, in the last several decades especially, we've been accommodating a new view of uniformitarianism that is broader and more accepting of things that are uncommon. Increasingly, rare and large-scale events and their driving forces are being added to the repertoire acceptable to uniformitarianists.

[With that, the panel discussion was adjourned.]

Endnotes

Endnotes

The notes that follow are all signaled in text by number; for easy reference, the text page is given in brackets following each note number below.

2.1 [p. 71]. The evolutionary history of the stepped-extinction anomaly, as documented here—which sees the perceived significance of the anomaly eventually driving the formulation of a new hypothesis—is in large part the reverse of what Alan Lightman and Owen Gingerich (1991) described for a different class of anomalies. They identified a group of anomalies that had been ignored, and were then "recognized as anomalies only after they were given compelling explanations within a new conceptual framework." Among the five examples with which they illustrated their class of anomalies was Wegener's fit of the continents, which Wegener held as anomalous in the absence of continental drift. Lightman and Gingerich believe that the fit of the continental edges was ignored as an anomaly until the advent of plate theory. Their point is mostly well taken, but the term "ignored" ought be qualified, since a small, vocal minority of highly qualified mooters did not relent in their efforts to argue Wegener's case prior to the confirmation of seafloor spreading. I would also point out that Wegener's drift hypothesis was essentially proved by the confirmation of seafloor spreading in 1966 (Hallam 1973; Glen 1975, 1982, 1987; Frankel 1982; Le Grand 1988), prior to the development of the plate-tectonics model. Lightman and Gingerich might also have used J. Tuzo Wilson's epochal transform-fault paper of 1965 to better illustrate their point: the transform article contained the germ seed of plate theory *but was ignored* until a year later, when the Eltanin 19 magnetic-anomaly profile was deciphered to prove seafloor spreading; and only then did that proof trigger the development of the plate-tectonics model.

3.1 [p. 96]. The trope linking childhood experience to scientific orientation is widely used in science pieces written for general audiences. In an essay written for the *New York Review of Books*, the biologist John Maynard Smith invoked biographical events to account for his own scientific development:

> When, as often happens, I find myself dissenting from something written by Stephen Jay Gould, I remind myself that we share a common childhood experience. We were both dinosaur nuts, at a time when to be interested in dinosaurs was to be an oddball. For both of us, that early passion has led to a life thinking about evolution, although in my case a world war, and a false start as an aircraft engineer, meant that I did not become a professional biologist until I was thirty. However, our paths diverged earlier. When I was eight, my father died, and we moved away from London to the country: for me, visiting the Natural History Museum was replaced by watching birds. I suspect that this switch may help to explain the disagreements between us, to which I shall return below. It also explains why I can review two books about dinosaurs only as an interested outsider, and not as an expert. (1991: 5)

Of course, the fact that scientists frequently *attribute* their intellectual development to the presence of the artifacts and institutions of popular science cannot be taken as proof of a causal connection. Nevertheless, the frequency of such accounts suggests that the relationships between the systems of popular and professional science merit serious attention.

3.2 [p. 100]. These different rates of coverage reflect, in part, the different institutional settings of the disciplines. Space studies are framed by events that have to be news today—launches, missions, approaching comets, etc. Paleontology, by comparison, is located in museums where the only "events" worthy of note appear to be open houses and identification days. Research "findings," as opposed to "events," can always wait until the next day's papers and, consequently, run a greater risk of never appearing at all.

3.3 [p. 103]. One important exception is the essentially metaphorical parallel that Stephen Jay Gould has drawn between catastrophic models of extinction and punctuated-equilibrium models of evolution. This connection rests at the level of fundamental assumptions about the shape and pace of change in the history of life, rather than as a claim for the efficacy of a common set of causal mechanisms. The lack of articulation between debates in evolutionary biology and the impact debates may reflect, in large part, the narrow focus on genetics that has increasingly characterized the field of evolutionary biology over the course of this century. This is only one example of the "segmentation processes" that define "lines of work" in science (Gerson 1983).

3.4 [p. 105]. In a citation analysis I began in the mid-1980's, this article proved to be the only citation shared by many of the recognized contributions to the impact debate. As an exercise in structural-equivalence analysis, my project was, therefore, an abject failure. As an indication that structures other than cohesive research programs and theoretical literatures may organize scientific activity, however, this failure underscored an important lesson.

3.5 [p. 105]. This by no means represents the entire population of publications addressed to the impact debate. A recent paper estimated that over 2,000 books and papers have addressed the topic (Glen 1990: 354 and this volume; see also Miller 1990: 382).

3.6 [p. 106]. In his discussion of thought communities or collectives, Fleck addressed the question of multiple membership, arguing that links among closely related fields were likely to be problematic: "If thought styles are very different, their isolation can be preserved even in one and the same person. But if they are related, such isolation is difficult. The conflict between closely allied thought styles makes their coexistence within the individual impossible and sentences the person involved either to lack of productivity or to the creation of a special style on the borderline of the field." (1979: 110)

If this mode of psychological explanation were to be extended, it would suggest that the rancor and name-calling of the impact debate is due, in part, to the repeated encounter of fundamental differences among disciplines that are initially masked by the shared label of "science." The reaction in a number of cases has been to attempt to expel those with different assumptions by labeling them "not scientists" or, according to some allegations, to sabotage their careers within the academy (Browne 1988).

3.7 [p. 109]. These figures need to be qualified in two ways. First, the levels of participation by different disciplines inevitably reflect size effects; there are fewer paleontologists than geologists or physicists. Consequently, an explanation in terms of the motivations of members of different disciplines may well not be warranted, but to the extent that levels of participation translate into levels of coverage in the mass media, the different levels of participation still have important effects in terms of shaping popular knowledge about the impact debate. Second, much of the publication has taken place in conference and symposia volumes. Nevertheless, consider that a scientific debate that warranted 32 mentions in *The New York Times* abstracts was explicitly addressed only six times in the *Journal of Paleontology*, eleven times in *Palaeogeography, Palaeoclimatology, Palaeoecology*, and twelve times in *Paleobiology* between 1980 and 1990.

3.8 [p. 109]. The lack of attention to debates within paleontology has characterized the impact debate from the start. In the original 1980 article, Alvarez et al. cited only one paleontologist in support of their characterization of biological changes at the K/T as sudden and catastrophic. Regardless of the current level of agreement with this characterization (which remains low, according to a number of informal polls and formal surveys), as of 1980 there were a number of positions on this point within paleontology, yet this debate was not acknowledged in the question as framed by geologists and physicists. This attitude toward evidence of the complexity of biological change clearly contrasts with the catastrophic hypothesis framed by Digby McLaren (1970) a decade earlier.

3.9 [p. 112]. The decomposition of a theory into elements that enjoy different levels of acceptance is not without precedent. In tracing the reception of Darwin's argument by Victorian England, Alvar Ellegard observed that "though it is practically certain that the evolution theory would not have been established at all if Darwin had not been able to support it by means of the naturalistic theory of Natural Selection, yet the majority of the general public, and a good many scientists, refused to accept the Natural Selection theory, while allowing themselves to be converted to evolutionism" (1990 [1958]: 17).

3.10 [p. 118]. This neighborhood was Hyde Park, home of the University of Chicago, and, consequently, home to a number of prominent impactors, including David Raup, Jack Sepkoski, and a research group that has published findings of soot at the K/T, suggesting worldwide wildfires following the alleged impact (Wolbach 1985). To the extent that there are any adjacency effects in the selection of this bookstore, therefore, they should favor impact accounts.

8.1 [p. 184]. The preceding "case histories" of Kelvin and Jeffreys are conspicuous, but mild, examples of how the creative potential at the high school, college, and graduate-university levels gets steamrollered by the righteous application of Newton's mechanics (as an arbitrary, but obvious, example of the canonical view of scientific principles)—an effect felt in other ways by Einstein and by other well-known "thinkers" who survived the indoctrination process with their cross-disciplinary enthusiasms more or less intact.

8.2 [p. 193]. As a contrived example, two parties connected by a transmission line might agree in advance that any sample of English text with a relative redundancy of X percent, with Y precision, will activate an electronic key that gives them access to a shared electronic data bank. The text itself, whatever it might appear to say, is meaningless in this context, which is known only to the transmitter and receiver. Unfortunately, this particular example would be a poorly kept secret, because the first thing a cryptographer might do in attempting to break such a code would be to search for patterns of Shannon's relative redundancy, given an alphabet and a dictionary of words—or an ability to generate one (cf. Searls 1992).

8.3 [p. 195]. A mechanical clock is made up, in essence, of a pendulum, a spring, and an escapement (gear and cog) mechanism, which, taken together, form a complex feedback system that regulates the meting out of equal increments of "time" (cf. Morrison 1991, p. 95). In this context, therefore, the ordinary clock mechanism, rather than being the stereotype of linear periodic motion, actually is an example of *controlled chaos* (cf. Moon 1992).

8.4 [p. 196]. The ongoing debate on the statistical validity of the initial conclusion by Raup and Sepkoski (1984), that the intermittent extinctions of the latest 250 million years of the fossil record have constituted a linear (harmonic) periodic pattern in biostratigraphy, is irrelevant to the eventual recognition of a familiar nonlinear-dynamical style of geometronumeric evolutionary progression recorded by that record. Or, it might be said that one person's dream is another's nightmare—or, for better or worse, pursuit of the scientific method in the absence of an open-system context and the universality essential to language is tantamount to Frankensteinian madness.

8.5 [p. 199]. The question mark closing the concluding sentence expresses the *uncertainty-of-precedence principle* in science, a principle that I have rediscovered repeatedly since my first scientific report—viz. that every time I have attributed priority for an idea to a particular person and/or date in history, I have later found an earlier example of it. The complexity of nonlinear-dynamical evolution, wherein feedback and recursion are prevalent (meaning repeated folding and stretching of the phase-space trajectories of the motion, hence kneading and fractal implication of information)—thereby implicating evolution in general—is such that evidence of a unique initial state in the fabric of the evolved states does not exist (or, alternatively, there is evidence of infinitely many initial states). Like the Gordian knot, there is no trace of a beginning or of an ending, an observation that, in geology, is attributed—together with the *principle of uniformitarianism*—to James Hutton, working in the late eighteenth century. Another way of saying the same thing is that the initial state of a nonlinear-dynamical evolutionary process is not knowable by any future complex state, or set of complex states, that evolves from it (cf. Spencer-Brown 1972, p. 105), an adage that applies to the notion of the "origin" of the universe (e.g. the "Big Bang"; cf. Alfvén 1980, 1981; Lerner 1991) as well as to the "origin" of a paradigm.

These observations pose questions for Darwinian and neo-Darwinian concepts of evolution—if those concepts are predicated on the idea that "family trees" can be traced back to unique sets of common ancestors (cf. Eldredge 1989). According to the uncertainty-of-precedence principle, the idea of a unique ancestral individual (asexual reproduction) or pair of individuals (sexual reproduction) is, at best, ambiguous, and, at worst, fallacious. However, a qualified view of organic evolution—constructed in parallel with the above remarks on recursive complexity (nonlinear-periodic/chaotic complexity)—would see the dendritic structure of macroscopic biologic diversity as "surficial" in character, analogous to dendritically branching streams traced upon an infinitely complex landscape. In this sense, macroscopic organic evolution is analogous to the dendritic bifurcations of *systems* of chaotic attractor fields, wherein any particular "dendrite," if geometrically abstracted from its chaotic microstructure (but note the dynamical absurdity of such an abstraction), can be exceedingly simple (algorithmically reduced, or compressed) relative to the fractal geometry of the particle trajectories of that same form within the context (or "dynamical matrix") of the interdependent system as a whole.

Striking nonbiological examples of this figurative double standard are given by *bifurcations of ring-vortex fields*, wherein an initial vortex ring bifurcates, or "n-furcates" (*appears* to subdivide into two or more unique entities that *appear* to be macroscopic replicas of the original structure), into some integer multiple of rings—passing through stages analogous to gastrulation and fission—following which each new ring does the same, and so on (abstractly) to infinity (e.g. Bradley 1965; Shaw 1986). The integer number, n, of rings per new generation may be even or odd and tends to be constant for a relatively invariant set of flow parameters, but it is highly sensitive to variations (*variable tuning*) of these parameters and would appear to have no upper numerical limit (i.e., the numerical base of the possible geometric progressions appears to be unconstrained except by factors that are extremely sensitive to the diversity of local accelerations in the matrix fluid relative to the trajectories of marker particles that trace out the forms of the vortices; cf. Shaw 1986). In the terminology of nonlinear dynamics, the fundamental bifurcation sequence is called a *period-doubling cascade*, or *period-doubling route to chaos* (i.e. where chaos, in this case, is a form of turbulence that is characteristic of doubling progressions; cf. Feigenbaum 1980), and therefore the ring-vortex cascades just described would have to be called *period-tripling, -quadrupling, -quintupling*, etc., *routes* to forms of turbulence that differ subtly from each other in the vicinity of the periodic-chaotic transition (thus testifying to the exquisite intricacy of nonlinear-dynamical regimes that tend to be lumped under the unqualified rubric "chaos"; cf. examples discussed by Shaw 1987a). The *microscopic* flow is both locally and globally chaotic (e.g., each ring resembles a strange attractor of the type illustrated by Ruelle 1980), while the *macroscopic* structure consists of a system of fixed-point attractors (the sets of rings). Such macroscopic attractors are subject to annihilations (global catastrophes), particularly given a punctuational change in tuning parameters, whereas the many-dimensional chaotic-attractor field—being a *pumped dissipative system operating far from equilibrium*—consists of globally persistent, though intermittent, regions of *transient chaos* (cf. Shaw 1994, Chap. 5).

The nonlinear-dynamical paradigm just described immediately reconciles the age-old controversy in geology, and now in the impact-extinction debates, concerning *uniformitarianism* vs. *catastrophism* as alternative dynamical regimes in nature—as well as controversies concerning the independence of the micro-states and macro-states involved in biological change. It is obvious, in the general context of natural processes, that these are not dynamically separable concepts (the "destruction" of a ring vortex is catastrophic only to a particular geometric form [ring vortex], not to the system of dynamical states that is essential to the existence of the form). Of greater relevance than the destruction of a particular macroscopic form, in the above dynamics, is the event, process, or processes that can change the *tuning* of the flow regime. Episodic or punctuational changes in tuning parameters are likely to correlate with major changes in macroscopic forms of fixed-point systems of vortex trees (attractors), but abrupt changes and/or "destructions" of fixed-point attractors *can* occur with arbitrarily gradual changes in tuning parameters—hence the nonlinear-dynamical paradigm offers no definite resolution to the chronic debate of gradual vs. punctuational change in macroscopic Darwinian evolution, nor does it support the common belief that wholesale macroscopic extinctions are necessarily correlated with wholesale physical cataclysms. What it *does* support is the view that *patterns* of punctuational changes in the *environment* of macroscopic forms (as expressed in the underlying dynamics at all scales from the microscopic to the global) accompany *patterns* of punctuational changes in the *diversity* of those macroscopic forms.

This is not good news to the advocates of the impact-extinction hypothesis (IEH) who would argue their case *only* on the basis of cause-effect correlations, because, in that case, the before/after, true/false nature of the evidence on the sequential timing and precision of events—which has been of such paramount importance to the gaining of impetus by the IEH—loses its power of absolute falsification, hence its power of *implicit* circumstantial verification (as in scientific hearsay with legalistic overtones). The importance attached to such evidence can be seen in the volume of the literature that argues endlessly about the before-or-after issue in boundary-event correlations; cf. papers in Sharpton & Ward (1990). But, by the argument urged above, whereby the nonlinear-dynamical paradigm ironically resolves the paradox of the dichotomous view of uniformity vs. catastrophe in nature—i.e., by replacing a *logic* paradigm of mutual exclusion with a *coupled-dynamical-action* paradigm of holistic inclusion—the abandonment of absolutism may turn out to be the salvation of the IEH in a future where the sharpness of the tools of stratigraphic correlation (and the instances of false correlations) is likely to increase dramatically. For example, a report I came upon recently demonstrated convincingly that a microtektite horizon in a sedimentary sequence definitely occurred within, rather than at the base of, a geomagnetic polarity chron, thereby "proving" that the related impact event or events could not be causally related to the dynamical mechanism or mechanisms of polarity reversals. According to the revised paradigm just described, however, evidence this self-constrained loses its power of absolute falsification—yet it *retains* the stratigraphic property of relative-pattern correlation (self-similarity and/or universality) between comparative sequences.

This discussion implies a *universal principle of evolutionary change,* the universality of which lies in the existence of invariances in numerical scaling that are independent of material states, size, and time (cf. Shaw 1994, Preface and Chap. 5). The role of universality parameters in global (holistic) systems helps to explain some of the seemingly dichotomous aspects of evolutionary change as evidenced by the fossil record, such as (1) the hierarchical classification of evolutionary principles (e.g. microevolution and macroevolution; cf. Burnet 1976; Gould 1982; Edelman 1985, 1987; Shaw 1987a; Eldredge 1989), (2) distinctions between mass extinctions and background extinctions (cf. Jablonski 1986a, b; Raup 1986, 1990), (3) the relativity of multiple—and seemingly independent or even antithetical—contemporaneous outcomes ("parallel universes"), for instance as they might be compared in terms of Sewell Wright's concepts of evolutionary landscapes (cf. Provine 1986), and (4) the reconciliation of the "tree of life" (taxonomically as well as dynamically) with the emerging biologic syntheses that deal with the implications of *symbiotic cohabitations of microbial fields* (cf. Sonea & Panisset 1983; Margulis & Sagan 1986; Margulis 1992). The type of systematics that would seem to be implied by aspect 4 is remarkably analogous to the hierarchical vortex-tree model by which I have attempted to illustrate the nature of macroscopic heterogeneity in terms of structural dynamics and the emerging diversity of coexisting forms—forms that are not just geometrically, or taxonomically, *possible* but are dynamically *compatible*—with differing relative simplicities within the global complexity.

The statements made in the article by Herbert R. Shaw are not to be construed as representing any viewpoint sanctioned by the U.S. Geological Survey, neither are they to be construed as being officially approved for publication by that agency.

10.1 [p. 220]. The three names introduced by Lyell are still used, but the time spans they represent have changed, because further studies of fossil assemblages have come up with finer distinctions. The Eocene is now preceded by the Paleocene [most remotely recent], and followed by the Oligocene [slightly recent], and the Miocene and Pliocene are followed by two more subdivisions, the Pleistocene [most recent] and the Holocene [wholly recent]. Lyell's was, by the way, the first subdivision of a geologic Period, and the first use of statistics in biostratigraphy.

References Cited

References Cited

Full citations of all sources cited in the text are supplied here, chapter by chapter. Information on the month and even the day of publication (or submission or acceptance), where known, is given for many key works, particularly in the references for the first two chapters. (No sources are cited in the Introduction.)

1. Glen, What the Debates Are About

Ahrens, T. J., & J. D. O'Keefe. 15 February 1983. Impact of an asteroid or comet in the ocean and extinction of terrestrial life. *Proc. Thirteenth Lunar & Planetary Sci. Conf.*, Part 2, *J. Geophys. Res.* 88, Supplement pp. A799–A806.

Alt, D., J. M. Sears & D. W. Hyndman. November 1988. Terrestrial maria: The origins of large basalt plateaus, hotspot tracks, and spreading ridges. *J. Geol.* 96(6): 647–62.

Alvarez, L. W., W. Alvarez, F. Asaro & H. V. Michel. June 1980. Extraterrestrial cause for the Cretaceous-Tertiary extinction. *Science* 208: 1095–1108.

Alvarez, W. 1986. Toward a theory of impact crises. *Eos* 67: 649–58.

Alvarez, W., & F. Asaro. October 1990. An extraterrestrial impact. *Scientific American* 263: 78–84.

Alvarez, W., & R. A. Muller. 1984. Evidence from crater ages for periodic impacts on the Earth. *Nature* 308: 718–20.

Alvarez, W., J. Smit, W. Lowrie, F. Asaro, S. V. Margolis, P. Claeys, M. Kastner & A. R. Hildebrand. August 1992. Proximal impact deposits at the Cretaceous-Tertiary boundary in the Gulf of Mexico: A restudy of DSDP Leg 77 Sites 536 and 540. *Geology* 20: 697–700.

Archibald, J. D., & L. J. Bryant. 1990. Differential Cretaceous/Tertiary extinctions of nonmarine vertebrates: Evidence from northeastern Montana. *In*

V. L. Sharpton & P. D. Ward, eds., *Global Catastrophes in Earth History, Geol. Soc. Amer. Special Paper* 247: 549–62.
Archibald, J. D., & W. A. Clemens. 1982. Late Cretaceous extinctions. *American Scientist* 70: 377–85.
Archibald, J. D., & D. L. Lofgren. 1990. Mammalian zonation near the Cretaceous-Tertiary boundary. *Geol. Soc. Amer. Special Paper* 243: 31–50.
Asaro, F., H. V. Michel, L. W. Alvarez, W. Alvarez & A. Montanari. 1988. Impacts and multiple iridium anomalies. *Eos* 69(16): 301–2.
Baksi, A. K. 4 December 1990. Timing and duration of Mesozoic-Tertiary flood-basalt volcanism. *Eos* 71(49): 1835–36.
Baksi, A. K., & E. Farrar. May 1991. 40 Argon/39 Argon dating of the Siberian Traps, USSR: Evaluation of the ages of the two major extinction events relative to episodes of flood-basalt volcanism in the USSR and the Deccan Traps, India. *Geology* 19: 461–64.
Bercovici, D., G. A. Glatzmaier & G. Schubert. 26 May 1989. Three-dimensional spherical models of convection in the Earth's mantle. Abstract: *Eos* 70(43): 1358.
Blum, J. D., & C. P. Chamberlain. 21 August 1992. Oxygen isotope constraints on the origin of impact glasses from the Cretaceous-Tertiary boundary. *Science* 257: 1104–7.
Bohor, B. F., E. E. Foord, P. J. Modreski & D. M. Triplehorn. 1984. Mineralogic evidence for an impact event at the Cretaceous-Tertiary boundary. *Science* 224: 867–69.
Bohor, B. F., P. J. Modreski & E. E. Foord. 8 May 1987. Shocked quartz in the Cretaceous-Tertiary boundary clays: Evidence for a global distribution. *Science* 236: 705–8.
Bohor, B. F., & R. Seitz. 1990. Cuban K/T catastrophe. *Nature* 344: 593.
Bourgeois, J., T. A. Hansen, P. L. Wiberg & E. G. Kauffman. 29 July 1988. A tsunami deposit at the Cretaceous-Tertiary boundary in Texas. *Science* 24: 567–70.
Briggs, J. C. 1990. Global extinctions, recoveries and evolutionary consequences. *Evolutionary Monographs* 13: 1–47. Chicago: Univ. of Chicago.
Carlisle, D. B. 14 May 1992. Diamonds at the K/T boundary. *Nature* 357: 119–20.
Carlisle, D. B., & D. R. Braman. 22 August 1991. Nanometre-size diamonds in the Cretaceous/Tertiary boundary clay of Alberta. *Nature* 352: 708–9.
———. 1993a. Extra-terrestrial amino acids at the Cretaceous-Tertiary boundary of Alberta. [In press, *Nature*, vol. 361.[
———. 1993b. Amino acids at the Cretaceous-Tertiary boundary, Red Deer Valley, Alberta, Canada. [In press, *Canadian J. Earth Sciences*.]
Carter, N. L., C. B. Officer & C. L. Drake. January 1990. Dynamic deformation of quartz and feldspar: Clues to causes of some natural crises. *Tectonophysics* 171: 373–91.
Claeys, P., S. V. Margolis & J.-G. Casier. 1992. Microtektites and mass extinctions from the Late Devonian of Belgium: Evidence for a +367 Ma aster-

oid impact. *Amer. & Canadian Geophys. Union Spring Meeting Abstracts Volume,* supplement to *Eos,* p. 328.

Clemens, W. A. 1982. Patterns of survival of the terrestrial biota during the Cretaceous/Tertiary transition. *Geol. Soc. Amer. Special Paper* 190: 407–13.

Clube, S. V. M. 1978. Does our Galaxy have a violent history? *Vistas in Astron.* 22: 77–118.

Clube, S. V. M., & W. M. Napier. 1984. The microstructure of terrestrial catastrophism. *Monthly Notices Royal Astronomical Soc.* 211: 953–68.

Colodner, D. C., E. A. Boyle, J. M. Edmond & J. Thomson. 30 July 1992. Post-depositional mobility of platinum, iridium, and rhenium in marine sediments. *Nature* 358: 402–4.

Courtillot, V. E. October 1990. What caused the mass extinction? A volcanic eruption. *Scientific American* 263: 85–92.

Courtillot, V. E., & J. Besse. 1987. Magnetic field reversals, polar wander, and core-mantle coupling. *Science* 237: 1140–47.

Covey, C., S. J. Ghan, J. J. Walton & P. R. Weissman. 1990. Global environmental effects of impact-generated aerosols: Results from a general circulation model. *In* V. L. Sharpton & P. D. Ward, eds., *Global Catastrophes in Earth History, Geol. Soc. Amer. Special Paper* 247: 263–70.

Crockett, J. H., C. B. Officer, F. C. Wezel & G. D. Johnson. January 1988. Distribution of noble metals across the Cretaceous/Tertiary boundary at Gubbio, Italy: Iridium variations as a constraint on the duration and nature of Cretaceous/Tertiary boundary events. *Geology* 16: 77–80.

Davenport, S. A., T. J. Wdowiak, D. D. Jones, & P. Wdowiak. 1990. Chondritic metal toxicity as a seed stock kill mechanism in impact-caused mass extinctions. *In* V. L. Sharpton & P. D. Ward, eds., *Global Catastrophes in Earth History, Geol. Soc. Amer. Special Paper* 247: 71–76.

Davis, M., P. Hut & R. A. Muller. 1984. Extinction of species by periodic comet showers. *Nature* 308: 715–17.

de Silva, S. L., J. A. Wolff & V. L. Sharpton. 1990. Explosive volcanism and associated features: Implications for models of endogenically shocked quartz. *In* V. L. Sharpton & P. D. Ward, eds., *Global Catastrophes in Earth History, Geol. Soc. Amer. Special Paper* 247: 139–45.

Emiliani, C. 1982. Extinctive evolution: Extinctive and competitive evolution combine into a unified model of evolution. *J. Theor. Biol.* 97: 13–33.

Emiliani, C., E. B. Kraus & E. M. Shoemaker. 1981. Sudden death at the end of the Mesozoic. *Earth and Planetary Science Letters* 55: 317–34.

Felitsyn, S. B., & P. A. Vaganov. 1988. Iridium in the ash of Kamchatkan volcanoes. *Internat. Geology Review* 30: 1288–91.

Finnegan, D. L., T. L. Miller & W. H. Zoller. 1990. Iridium and other trace-metal enrichments from Hawaiian volcanoes. 1990. *In* V. L. Sharpton & P. D. Ward, eds., *Global Catastrophes in Earth History, Geol. Soc. Amer. Special Paper* 247: 111–16.

Fischer, A. G., & M. A. Arthur. 1977. Secular variations in the pelagic realm. *In* H. E. Cook & P. Enos, eds., *Deep-water Carbonate Environments, Soc. of Econ. Paleontol. and Mineral. Spec. Publ.* 25: 19–50.

French, B. M. 1990. Twenty-five years of the impact-volcanic controversy: Is there anything new under the sun? Or inside the Earth? *Eos* 71(17): 411–14.

Ganapathy, R. August 1980. A major meteorite impact on Earth 65 million years ago: Evidence from the Cretaceous-Tertiary boundary clay. *Science* 209: 921–23.

Glen, W. 1982. *The Road to Jaramillo: Critical Years of the Revolution in Earth Science*. Stanford, Calif.: Stanford Univ. Press.

Gostin, V. A., R. R. Keays & M. W. Wallace. 17 August 1989. Iridium anomaly from the Acraman impact ejecta horizon: Impacts can produce sedimentary iridium peaks. *Nature* 340: 542–44.

Gould, S. J. March 1965. Is uniformitarianism necessary? *Amer. J. Science* 263: 223–28.

Gratz, A. J., W. J. Nellis & N. A. Hinsey. 1992. Laboratory simulation of explosive volcanic loading and implications for the cause of the K-T boundary. *Geophys. Res. Letters* 19(13); 1391–94.

Hallam, A. 1987. End-Cretaceous mass extinction event: Argument for terrestrial causation. *Science* 238: 1237–42.

———. 1988. A compound scenario for the end-Cretaceous mass extinctions. *In* M. Lamolda, E. G. Kauffman, O. H. Walliser, H. J. Hansen, K. L. Rasmussen, R. Gwodz & H. Kunzendorf, eds., *Revista Española de Paleontología*, No. Extraordinario, *Paleontology and Evolution: Extinction Events* 7–20.

———. 1989. Catastrophism in geology. In *Catastrophes and Evolution: Astronomical Foundations*, pp. 25–55. Cambridge: Cambridge Univ. Press. (Lecture delivered at a meeting of the Royal Astronomical Society at Oxford University, 6 Sept. 1988.)

Hambrey, M. J., & W. B. Harland. 1981. *Earth's Pre-Pleistocene Glacial Record*. Cambridge, England, & New York: Cambridge Univ. Press.

Hansen, H. J., K. L. Rasmussen, R. Gwodz & H. Kunzendorf. 1987. Iridium-bearing carbon black from the Cretaceous-Tertiary boundary. *Bull. Geol. Soc. Denmark* 36 (pts. 3–4): 308–14.

Hartung, J. B., & R. R. Anderson. 1988. A compilation of information and data on the Manson impact structure. *Lunar & Planetary Institute Rep.* 88-08.

Hazel, J. E. 1988. Chronostratigraphy of Upper Eocene microspherules. *Geol. Soc. Amer. Abstr. with Programs* 20: A177–78.

———. 1989. Chronostratigraphy of Upper Eocene microspherules. *Palaios* 4: 318–29.

Herring, J. R. 1990. Charcoal fluxes into sediments of the North Pacific Ocean: The Cenozoic record of burning. *In* E. T. Sundquist & W. S. Broeker, eds., *The carbon cycle and atmospheric* CO_2*: Natural variations, Archean to present*. Amer. Geophys. Union, Geophysical Monographs, 419–42.

Hickey, L. J., & L. J. McSweeny. 26 March 1992. Plants at the K/T boundary. *Nature* 356: 295–96.

Hildebrand, A. R., & W. V. Boynton. 18 May 1990. Proximal Cretaceous-Tertiary boundary impact deposits in the Caribbean. *Science* 248: 843–47.

Hildebrand, A. R., & G. T. Penfield. 4 December 1990. A buried 180-km-diameter probable impact crater on the Yucatan Peninsula, Mexico. *Eos* 71(43): 1425.

Hoffman, A. 1985. Patterns of family extinction depend on definition and geological time scale. *Nature* 315: 659–62.

Hoffman, A. & J. Ghiold. 1985. Randomness in the pattern of "mass extinction" and "waves of origination." *Geology* 122: 1–4.

Hsü, K. J. 1986. *The Great Dying: Cosmic Catastrophe*. New York: Harcourt Brace Jovanovich.

Hsü, K. J., & J. A. McKenzie. 1990. Carbon-isotope anomalies at era boundaries: Global catastrophes and their ultimate cause. *In* V. L. Sharpton & P. D. Ward, eds., *Global Catastrophes in Earth History, Geol. Soc. Amer. Special Paper* 247: 61–70.

Hsü, K. J., et al. 16 April 1982. Mass mortality and its environmental and evolutionary causes. *Science* 216: 249–56.

Huffman, A. R., J. M. Brown & N. L. Carter. 1990. Temperature dependence of shock-induced microstructures in tectonosilicates. *In* S. C. Schmidt, J. N. Johnson & L. W. Davison, eds., *Proc. Amer. Phys. Soc. Topical Conference on Shock*.

Hut, P., W. Alvarez, W. Elder, E. G. Kauffman, T. A. Hansen, G. Keller, E. M. Shoemaker & P. Weissman. 1985. Comet showers as possible causes of stepwise mass extinctions. *Eos Abstract*. 66: 813.

Irving, E. 1988. The paleomagnetic confirmation of continental drift. *Eos* 69(44): 994–1014.

Izett, G. A. 1987. Authigenic "spherules" in K-T boundary sediments at Caravaca, Spain, and Raton Basin, Colorado, New Mexico, may not be impact derived. *Geol. Soc. Amer. Bull.* 98: 78–86.

———. 25 November 1991. Tektites in Cretaceous-Tertiary boundary rocks on Haiti and their bearing on the Alvarez impact extinction hypothesis. *J. Geophys. Res.* 96(E4): 20,879–20,905.

Izett, G. A., W. A. Cobban, J. D. Obradovich & M. J. Kunk. 1993 [in press]. The Manson impact structure: $^{40}Ar/^{39}Ar$ age and its distal impact ejecta in the Pierre Shale in southeastern South Dakota. *Science*.

Izett, G. A., G. B. Dalrymple & L. W. Snee. 1991. ^{40}Ar-^{39}Ar age of Cretaceous-Tertiary boundary tektites from Haiti. *Science* 252: 1539, 1543.

Izett, G. A., F.J.-M.R. Maurrasse, F. E. Lichte, G. P. Meeker & R. Bates. November 1990. Tektites in Cretaceous-Tertiary boundary rocks on Haiti. *U.S. Geol. Survey Open File Report* 90-635: 1–31.

Izett, G. A., & W. R. Premo. 1992. Isotopic signatures of black tektites from the K-T boundary on Haiti: Implications for the age and type of source material. *Meteoritics* 27: 413–23. [Received Jan. 20, 1992; accepted June 12, 1992.]

Jablonski, D. 1986. Evolutionary consequences of mass extinctions. *In* D. M. Raup & D. Jablonski, eds., *Patterns and Processes in the History of Life: A Report of the Dahlem Workshop*. Berlin: Springer Verlag.

Jansa, L. F., G. Pe-Piper, B. P. Roberson & O. Friedenreich. 1989. Montagnais:

A submarine impact structure on the Scotian shelf, eastern Canada: *Geol. Soc. Amer. Bull.* 101: 450–63.
Jéhanno, C., D. Boclet, L. Froget, B. Lambert, E. Robin, R. Rocchia & L. Turpin. 1992. The Cretaceous-Tertiary boundary at Beloc, Haiti: No evidence for an impact in the Caribbean Area. *Earth & Planetary Sci. Lett.* 109: 229–41.
Johnson, K. R., & L. J. Hickey. 1990. Megafloral change across the Cretaceous/Tertiary boundary in the northern Great Plains, Rocky Mountains. *In* V. L. Sharpton & P. D. Ward, eds., *Global Catastrophes in Earth History, Geol. Soc. Amer. Special Paper* 247: 433–44.
Johnson, K. R., D. J. Nichols, M. Attrep & C. J. Orth. 31 August 1989. High resolution leaf-fossil record spanning the Cretaceous/Tertiary boundary. *Nature* 340: 708–11.
Kauffman, E. G. 1988. Stepwise mass extinctions. *In* M. Lamolda, E. G. Kauffman & O. H. Walliser, eds., *Revista Española de Paleontología*. No. Extraordinario, *Paleontology and Evolution: Extinction Events* 58–71.
Keller, G. June 1989. Extended Cretaceous/Tertiary boundary extinctions and delayed population changes in planktonic foraminifera from Brazos River, Texas. *Paleoceanography* 4(3): 287–332.
Keller, G., S. L. D'Hondt, C. J. Orth, J. S. Gilmore, P. Q. Oliver, E. M. Shoemaker & E. Molina. 1987. Late Eocene impact microspherules: Stratigraphy, age, and geochemistry. *Meteoritics* 22: 25–60.
Kirschner, C. E., A. Grantz & M. W. Mullen. May 1992. Impact origin of the Avak structure, Arctic Alaska, and genesis of the Barrow Gas Fields, *Amer. Assoc. Petrol. Geologists* 76: 651–79.
Kitchell, J. A., D. L. Clark & A. M. Gombos, Jr. 1986. Biological selectivity of extinction: A link between background and mass extinction. *Palaios* 1: 501–11.
Kitchell, J. A., & D. Pena. 1984. Periodicity of extinctions in the geologic past: Deterministic versus stochastic explanations. *Science* 226: 689–92.
Koeberl, C. 1989. Iridium enrichment in volcanic dust from blue icefields, Antarctica, and possible relevance to the K/T boundary event. *Earth & Planetary Sci. Letters* 92: 317–22.
Kring, D. A., & W. V. Boynton. 9 July 1992. Petrogenesis of an augite-bearing melt rock in the Chicxulub structure and its relationship to K/T impact spherules in Haiti. *Nature* 358: 141–43.
Kyte, F. T., & J. Smit. 1986. Regional variations in spinel compositions: An important key to the Cretaceous-Tertiary event. *Geology* 14(6): 485–87.
Kyte, F. T., & J. T. Wasson. 6 June 1986. Accretion rate of extra-terrestrial matter: Iridium deposited 33 to 67 million years ago. *Science* 232: 1225–29.
Kyte, F. T., Z. Zhou & J. T. Wasson. 18/25 December 1980. Siderophile-enriched sediments from the Cretaceous-Tertiary boundary. *Nature* 288: 651–56.
Loper, D. E., & K. McCartney. December 1986. Mantle plumes and periodicity of magnetic field reversals. *Geophys. Res. Letters* 13(13): 1525–28.

———. 18 October 1988. Shocked quartz found at the K/T boundary. *Eos* 69(42): 961, 971–72.

———. 1990. On impacts as a cause of geomagnetic field reversals or flood basalts. *In* V. L. Sharpton & P. D. Ward, *Global Catastrophes in Earth History, Geol. Soc. Amer. Special Paper* 247: 19–25.

Lowe, D. R., G. R. Byerly, F. Asaro & F. J. [sic] Kyte. 1 September 1989. Geological and geochemical record of 3,400-million-year-old terrestrial meteorite impacts. *Science* 245: 959–62.

Luck, J. M., & K. K. Turekian. 11 November 1983. Osmium-187/Osmium-186 in manganese nodules and the Cretaceous-Tertiary boundary. *Science* 222: 613–15.

Lutz, T. M. 1985. The magnetic reversal record is not periodic. *Nature* 308: 404–7.

———. 10 November 1986. Evaluating periodic, episodic, and poisson models: Reversals and meteorite impacts. *Geol. Soc. Amer. Abstr. with Programs* 18(6): 677.

Lyons, J. B., & C. B. Officer. 1992. Mineralogy and petrology of the Haiti Cretaceous/Tertiary section. *Earth & Planetary Sci. Letters* 109: 205–24.

Matese J. J., & D. P. Whitmire. January 1986. Planet X and the origins of the shower and steady-state flux of short-period comets. *Icarus* 65(1): 37–50.

McHone, J. F., & R. L. Nieman. November 1989. K/T boundary stishovite: Detection by solid-state nuclear magnetic resonance and powder x-ray diffraction. *Geol. Soc. Amer. Abstr.*, A120.

McLaren, D. J. 1970. Time, life and boundaries. *J. Paleontology* 44: 801–15.

———. 1989. Detection and significance of mass killings. *In* K. J. Hsü et al., eds., *Rare Events, Mass Extinction and Evolution: Second Workshop on Rare Events in Geology*, CIGCP1997, Beijing, China, March 1987, special issue of *Historical Biology* 2(1): 5–15.

McLean, D. M. 1981. Cretaceous-Tertiary extinctions and possible terrestrial and extraterrestrial causes. Proc. of Workshop held in Ottawa, Canada, 19–20 May 1981, by K-TEC group. *Syllogeous* 39. Ottawa: National Museums of Canada.

Melosh, H. J., N. M. Schneider, K. J. Zahnle & D. Latham. 1990. Ignition of global wildfires at the Cretaceous/Tertiary boundary. *Nature* 343: 251–54.

Michel, H., F. Asaro & W. Alvarez. 1991. Geochemical study of the Cretaceous-Tertiary boundary region at Hole 752b. *Proc. Ocean Drilling Program, Scientific Results* No. 121: 415–22. Joint Oceanographic Institutions for Deep Earth Sampling.

Montanari, A. December 1986. Spherules from the Cretaceous/Tertiary boundary clay at Gubbio, Italy: The problem of outcrop contamination. *Geology* 14: 1024–26.

———. 1990. Geochronology of the terminal Eocene impacts: An update. *In* V. L. Sharpton & P. D. Ward, eds., *Global Catastrophes in Earth History, Geol. Soc. Amer. Special Paper* 247: 607–16.

Montanari, A., R. L. Hay, W. Alvarez, F. Asaro, H. V. Michel, L. W. Alvarez

& J. Smit. 1983. Spheroids at the Cretaceous-Tertiary boundary are altered impact droplets of basaltic composition. *Geology* 11: 668–71.

Mount, J. F., S. V. Margolis, W. Showers, P. Ward & E. Doehne. 1986. Carbon and oxygen isotope stratigraphy of the Upper Maastrichtian, Zumaya, Spain: A record of oceanographic and biologic changes at the end of the Cretaceous Period. *Palaios* 1: 87–92.

Muller, R. A., & D. E. Morris. 1986. Geomagnetic reversals from impacts on the Earth. *Geophys. Res. Letters* 13(11): 1177–80.

———. 1989. Geomagnetic reversals driven by sudden climate changes. *Eos* 70(15): 276.

Naslund, H. R., C. B. Officer & G. D. Johnson. 1986. Microspherules in Upper Cretaceous and Lower Tertiary clay layers at Gubbio, Italy. *Geology* 14: 923–26.

Negi, J. G., & R. K. Tiwari. August 1983. Matching long-term periodicities of geomagnetic reversals and galactic motions of the Solar System. *Geophys. Res. Letters* 10(8): 713–16.

Nichols, D. J. 26 March 1992. Plants at the K/T boundary. *Nature* 356: 295.

Oberbeck, V. R. 1975. The role of ballistic erosion and sedimentation in lunar stratigraphy. *Rev. Geophys. Space Phys.* 13: 337–62.

Oberbeck, V. R., & H. Aggarwal. March 1992. Impact crater deposit production on Earth. *Proc. Lunar & Planetary Sci., Abstr. XXIII Conference,* pp. 1011–12.

Oberbeck, V. R., & J. R. Marshall. March 1992. Impacts, flood basalts, and continental breakup. *Lunar & Planetary Sci. Abstr. XXIII Conference,* pp. 1013–14.

Oberbeck, V. R., J. R. Marshall & H. Aggarwal. October 1992. Ancient "glacial" deposits are ejecta of large impacts: The ice age paradox explained. *Eos, Abstr. Vol.,* 1992 Fall Meeting, p. 99.

———. 1993. Impacts, tillites, and the breakup of Gondwanaland. *J. Geology* 101: 1–19.

Officer, C. B., & N. L. Carter. 1991. A review of the structure, petrology, and dynamic deformation characteristics of some enigmatic terrestrial structures. *Earth-Science Reviews* 30: 1–49, Amsterdam: Elsevier Science.

Officer, C. B., & C. L. Drake. 25 March 1983. The Cretaceous-Tertiary transition. *Science* 219: 1383–90.

———. 1985. Terminal Cretaceous environmental events. *Science* 227: 1161–67.

———. 20 June 1989. Cretaceous/Tertiary extinctions: We know the answer, but what is the question? *Eos* 70(25): 659–60.

Officer, C. B., C. L. Drake, J. L. Pindell & A. A. Meyerhoff. April 1992. Cretaceous-Tertiary events and the Caribbean caper. *GSA Today* 2(4): 69–70, 73–76.

Officer, C. B., A. Hallam, C. L. Drake & J. C. Devine. 1987. Late Cretaceous and paroxysmal Cretaceous/Tertiary extinctions. *Nature* 326: 143–49.

O'Keefe, J. D., & J. T. Ahrens. 1989. Impact production of CO_2 by the Cretaceous/Tertiary extinction bolide and the resultant heating of the Earth. *Nature* 338: 247–49.

Orth, C. J., M. Attrep, Jr., X. Y. Mao, E. G. Kauffman & R. Diner. October 1987. Iridium abundance peaks at upper Cenomanian stepwise extinction boundaries. *Geol. Soc. Amer. Abstr. with Programs* 19(7): 796.

Orth, C. J., M. Attrep & L. R. Quintana. 1990. Iridium abundance patterns across bio-event horizons in the fossil record. *In* V. L. Sharpton & P. D. Ward, eds., *Global Catastrophes in Earth History, Geol. Soc. Amer. Special Paper* 247: 45–59.

Orth, C. J., M. Attrep, L. R. Quintana, W. P. Elder, E. G. Kauffman, R. Diner & T. Villamil. 1993 [in press]. Elemental abundance anomalies in the Late Cenomanian extinction interval: A search for the source(s). *Earth & Planetary Sci. Letters*.

Orth, C. J., J. S. Gilmore, J. D. Knight, C. L. Pillmore, R. H. Tschudy & J. E. Fassett. December 1981. An iridium abundance anomaly at the palynological Cretaceous-Tertiary boundary in northern New Mexico. *Science* 214: 1341–43.

Pantin, C. F. A. 1968. *The Relations among the Sciences*. New York: Cambridge Univ. Press.

Penfield, G. T., & Z. A. Camargo. 1981. Definition of a major igneous zone in the central Yucatan platform with aeromagnetics and gravity. *Soc. Explor. Geophys., 51st Ann. Meeting Tech. Program*: 37 [Abstr.].

Perlmutter, S., R. A. Muller, C. A. Pennypacker, C. K. Smith, L. P. Wang, S. White & H. S. Yang. 1990. A search for Nemesis: Current status and review of theory. *In* V. L. Sharpton & P. D. Ward, eds., *Global Catastrophes in Earth History, Geol. Soc. Amer. Special Paper* 247: 87–91.

Pollastro, R. M., & C. L. Pillmore. 1987. Mineralogy and petrology of the Cretaceous-Tertiary boundary clay bed and adjacent clay-rich rocks, Raton Basin, New Mexico, and Colorado. *J. Sed. Petrol.* 57(3): 456–66.

Premoli-Silva, I., R. Coccioni & A. Montanari, eds. 1988. *The Eocene-Oligocene Boundary in the Marche-Umbria Basin (Italy)* (I.U.G.S. Memoir). Ancona: Aniballdi.

Prinn, R. G., & B. F. Fegley, Jr. 1987. Bolide impacts, acid rain, and biospheric traumas at the Cretaceous-Tertiary boundary. *Earth and Planetary Sci. Letters* 83: 1–15.

Rampino, M. R. 1987. Impact cratering and flood basalt volcanism. *Nature* 327: 468.

———. 27 October 1992. Ancient "glacial" deposits are ejecta of large impacts: The ice age paradox explained. *Eos*, AGU Fall Meeting Abstracts vol.: 99.

Rampino, M. R., & R. B. Stothers. 1984a. Terrestrial mass extinctions, cometary impacts and the Sun's motion perpendicular to the galactic plane. *Nature* 308: 709–12.

———. 21 December 1984b. Geological rhythms and cometary impacts. *Science* 226: 1427.

Rampino, M. R., & T. Volk. 1988. Mass extinctions, atmospheric sulphur and climatic warming at the K/T boundary. *Nature* 332: 63–65.

Raup, D. M. 1985a. Magnetic reversals and mass extinctions. *Nature* 314: 341–43.

———. 3 October 1985b. Rise and fall of periodicity. *Nature* 317: 384–85.

———. 1991. Periodicity of extinctions: A review. *In* D. W. Muller, J. A. McKenzie & H. Weissert, eds., *Controversies in Modern Geology*, pp. 193–208. New York: Academic Press.

Raup, D. M., & J. J. Sepkoski, Jr. 1984. Periodicity of extinctions in the geologic past. *Proc. Nat. Acad. Sciences* 81: 801–5.

———. 1986. Periodic extinctions of families and genera. *Science* 231: 833–36.

———. 1988. Testing for periodicity of extinction. *Science* 241: 94–96.

Retallack, G. J., G. D. Leahy & M. D. Spoon. 1987. Evidence from paleosols for ecosystem changes across the Cretaceous/Tertiary boundary in eastern Montana. *Geology* 15: 1090–93.

Rezanov, I. A. 1980. *Veliskiye Katastrofy Istorii Zemli*. Moscow: Nauka. [Transl. 1984 by H. C. Creighton as *Catastrophes in the Earth's History*. Moscow: Mir Publishers.]

Rice, A. 1987. Shocked minerals at the K/T boundary: Explosive volcanism as a source. *Phys. Earth & Planetary Interiors* 48: 167–74.

Richards, M. A. & R. A. Duncan. 24 October 1989. Flood basalts as plume initiation events: Active vs. passive rifting. *Eos, Amer. Geophys. Union Fall Meeting Abstracts*, p. 1357.

Roddy, D. J., S. H. Schuster, M. Grant Rosenblatt, L. B. Grant, P. J. Hassig, & K. N. Kreyenhagen. 1987. Computer simulations of ten-km-diameter asteroid impacts into oceanic and continental sites: Preliminary results on atmospheric passage. *NASA Technical Memorandum* No. 89810: 377–79.

Russel, D. A. 1979. The enigma of the extinction of the dinosaurs. *Ann. Rev. Earth & Planetary Sciences* 7: 163–82.

Saito, T., T. Yamanoi & K. Kaiho. 1986. End-Cretaceous devastation of terrestrial flora in the boreal Far East. *Nature* 323: 253–55.

Schubert, G. 1989. Mantle convection and the cooling of the Earth. *Trans. Amer. Geophys. Union* 70(43): 999. [Abstracts for the fall meeting.]

Schultz, P. H. 1982. Geological implications of impacts of large asteroids and comets on the Earth. *Geol. Soc. Amer. Special Paper* 190: 291–96.

Schultz, P. H., & D. E. Gault. 1990. Prolonged global catastrophes from oblique impacts. *In* V. L. Sharpton & P. D. Ward, eds., *Global Catastrophes in Earth History, Geol. Soc. Amer. Special Paper* 247: 239–61.

Sears, J. W., & D. Alt. 1992. Impact origin of large intracratonic basins, the stationary Proterozoic crust, and the transition to modern plate tectonics. *In* M. J. Bartholomew, D. W. Hyndman, D. W. Mogk & R. Mason eds. *Basement Tectonics 8: Characterization and Comparison of Ancient and Mesozoic Continental Margins, Proc. 8th International Conference on Basement Tectonics* (Butte, Montana, 1988): 385–92. Dordrecht, The Netherlands: Kluwer Academic. [Received 12 August 1988; revision accepted 1 March 1989.]

Sepkoski, J. J., Jr. 1990. The taxonomic structure of periodic extinction. In

V. L. Sharpton & P. D. Ward, eds., *Global Catastrophes in Earth History, Geol. Soc. Amer. Special Paper* 247: 33–44.

Sharpton, V. L., & K. Burke. 1987. Cretaceous-Tertiary impacts. *Meteoritics* 22: 499–500.

Sharpton, V. L., G. B. Dalrymple, L. E. Marin, G. Ryder, B. C. Schuraytz & J. Urrutia-Fucugauchi. 29 October 1992. New links between the Chicxulub impact structure and the Cretaceous/Tertiary boundary. *Nature* 359: 819–21. [Submitted 7 August 1992.]

Shaw, H. R. 1987. The periodic structure of the natural record and nonlinear dynamics. *Eos* 68(50): 1651–65.

———. 1994 [forthcoming]. *Craters, Cosmos, and Chronicles: A New Theory of Earth.* Stanford, Calif.: Stanford Univ. Press.

Sheehan, P. M., D. E. Fastovsky, R. G. Hoffman, C. B. Berghaus & D. L. Gabriel. 8 November 1991. Sudden extinction of the dinosaurs: Latest Cretaceous, Upper Great Plains, U.S.A. *Science* 254: 835–39.

Sheehan, P. M., & T. A. Hansen. October 1986. Detritus feeding as a buffer to extinction at the end of the Cretaceous. *Geology* 14: 868–70.

Signor, P. W., III. July 1992. Taxonomic diversity and faunal turnover in the early Cambrian: Did the most severe mass extinction of the Phanerozoic occur in the Botomian Stage? *Fifth North American Paleontological Convention, Abstr. & Prog.* The Paleontological Society, Special Publ. 6: 272.

Signor, P. W., III, & J. H. Lipps. 1982. Sampling bias, gradual extinction patterns and catastrophes in the fossil record. *In* L. T. Silver & P. H. Schultz, eds., *Geological Implications of Impacts of Large Asteroids and Comets on the Earth, Geol. Soc. Amer. Special Paper* 190: 291–96.

Sigurdsson, H., S. D'Hondt, M. A. Arthur, T. J. Bralower, J. C. Zachos, M. Van Fossen & J. E. T. Channell. 1991. Glass from the Cretaceous-Tertiary boundary in Haiti. *Nature* 349: 482–87.

Sloan, R. E., J. K. Rigby, L. M. Van Valen & D. Gabriel. 1986. Gradual dinosaur extinction and simultaneous ungulate radiation in the Hell Creek Formation. *Science* 232: 629–33.

Sloan, R. E., & L. Van Valen. 9 April 1965. Cretaceous mammals from Montana. *Science* 148: 220–27.

Smit, J., & J. Hertogen. 1980. An extraterrestrial event at the Cretaceous-Tertiary boundary. *Nature* 285: 198–200.

Smit, J., & F. T. Kyte. 1984. Siderophile-rich magnetic spheroids from the Cretaceous boundary in Umbria, Italy. *Nature* 310: 304–5.

Smit, J., A. Montanari, N. H. M. Swinburne, W. Alvarez, A. R. Hildebrand, S. V. Margolis, P. Claeys, W. Lowrie & F. Asaro. February 1992. Tektite-bearing deep-water clastic unit at the Cretaceous-Tertiary boundary in northeastern Mexico. *Geology* 20: 99–103.

Stanley, S. M. April 1984. Temperature and biotic crises in the marine realm. *Geology* 12: 205–8.

———. 1987. *Extinction.* New York: Scientific American Library/W. H. Freeman.

Stigler, S. M., & M. J. Wagner. 1988. Response to D. M. Raup, and J. J. Sepkoski, Jr., "Testing for periodicity of extinction." *Science* 241: 96–98.

Strothers, R. B. 1985. Terrestrial record of the solar system's oscillation about the galactic plane. *Nature* 317: 338–41.

———. 1986. Periodicity of the Earth's magnetic field. *Nature* 322: 444–46.

Strong, C. P., R. R. Brooks, S. M. Wilson, R. O. Reeves, C. J. Orth, X.Y. Mao, L. R. Quintana & E. Anders. 1987. A new Cretaceous-Tertiary boundary site at Flaxbourne River, New Zealand: Biostratigraphy and geochemistry. *Geochem. et Cosmochem. Acta* 51: 2769–77.

Surlyk, F. & M. B. Johansen. 16 March 1984. End-Cretaceous brachiopod extinctions in the chalk of Denmark. *Science* 223: 1174.

Sutherland, F. L. 1988. Demise of the dinosaurs and other denizens—by cosmic clout, volcanic vapours or other means. *J. & Proc., Royal Soc. New South Wales* 121: 123–64.

Swinburne, N. H. M., A. Montanari, W. Alvarez, J. Smit, A. R. Hildebrand, S. V. Margolis & P. Claeys. 1991. Evidence from the Arroyo el Mimbral (Northeastern Mexico) for the nearby impact of the K/T bolide. *Geol. Soc. Amer. Annual Meeting, Abstr. with Programs* 23(5): A421.

Swisher, C. C. III, J. M. Grahales-Nishimura, A. Montanari, S. V. Margolis, P. Claeys, W. Alvarez, P. Renne, E. Cedillo-Pardo, F. J.-M. R. Maurrasse, G. H. Curtis, J. Smit & M. O. McWilliams. 14 August 1992. Coeval ^{40}Ar/^{39}Ar ages of 65.0 million years ago from Chicxulub crater melt rock and Cretaceous-Tertiary boundary tektites. *Science* 257: 954–58.

Teichert, C. Dec. 1986. Times of crisis in the evolution of the Cephalapoda. *Paleontologische Zeitschrift* 60(3&4): 227–43.

Thierstein, H. R. October 1981. Late Cretaceous nannoplankton and the change at the Cretaceous-Tertiary boundary. *Soc. of Econ. Paleontol. and Mineral. Spec. Publ.* 32: 355–94.

———. 1982. Terminal Cretaceous plankton extinctions: A critical assessment. *In* L. T. Silver & P. E. Schultz, eds., *Geological Implications of Impacts of Large Asteroids and Comets on the Earth, Geol. Soc. Amer. Special Paper* 190: 385–99.

Tinus, R. W., & D. J. Roddy. 1990. Effects of global atmospheric perturbations on forest ecosystems in the northern temperate zone: Predictions of seasonal depressed-temperature kill mechanisms, biomass production, and wildfire soot emissions. *In* V. L. Sharpton & P. D. Ward, eds., *Global Catastrophes in Earth History, Geol. Soc. Amer. Special Paper* 247: 77–86.

Toutain, J.-P., & G. Meyer. December 1989. Iridium-bearing sublimates at a hot-spot volcano (Piton de la Fournaise, Indian Ocean). *Geophys. Res. Letters* 16(12): 1391–94.

Tschudy, R. H., C. L. Pillmore, C. J. Orth, J. S. Gilmore & J. D. Knight. 7 September 1984. Disruption of the terrestrial plant ecosystem at the Cretaceous-Tertiary boundary, Western Interior. *Science* 225: 1030–32.

Tschudy, R. H., & B. D. Tschudy. 1986. Extinction and survival of plant life following the Cretaceous/Tertiary boundary event, Western Interior, North America. *Geology* 14: 667–70.

Van Valen, L., & R. E. Sloan, 1977. Ecology and the extinction of the dinosaurs. *Evolutionary Theory* 2: 37–64.

Vickery, A. M., & H. J. Melosh. 1990. Atmospheric erosion and impactor retention in large impacts, with application to mass extinctions. *In* V. L. Sharpton & P. D. Ward, eds., *Global Catastrophes in Earth History, Geol. Soc. Amer. Special Paper* 247: 289–300.

Vogt, P. R. 1972. Evidence for global synchronism in mantle plume convection, and possible significance for geology. *Nature* 240: 338–42.

Wallace, M. W., G. E. Williams, V. A. Gostin & R. R. Keays. 1990. The Late Proterozoic Acraman impact: Toward an understanding of impact events in the sedimentary record. *Mines and Energy Reviews, South Australia* 157: 29–36.

Wang, K., & H. H. J. Geldsetzer. 1993 [in press]. A Late Devonian impact event (about 1.5 Ma after the Frasnian/Famennian crisis) in South China and Western Australia and its association with a possible mass extinction event.

Wang, K., B. D. E. Chatterton, M. Attrep & C. J. Orth. 1992. Iridium abundance maxima at the latest Ordovician mass-extinction horizon, Yangtze Basin, China: Terrestrial or extraterrestrial? *Geology* 20(1): 39–42.

Wang, K., C. J. Orth, M. Attrep, B. D. E. Chatterton, H. G. Hou & H. H. J. Geldsetzer. August 1991. Geochemical evidence for a catastrophic biotic event at the Frasnian-Famennian boundary in South China. *Geology*. 19(8): 776–79.

Ward, P. D. 1983. The extinction of the ammonites. *Scientific American* 249: 136–47.

———. 1990. A review of Maastrichtian ammonite ranges. *In* V. L. Sharpton & P. D. Ward, eds., *Global Catastrophes in Earth History, Geol. Soc. Amer. Special Paper* 247: 519–30.

Ward, P. D., J. Wiedmann & J. F. Mount. November 1986. Maastrichtian molluscan biostratigraphy and extinction patterns in a Cretaceous/Tertiary boundary section exposed at Zumaya, Spain. *Geology* 14: 899–903.

Whitmire, D. P., & A. A. Jackson. 1984. Are periodic mass extinctions driven by a distant solar companion? *Nature* 308: 713–15.

Whitmire, D. P., & J. J. Matese. 1985. Periodic comet showers and Comet X. *Nature* 313: 36–38.

Wolbach, W. S., I. Gilmour & E. Anders. 1990. Major wildfires at the Cretaceous/Tertiary boundary. *In* V. L. Sharpton & P. D. Ward, eds., *Global Catastrophes in Earth History, Geol. Soc. Amer. Special Paper* 247: 391–400.

Wolbach, W. S., I. Gilmour, E. Anders, C. J. Orth & R. R. Brooks. 1988. Global fire at the Cretaceous-Tertiary boundary. *Nature* 334: 665–69.

Wolbach, W. S., R. S. Lewis & E. Anders. 1985. Cretaceous extinctions: Evidence for wildfires and search for meteoritic material. *Science* 230: 167–70.

Wolfe, J. A. 1990. Palaeobotanical evidence for a marked temperature increase following the Cretaceous/Tertiary boundary. *Nature* 343: 153–56.

———. 1991. Paleobotanical evidence for a June impact winter at the Cretaceous/Tertiary boundary. *Nature* 352: 420–23.

———. 26 March 1992. Plants at the K/T boundary. *Nature* 356: 296.

Zahnle, K. J. 1990. Atmospheric chemistry by large impacts. *In* V. L. Sharpton & P. D. Ward, eds., *Global Catastrophes in Earth History, Geol. Soc. Amer. Special Paper* 247: 271–88.

Zhao, M., & J. L. Bada. 8 June 1989. Extraterrestrial amino acids in Cretaceous/Tertiary boundary sediments at Stevns Klint, Denmark. *Nature* 339: 463–65.

Zoller, N. H., J. R. Parrington & J. M. Phelan Kotra. 1983. Iridium enrichment in airborne particles from Kilauea volcano. *Science* 222: 1118.

2. Glen, How Science Works in the Debates

Ahrens, T. J., & J. D. O'Keefe. 15 February 1983. Impact of an asteroid or comet in the ocean and extinction of terrestrial life. *Proc. Lunar & Planetary Sci. Conf., Abstr. XIII*, Part 2, *J. Geophys. Res.* 88: Supplement pp. A799–A806.

Alt, D., J. M. Sears & D. W. Hyndman. November 1988. Terrestrial maria: The origins of large basalt plateaus, hotspot tracks and spreading ridges. *J. Geol.* 96(6): 647–62.

Alvarez, L. W., W. Alvarez, F. Asaro & H. V. Michel. June 1980. Extraterrestrial cause for the Cretaceous-Tertiary extinction. *Science* 208: 1095–1108.

Alvarez, W., & F. Asaro. October 1990. An extraterrestrial impact. *Scientific American* 263: 78–84.

Asaro, F., H. V. Michel, L. W. Alvarez, W. Alvarez & A. Montanari. 19 April 1988. Impacts and multiple iridium anomalies. *Eos* 69(16): 301–2.

Bacon, F. 1620. Novum organum, Aphorism xlix. *In* anon., *The Oxford Book of Quotations*. Oxford: Oxford Univ. Press. [Transl. by Spedding.]

Baksi, A. K. 4 December 1990. Timing and duration of Mesozoic-Tertiary flood-basalt volcanism. *Eos* 71(49): 1835–36.

Bauer, H. H. 1992. *Scientific Literacy and the Myth of the Scientific Method*. Champaign, Ill.: Univ. of Illinois Press.

Ben-David, J. 1971. *The Scientist's Role in Society*. New York: Prentice-Hall.

Blum, J. D., & C. P. Chamberlain. 21 August 1992. Oxygen isotope constraints on the origin of impact glasses from the Cretaceous-Tertiary boundary. *Science* 257: 1104–7.

Bohm, D., & F. D. Peat. 1987. *Science, Order, and Creativity*. New York: Bantam Books.

Bohor, B. F., E. E. Foord, P. J. Modreski & D. M. Triplehorn. 1984. Mineralogic evidence for an impact event at the Cretaceous-Tertiary boundary. *Science* 224: 867–69.

Bohor, B. F., P. J. Modreski & E. E. Foord. 8 May 1987. Shocked quartz in the Cretaceous-Tertiary boundary clays: Evidence for a global distribution. *Science* 236: 705–8.

Briggs, J. C. 1990. Global extinctions, recoveries and evolutionary consequences. *Evolutionary Monographs* 13: 1–47. Chicago: Univ. of Chicago.

———. October 1991a. A Cretaceous-Tertiary mass extinction? *Bioscience* 41(9): 619–24.
———. 1991b. Global species diversity. *J. Natural History* 25: 1403–6.
Carlisle, D. B., & D. R. Braman. 22 August 1991. Nanometre-size diamonds in the Cretaceous/Tertiary boundary clay of Alberta. *Nature* 352: 708–9.
———. 1993. Extra-terrestrial amino acids at the Cretaceous-Tertiary boundary of Alberta. [In press, *Nature*, vol. 361.]
Carter, N. L., & C. B. Officer. May 1989. Comment and reply on "Microscopic lamellar deformation features in quartz: Discriminative characteristics of shock-generated varieties." *Geology* 17(5): 477–80.
Chamberlain, T. C. 1890. The method of multiple working hypotheses. *Science* (old series) 15: 92–96 [Reprinted 1965, 148: 754–59].
Clemens, E. S. 1986. Of asteroids and dinosaurs: The role of the press in the shaping of scientific debate. *Social Studies of Sci.* 16: 421–56.
Clemens, W. A. 1982. Patterns of survival of the terrestrial biota during the Cretaceous/Tertiary transition. *Geol. Soc. Amer. Special Paper* 190: 407–13.
Clube, S. V. M., & W. M. Napier. 1984. The microstructure of terrestrial catastrophism. *Monthly Notices Royal Astronomical Soc.* 211: 953–68.
Collins, H. M. 1974. The TEA set: Tacit knowledge and scientific networks. *Science Studies* 4: 165–86.
———. 1985. *Changing order: Replication and Induction in Scientific Practice*. Beverly Hills: Sage Publications.
Courtillot, V. E., & J. Besse. 1987. Magnetic field reversals, polar wander, and core-mantle coupling. *Science* 237: 1140–47.
Crockett, J. H., C. B. Officer, F. C. Wezel & G. D. Johnson. January 1988. Distribution of noble metals across the Cretaceous/Tertiary boundary at Gubbio, Italy: Iridium variations as a constraint on the duration and nature of Cretaceous/Tertiary boundary events. *Geology* 16: 77–80.
Courtillot, V. E. October 1990. What caused the mass extinction? A volcanic eruption. *Scientific American* 263: 85–92.
De Laubenfels, M. W. January 1956. Dinosaur extinction: One more hypothesis. *J. Paleontology* 30(1): 207–12.
Douglas, M. 1986. *How Institutions Think*. Syracuse, N.Y.: Syracuse Univ. Press.
Eldredge, N., & S. J. Gould. 1972. Punctuated equilibria: An alternative to phyletic gradualism. *In* T. J. Schopf, ed., *Models in Paleobiology*, pp. 82–115. San Francisco: Freeman, Cooper.
Evernden, J. F., D. E. Savage, G. H. Curtis & G. T. James. 1964. Potassium-argon dates and the Cenozoic mammalian chronology of North America. *Amer. J. Sci.* 262: 145–98.
Feyerabend, P. K. 1989. Philosophy of science: A subject with a great past. In R. H. Stuewer, ed., *Historical and Philosophical Perspectives of Science*. New York: Gordon & Breach.
Fleck, L. 1935. *Entstehung und Entwicklung einer wissenschaftlichen Tatsache*. [Transl. 1979 by F. Bradley & T. J. Trenn as *The Genesis and Development of a Scientific Fact*. Chicago: Univ. of Chicago Press.]

Frankel, H. 1982. The development, reception, and acceptance of the Vine-Matthews-Morley Hypothesis. *Hist. Studies in Phys. Sci.* 13(2): 1–39.

Gallagher, K., & C. Hawkesworth. 2 July 1992. Dehydration melting and the generation of continental flood basalts. *Nature* 358: 57–59.

Gallant, R. 1964. *Bombarded Earth.* London: John Baker.

Garvey, W. D., & K. Tomita. 1972. Scientific communication in geophysics. *Eos, Trans. Amer. Geophys. Union* 53: 772–77.

Gault, D. E., & C. P. Sonnett. 1982. Laboratory simulation of pelagic asteroidal impact: Atmospheric injection, benthic topography, and the surface wave radiation field. *In* L. T. Silver & P. E. Schultz, eds., *Geological Implications of Impacts of Large Asteroids and Comets on the Earth, Geol. Soc. Amer. Special Paper* 190: 69–92.

Gerson, E. M. 1983. Scientific work and social worlds. *Knowledge: Creation, Diffusion, Utilization* 4(3): 357–77.

———. [Forthcoming.] *The American System of Research: Evolutionary Biology 1890–1950.* Berkeley, Calif.: Univ. of California Press.

Gilbert, G. K. 1886. The inculcation of scientific method by example, with an illustration drawn from the Quaternary geology of Utah. *Amer. J. Sci.* 31: 284–99.

Glen, W. 1975. *Continental Drift and Plate Tectonics.* Columbus, Ohio: Charles E. Merrill.

———. 1981. The first potassium-argon geomagnetic polarity reversal time scale: A premature start by Martin G. Rutten. *Centaurus* 25: 222–38.

———. 1982. *The Road to Jaramillo: Critical Years of the Revolution in Earth Science.* Stanford, Calif.: Stanford Univ. Press.

———. 1987. Heresiarchs, converts and disciples. *Palaios* 2(2): 199–201.

———. 1989. Musings on the review process. *Palaios* 4: 397–99.

———. July-Aug. 1990. What killed the dinosaurs. *Amer. Scientist* 78: 354–70.

———. 12 July 1991. The power of gestalt, standards of appraisal, and the polemic mode. *Abstracts, Biannual Meeting, International Society for the History, Philosophy and Social Studies of Biology,* p. 25.

———. 24 August–3 September 1992a. The history of the current impact/volcanism/mass-extinction debates, *Abstracts Volume, 29th International Geological Congress,* p. 330.

———. 24 August–3 September 1992b. Mindset, standards, and style in the mass-extinction debates. *Abstracts Volume, 29th International Geological Congress.*

Gostin, V. A., R. R. Keays & M. W. Wallace. 17 August 1989. Iridium anomaly from the Acraman impact ejecta horizon: Impacts can produce sedimentary iridium peaks. *Nature* 340: 542–44.

Gould, S. J. March 1965. Is uniformitarianism necessary? *Amer. J. Sci.* 263: 223–28.

———. February 1984. Ediacaran experiment. *Natural History*: 14–23.

———. October 1989. An asteroid to die for. *Discover*: 60–65.

———. March 1992. Dinosaurs in the haystack. *Natural History*: 2–12.

Gould, S. J., & N. Eldredge. 1977. Punctuated equilibria: The tempo and mode of evolution reconsidered. *Paleobiology* 3: 115–51.

Grieve, R. A. 1990. Shocked minerals and the K/T controversy. *Eos* 71(46): 1792.

Hallam, A. 1973. *A Revolution in the Earth Sciences: From Continental Drift to Plate Tectonics*. Oxford: Oxford Univ. Press.

———. 1988. A compound scenario for the end-Cretaceous mass extinctions. *In* M. Lamolda, E. G. Kauffman, O. H. Walliser, H. J. Hansen, K. L. Rasmussen, R. Gwodz & H. Kunzendorf, eds., *Revista Española de Paleontología*. No. Extraordinario, *Paleontology and Evolution: Extinction Events* 7–20.

Hildebrand, A. R., & W. V. Boynton. 1987. The K/T impact excavated oceanic mantle: evidence from REE abundances. *Lunar and Planetary Science* 18: 427–28.

Hildebrand, A. R., & G. T. Penfield. 4 December 1990. A buried 180-km-diameter probable impact crater on the Yucatan Peninsula, Mexico. *Eos* 71(43): 1425.

Hoffman, A., & M. H. Nitecki. December 1985. Reception of the asteroid hypothesis of terminal Cretaceous extinctions. *Geology* 13: 884–87.

Hut, P., W. Alvarez, W. Elder, E. G. Kauffman, T. A. Hansen, G. Keller, E. M. Shoemaker & P. Weissman. 1985. Comet showers as possible causes of stepwise mass extinctions. *Eos Abstract* 66: 813.

Hut, P., W. Alvarez, W. Elder, T. A. Hansen, E. G. Kauffman, G. Keller, E. M. Shoemaker & P. Weissman. 10 September 1987. Comet showers as a cause of mass extinctions. *Nature* 329: 118–26.

Irving, E. 1988. The paleomagnetic confirmation of continental drift. *Eos* 69(44): 994–1014.

Izett, G. A. 1990. The Cretaceous/Tertiary boundary interval, Raton Basin, Colorado and New Mexico, and its contents of shock-metamorphosed minerals: Evidence relevant to the K-T boundary impact extinction theory. *Geol. Soc. Amer. Special Paper* 249: 1–100.

Izett, G. A., F. J.-M. R. Maurrasse, F. E. Lichte, G. P. Meeker & R. Bates. November 1990. Tektites in Cretaceous-Tertiary boundary rocks on Haiti. *U.S. Geol. Survey Open File Report* 90-365: 1–31.

Jablonski, D. 10 January 1986. Background and mass extinctions: The alternation of macroevolutionary regimes. *Science* 231: 129.

Johnson, J. G. 1990. Method of multiple working hypotheses: A chimera. *Geology* 18: 44–45.

Katz, E., & P. F. Lazersfeld. 1955. *Personal Influence*. Glencoe, Ill.: Free Press.

Kaufmann, E. G. 1984. The fabric of Cretaceous marine extinctions. *In* W. A. Berggren & J. Van Couvering, eds. *Catastrophes in Earth History: The New Uniformitarianism*, pp. 151–246. Princeton, N.J.: Princeton Univ. Press.

———. 1988. Stepwise mass extinctions. *In* M. Lamolda, E. G. Kauffman & O. H. Walliser, eds., *Revista Española de Paleontología*, No. Extraordinario, *Paleontology and Evolution: Extinction Events* 58–71.

Keller, G. June 1989. Extended Cretaceous-Tertiary boundary extinctions and delayed population changes in planktonic foraminifera from Brazos River, Texas. *Paleoceanography* 4(3): 287–332.

Kerr, R. A. February 1988. Was there a prelude to the Dinosaurs' demise? *Science* 239: 729–30.

Kring, D. A., & W. V. Boynton. 9 July 1992. Petrogenesis of an augite-bearing melt rock in the Chicxulub structure and its relationship to K/T impact spherules in Haiti. *Nature* 358: 141–43.

Kuhn, T. S. 1970. *The Structure of Scientific Revolutions*, 2d ed., enl. Chicago: Univ. of Chicago Press.

Kyte, F. T., and J. T. Wasson. 6 June 1986. Accretion rate of extra-terrestrial matter: Iridium deposited 33 to 67 million years ago. *Science* 232: 1225–29.

Kyte, F. T., Z. Zhou & J. T. Wasson 18/25 December 1980. Siderophile-enriched sediments from the Cretaceous-Tertiary boundary. *Nature* 288: 651–56.

Latour, B. 1987. *Science in Action*. Cambridge, Mass.: Harvard Univ. Press.

Laudan, R., & L. Laudan. 1989. Dominance and the disunity of method: Solving the problems of innovation and consensus. *Phil. Science* 56: 221–37.

Le Grand, H. E. 1988. *Drifting Continents and Shifting Theories: The Modern Revolution in Geology and Scientific Change*. Cambridge: Cambridge Univ. Press.

Le Grand, H. E., & W. Glen. 1993. Chokeholds, radiolarian cherts, and Davey Jones's locker. *Perspectives on Science: Historical/Philosophical/Social* 1(1): 24–65.

Lemke, J. L., M. H. Nitecki & H. Pullman. September 1980. Studies of the acceptance of plate tectonics. In M. Sears & D. Merriman, eds., *Oceanography: The Past, Proc. 3d Internat. Congress on History of Oceanography*, pp. 614–21. New York: Springer-Verlag.

Lightman, A., & O. Gingerich. 1991. When do anomalies begin? *Science* 255: 690–95.

Lipps, J. H. 1981. What, if anything, is micropaleontology? *Paleobiology* 7(2): 167–99.

Locke, W. W. 1990. Comments and reply on "Method of multiple working hypotheses: A chimera." *Geology* 18: 918.

Loper, D. E., & K. McCartney. December 1986. Mantle plumes and periodicity of magnetic field reversals. *Geophys. Res. Letters* 13(13): 1525–28.

———. 18 October 1988. Shocked quartz found at the K/T boundary. *Eos* 69(42): 961, 971–72.

———. 1989. Emergence of a rival paradigm to account for the Cretaceous/Tertiary event. *J. Geol. Educ.* 37: 1–13.

Luck, J. M., & K. K. Turekian. 11 November 1983. Osmium 187/osmium 186 in manganese nodules and the Cretaceous-Tertiary boundary. *Science* 222: 613.

Margolis, H. 1987. *Patterns, Thinking and Cognition: A Theory of Judgment*. Chicago: Univ. of Chicago Press.

Marvin, U. B. 1990. Impact and its revolutionary implications for geology. *In* V. L. Sharpton & P. D. Ward, eds., *Global Catastrophes in Earth History, Geol. Soc. Amer. Special Paper* 247: 87–91.

McLaren, D. J. 1970. Time, life and boundaries. *J. Paleontology* 44: 801–15.

McLean, D. M. 1981. Cretaceous-Tertiary extinctions and possible terrestrial and extraterrestrial causes. Proc. of workshop held in Ottawa, Canada, 19–20 May 1981, by K-TEC group. *Syllogeous* 39. Ottawa: National Museums of Canada.

Melosh, H. J., N. M. Schneider, K. J. Zahnle & D. Latham. 1990. Ignition of global wildfires at the Cretaceous/Tertiary boundary. *Nature* 343: 251–54.

Merton, R. K. 1951. Social structure and anomie. *In* R. K. Merton, *Social Theory and Social Structure*. [2nd ed., 1957, pp. 131–94.] New York: Free Press.

Montanari, A. 1990. Geochronology of the terminal Eocene impacts: An update. *In* V. L. Sharpton & P. D. Ward, eds., *Global Catastrophes in Earth History, Geol. Soc. Amer. Special Paper* 247: 607–16.

Morgan, W. J. 1971. Convection plumes in the lower mantle. *Nature* 230: 42–43.

Mount, J. F., S. V. Margolis, W. Showers, P. Ward & E. Doehne. 1986. Carbon and oxygen isotope stratigraphy of the Upper Maastrichtian, Zumaya, Spain: A record of oceanographic and biologic changes at the end of the Cretaceous Period. *Palaios* 1: 87–92.

Nelkin, D. 1984. Background paper for Science in the Streets. *In* anon., ed., *Report of the Twentieth Century Fund Task Force on the Communication of Scientific Risk*. New York: Priority Press.

Newell, N. D. 1982. Mass extinctions: Illusions or realities? *In* L. T. Silver & P. H. Schultz, eds., *Geological Implications of Impacts of Large Asteroids and Comets on the Earth, Geol. Soc. Amer. Special Paper* 190: 257–63.

Nitecki, M. H., J. L. Lemke, H. W. Pullman & M. E. Johnson. 1978. Acceptance of plate tectonic theory by geologists. *Geology* 6(11): 661–64.

Officer, C. B., & C. L. Drake. 25 March 1983. The Cretaceous-Tertiary transition. *Science* 219: 1383–90.

———. 1985. Terminal Cretaceous environmental events. *Science* 227: 1161–67.

———. 1989. Cretaceous/Tertiary extinctions. *Eos* 70: 25.

O'Keefe, J. D., & T. Ahrens. 8 July 1982. Impact mechanics of the Cretaceous-Tertiary extinction bolide. *Nature* 298: 123–27.

Olmez, I., D. L. Finnegan & W. H. Zoller. 1986. Iridium emissions from Kilauea volcano. *J. Geophys. Res.* 91: 653–63.

Orth, C. J., M. Attrep, Jr., X. Y. Mao, E. G. Kauffman & R. Diner. October 1987. Iridium abundance peaks at upper Cenomanian stepwise extinction boundaries. *Geol. Soc. Amer. Abstr. with Programs* 19(7): 796.

Ostry, B. 1977. The illusion of understanding: Making the ambiguous intelligible. *Oral History Review*, p. 10.

Overstreet, H. A. 1949. *The Mature Mind*. New York: W. W. Norton.

Pantin, C. F. A. 1968. *The Relations among the Sciences*. New York: Cambridge Univ. Press.

Penfield, G. T., & Z. A. Camargo. 1981. Definition of a major igneous zone in the central Yucatan platform with aeromagnetics and gravity. *Soc. Explor. Geophys., 51st Ann. Meeting Tech. Program:* 37 [Abstr.].

Posey, C. J. 1989. A future for models? *Eos* 70: 890.

Rampino, M. R. 11 June 1987. Impact cratering and flood basalt volcanism. *Nature* 327: 468.

Rice, A. 1987. Shocked minerals at the K/T boundary: Explosive volcanism as a source. *Phys. Earth & Planetary Interiors* 48: 167–74. [Received 22 April 1986; revision accepted 12 December 1987.]

Richards, M. A., & R. A. Duncan. 24 October 1989. Flood basalts as plume initiation events: Active vs. passive rifting. *Eos, Amer. Geophys. Union Fall Meeting Abstracts*, p. 1357.

Rogers, E. P. 1982. *Diffusion of Innovation*. New York: Free Press.

Rubin, D. C., ed. 1986. *Autobiographical Memory*. New York: Cambridge Univ. Press.

Russel, D. A. 1975. Reptilian diversity and the Cretaceous-Tertiary transition in North America. *Geol. Assoc. Canada, Special Paper* 13: 119–36.

Russel, D. A., & W. Tucker. 19 February 1971. Supernovae and the extinction of the dinosaurs. *Nature* 229: 553–54.

Saito, T., T. Yamanoi & K. Kaiho. 1986. End-Cretaceous devastation of terrestrial flora in the boreal Far East. *Nature* 323: 253–55.

Schubert, G. 1989. Mantle convection and the cooling of the Earth. *Trans., Am. Geophys. Union* 70(43): 999. [Abstracts for the Fall Meeting, December 4–8.]

Sharpton, V. L., G. B. Dalrymple, L. E. Marin, G. Ryder, B. C. Schuraytz, & J. Urrutia-Fucugauchi. 29 October 1992. New links between the Chicxulub impact structure and the Cretaceous/Tertiary boundary. *Nature* 359: 819–21. [Submitted 7 August 1992.]

Sharpton, V. L., & P. D. Ward, eds. 1991. *Global Catastrophes in Earth History: An Interdisciplinary Conference on Impacts, Volcanism, and Mass Mortality, Geol. Soc. Amer. Special Paper* 247.

Shaw, H. R. 1987. The periodic structure of the natural record and nonlinear dynamics. *Eos* 68(50): 1651–65.

———. 1994 [forthcoming]. *Craters, Cosmos, and Chronicles: A New Theory of Earth*. Stanford, Calif.: Stanford University Press.

Sheehan, P. M., D. E. Fastovsky, R. G. Hoffman, C. B. Berghaus & D. L. Gabriel. 8 November 1991. Sudden extinction of the dinosaurs: Latest Cretaceous, Upper Great Plains, U.S.A. *Science* 254: 835–39.

Shoemaker, E. M. 1977. Astronomically observable crater-forming projectiles. *In* D. J. Roddy, R. O. Pepin & R. B. Merrill, eds., *Impact and Explosion Cratering: Planetary and Terrestrial Implications*, pp. 617–28. New York: Pergamon Press.

Shoemaker, E. M., J. G. Williams, E. F. Helin & R. F. Wolfe. 1979. Earth-crossing asteroids: Orbital classes, collision rates with Earth, and origin. *In* T. Gehrels, ed., *Asteroids*, pp. 253–82. Tucson: Univ. of Arizona Press.

Schubert, G. October 1989. Mantle convection and the cooling of the Earth. *Eos, Amer. Geophys. Union Fall Meeting Abstracts*, 70: 1358.

Signor, P. W., III, & J. H. Lipps. 1982. Sampling bias, gradual extinction patterns and catastrophes in the fossil record. *In* L. T. Silver & P. H. Schultz, eds., *Geological Implications of Impacts of Large Asteroids and Comets on the Earth, Geol. Soc. Amer. Special Paper* 190: 291–96.

Silver, L. T., & P. H. Schultz, eds. 1982. *Geological Implications of Impacts of*

Large Asteroids and Comets on the Earth, *Geol. Soc. Amer. Special Paper* 190: 528.
Spence, D. P. 1982. *Narrative truth and historical truth: Meaning and interpretation in psychoanalysis*. New York: W. W. Norton.
Strauss, A. L. 1978. A social worlds perspective. *Studies in Symbolic Interaction* 1: 119–28.
Sutherland, F. L. 1988. Demise of the dinosaurs and other denizens—by cosmic clout, volcanic vapours or other means. *J. & Proc., Royal Soc. New South Wales* 121: 123–64.
Toutain, J.-P., & G. Meyer. December 1989. Iridium-bearing sublimates at a hot-spot volcano (Piton de la Fournaise, Indian Ocean). *Geophys. Res. Letters* 16(12): 1391–94.
Tschudy, R. H., C. L. Pillmore, C. J. Orth, J. S. Gilmore & J. D. Knight. 7 September 1984. Disruption of the terrestrial plant ecosystem at the Cretaceous-Tertiary boundary, Western Interior. *Science* 225: 1030–32.
Valentine, J. W., & M. M. Eldridge. 1972. Global tectonics and the fossil record. *J. Geology* 80: 167–84.
Valentine, J. W., & E. M. Moores. 1970. Plate tectonics regulation of faunal diversity and sea level: A model. *Nature* 228: 657–59.
Vogt, P. R. 1972. Evidence for global synchronism in mantle plume convection, and possible significance for geology. *Nature* 240: 338–42.
Westrum, R. 1978. Science and social intelligence about anomalies: The case of the meteorites. *Soc. Studies of Sci.* 8: 461–93.
Wilhelms, D. E. 1987. The geological history of the Moon. *U.S. Geol. Survey Prof. Paper* 1348.
———. 1993. *To a Rocky Moon*. Tucson: Univ. of Arizona Press.
Wilson, J. T. 1963. A possible origin of the Hawaiian Islands. *Canadian J. Phys.* 41: 863–70.
———. 1965. A new class of faults. *Nature* 207: 343–47.
Yuen, D. A., & G. Schubert. 1976. Mantle plumes: A boundary layer approach for Newtonian and non-Newtonian temperature-dependent rheologies. *J. Geophys. Res.* 81: 2499.
Zindler, A., & S. Hart. 1992. Chemical geodynamics. *Ann. Rev. Earth & Planetary Sciences* 14: 493–570.
Zoller, N. H., J. R. Parrington & J. M. Phelan Kotra. 1983. Iridium enrichment in airborne particles from Kilauea volcano. *Science* 222: 1118.

3. E. Clemens, The Impact Hypothesis and Popular Science

Alvarez, L. W. 1983. Experimental evidence that an asteroid impact led to the extinction of many species 65 million years ago. *Proc. Nat. Acad. Sci.* 80(January): 627–42.
Alvarez, L. W., W. Alvarez, F. Asaro & H. V. Michel. 1980. Extraterrestrial cause for the Cretaceous-Tertiary extinction: Experimental results and theoretical interpretation. *Science* 208(June 6): 1095–1108.
Alvarez, W. 1991a. Apocalypse then: Massive impact and the death of the

dinosaurs. Talk presented at the Institute of Human Origins, Berkeley, Calif., October 14.

———. 1991b. The gentle art of scientific trespassing. *GSA Today* 1(2): 29–31, 34.

Alvarez, W., & F. Asaro. October 1990. An extraterrestrial impact. *Scientific American* 263(4): 78–84.

Alvarez, W., F. Asaro & A. Montanari. 1990. Iridium profile for 10 million years across the Cretaceous-Tertiary boundary at Gubbio (Italy). *Science* 250: 1700–1702.

Andrews, R. C. 1943. *Under a Lucky Star: A Lifetime of Adventure*. New York: Viking.

Anon. 1979. Death of the dinosaurs. *Astronomy* 7(September): 64.

Aubry, M.-P., F. M. Gradstein & L. F. Jansa. 1990. The late early Eocene Montagnais bolide: No impact on biotic diversity. *Micropaleontology* 36(2): 164–72.

Ausubel, J. H. 1991. A second look at the impacts of climate change. *American Scientist* 79(May-June): 210–21.

Bakker, R. T. 1986. *The Dinosaur Heresies: New Theories Unlocking the Mystery of the Dinosaurs and Their Extinction*. New York: William Morrow.

Barry, D. [n.d.] Daddies and dinosaurs. *Washington Post*: K1.

Barton, B. 1989. *Dinosaurs, Dinosaurs*. New York: Thomas Y. Crowell.

Beatty, J. K. 1991. Killer crater in the Yucatan? *Sky and Telescope* (July): 38–40.

Benton, M. J. 1987. *Prehistoric World*. New York: Simon & Schuster.

———. 1990. Scientific methodologies in collision: The history of the study of the extinction of the dinosaurs. *Evolutionary Biology* 24: 371–400.

Bigelow, B. V. 1991. Scientists not amused by parody of dinosaur study. *San Diego Tribune* (November 1): B-12.

Bohor, B. F. 1990. Shock-induced microdeformations in quartz and other mineralogical indications of an impact event at the Cretaceous-Tertiary boundary. *Tectonophysics* 171: 359–72.

Brandenburg, A. 1985. *My Visit to the Dinosaurs*. New York: Harper & Row.

———. 1988. *Digging up Dinosaurs*. New York: Harper Trophy.

Broad, W. J. 1991. Asteroids, a menace to early life, could still destroy Earth: There's a "doomsday rock." But when will it strike? *The New York Times* (June 18): B5, B10.

Brouwers, E. M., W. A. Clemens, R. A. Spicer, T. A. Ager, L. D. Carter & W. V. Sliter. 1987. Dinosaurs on the North Slope: High latitude, latest Cretaceous environments. *Science* 237: 1608–10.

Browne, M. W. 1985. Dinosaur experts resist meteor extinction idea. *The New York Times* (October 29): 21–22.

———. 1988. Debate over dinosaur extinction takes an unusually rancorous turn. *The New York Times* (January 19): 19 et seq.

Buffetaut, E. 1990. Vertebrate extinctions and survival across the Cretaceous-Tertiary boundary. *Tectonophysics* 171: 337–45.

Chapman, C. R., & D. Morrison. 1991. Chicken Little was right. *Discover* (May): 40–43.

Cisowski, S. M. 1990. A critical review of the case for, and against, extraterrestrial impact at the K/T boundary. *Surveys in Geophysics* 11: 55–131.

Clemens, E. S. 1986. Of asteroids and dinosaurs: The role of the press in the shaping of scientific debate. *Social Studies of Science* 16: 421–56.

———. 1987. Changing texts and shifting audiences: The development of scientific publishing in the United States. Paper presented at the annual meetings of the Society for Social Studies of Science, Worcester, Mass.

Cloudsley-Thompson, J. L. 1978. *Why the Dinosaurs Became Extinct*. Shildon, England: Meadowfield.

Desmond, A. 1982. *Archetypes and Ancestors: Paleontology in Victorian London, 1850–1875*. Chicago: Univ. of Chicago Press.

Doukhan, J. C., O. Goltrant, P. Cordier, A. R. Huffman, N. L. Carter & C. B. Officer. 1990. Plane features in shocked quartz: A transmission electron microscope investigation. *Eos* 71(43): 1655.

Durrell, G. 1989. *The Fantastic Dinosaur Adventure*. New York: Simon & Schuster.

Ehrlich, P. R., J. Harte, M. A. Harwell, P. H. Raven, C. Sagan, G. M. Woodwell, J. Berry, E. S. Ayensu, A. H. Ehrlich, T. Eisner, S. J. Gould, H. D. Grover, R. Herrera, R. M. May, E. Mayr, C. P. McKay, H. A. Mooney, N. Myers, D. Pimentel & J. M. Teal. 1983. Long-term biological consequences of nuclear war. *Science* 222(December 23): 1293–1300.

Ellegard, A. 1958. [Rev. ed., with new Introduction 1990.] *Darwin and the General Reader: The Reception of Darwin's Theory of Evolution in the British Periodical Press, 1859–1872*. Chicago: Univ. of Chicago Press.

Feldmann, R. M. 1990. On impacts and extinction: Biological solutions to biological problems. *J. Paleontology* 64(1): 151–54.

Fleck, L. 1979. *The Genesis and Development of a Scientific Fact*. Chicago: Univ. of Chicago Press.

Gaster, B. 1990. Assimilation of scientific change: The introduction of molecular genetics into biology textbooks. *Social Studies of Science* 20(3): 431–54.

Gerson, E. M. 1983. Scientific work and social worlds. *Knowledge: Creation, Diffusion, Utilization* 4(3): 357–77.

Gieryn, T. F. 1983. Boundary-work and the demarcation of science from non-science: Strains and interests in professional ideologies of scientists. *Amer. Sociological Review* 48: 781–95.

Gitlin, T. 1980. *The Whole World Is Watching: Mass Media in the Making and Unmaking of the New Left*. Berkeley: Univ. of California Press.

Glen, W. 1990. What killed the dinosaurs? *American Scientist* 78(July-August): 354–70.

Gould, S. J. August 1989a. The dinosaur rip-off. *Natural History*: 14–18.

———. 1989b. *Wonderful Life: The Burgess Shale and the Nature of History*. New York: W. W. Norton.

Griffith, B. 1991. "Zippy." *Arizona Daily Star* (August 11).

Hallam, A., & K. Perch-Nielsen. 1990. The biotic record of events in the marine realm at the end of the Cretaceous: Calcareous, siliceous and organic-walled microfossils and macroinvertebrates. *Tectonophysics* 171: 347–57.

Heisler, J. 1990. Monte Carlo simulations of the Oort Comet Cloud. *Icarus* 88: 104–21.

Herbert, T. D., & S. L. D'Hondt. 1990. Precessional climate cyclicity in Late Cretaceous-Early Tertiary marine sediments: A high-resolution chronometer of Cretaceous-Tertiary boundary events. *Earth & Planetary Science Letters* 99: 263–75.

Hildebrand, A. R., & W. V. Boynton. 18 May 1990. Proximal Cretaceous-Tertiary boundary impact deposits in the Caribbean. *Science* 248: 843–47.

Hilgartner, S. 1990. The dominant view of popularization: Conceptual problems, political uses. *Social Studies of Science* 20(3): 519–40.

Hoffman, A., & M. H. Nitecki. December 1985. Reception of the asteroid hypothesis of terminal Cretaceous extinctions. *Geology* 13: 884–87.

Honan, W. H. 1990. Say goodbye to the stuffed elephants. *The New York Times Magazine* (January 14): 35–38.

Ingoglia, G. 1991. *Let's Look at Dinosaurs*. New York: Grosset & Dunlap.

Jastrow, R. 1983. The dinosaur massacre: A double-barreled mystery. *Science Digest* (September): 151–53 et seq.

Joyce, W. 1988. *Dinosaur Bob and His Adventures with the Family Lazardo*. New York: Harper & Row.

Kerr, R. A. 11 January 1991. Dinosaurs and friends snuffed out? *Science* 251: 160–62.

Kielan-Jawarowska, Z. 1969. *Hunting for Dinosaurs*. Cambridge, Mass.: MIT Press.

Lauber, P., & D. Henderson. 1991. *Living With Dinosaurs*. New York: Bradbury Press.

Lessem, D. 19 May 1991. The great dinosaur rip-off. *The New York Times Book Review*: 1 et seq.

Lowrie, W., W. Alvarez & F. Asaro. 1990. The origin of the white beds below the Cretaceous-Tertiary boundary in the Gubbio section, Italy. *Earth & Planetary Science Letters* 98: 303–12.

Maddox, J. 19 April 1984. Extinctions by catastrophe? *Nature* 308: 685.

Martin, P. S. 1990. 40,000 years of extinctions on the "planet of doom." *Palaeogeography, Palaeoclimatology, Palaeoecology* 82: 187–201.

McGhee, G. R., Jr. 1989. The Frasnian-Famennian extinction event. *In* S. K. Donovan, ed., *Mass Extinctions: Processes and Evidence*, pp. 133–51. London: Bellhaven.

McLaren, D. J. 5 September 1970. Presidential address: Time, life, and boundaries. *J. Paleontology* 44: 801–15.

Merrill, R. T., & P. L. McFadden. 1990. Paleomagnetism and the nature of the geodynamo. *Science* 248: 345–50.

Mervis, J. 1991. For fun, Los Alamos team goes digging for dinosaurs. *The Scientist* 5(17; September 2): 1, 8.

Miller, A. 1990. Review of S. K. Donovan, ed., *Mass Extinctions: Processes and Evidence*. *Palaios*: 382–84.

Most, B. 1987. *Dinosaur Cousins?* New York: Harcourt, Brace, Jovanovich.

Muller, R. A. 24 March 1985. An adventure in science: The pleasures of being an astrophysicist. *The New York Times Magazine*: 34–46 et seq.

Nelkin, D. 1987. *Selling Science: How the Press Covers Science and Technology*. New York: W. H. Freeman.

Officer, C. B. 1990. Extinctions, iridium, and shocked minerals associated with the Cretaceous/Tertiary transition. *J. Geol. Educ.* 38: 402–25.

Oram, H., & S. Kitamura. 1990. *A Boy Wants a Dinosaur*. New York: Farrar, Strauss & Giroux.

Otto, C. 1991. *Dinosaur Chase*. New York: Harper Collins.

Prelutsky, J. 1988. *Tyrannosaurus Was a Beast*. New York: Greenwillow Books.

Rainger, R. 1988. Vertebrate paleontology as biology: Henry Fairfield Osborn and the American Museum of Natural History. *In* R. Rainger, K. R. Benson, & J. Maienschein, eds., *The American Development of Biology*, pp. 219–56. Philadelphia: Univ. of Pennsylvania Press.

———. 1991. Public education and the limits of museum research: American vertebrate paleontology circa 1930. Paper presented at the meetings of the International Society for the History, Philosophy and Social Studies of Science, Evanston, Ill.

Raup, D. M. 1986. *The Nemesis Affair: A Story of the Death of Dinosaurs and the Ways of Science*. New York: W. W. Norton.

Robbins, J. 3 March 1991. The real Jurassic Park. *Discover* 12: 52–59.

Rocchia, R., D. Boclet, P. Bonte, C. Jehanno, Y. Chen, V. Courtillot, C. Mary & F. Wezel. 1990. The Cretaceous-Tertiary boundary at Gubbio revisited: Vertical extent of the Ir anomaly. *Earth & Planetary Science Letters* 99: 206–19.

Schlein, M. 1991. *Discovering Dinosaur Babies*. New York: Four Winds/Macmillan.

Schneider, D. A., & D. V. Kent. 1990. Ivory Coast microtektites and geomagnetic reversals. *Geophys. Res. Letters* 17(2): 163–66.

Schowalter, J. E. 1979. Children's fascination with dinosaurs. *Children Today* 8(May-June): 2–5.

Smith, J. M. 25 April 1991. Dinosaur Dilemmas. *The New York Review of Books*: 5–7.

Stevens, W. K. 20 August 1991. Species loss: Crisis or false alarm? *The New York Times*: B5–B6.

Thomson, K. S. January-February 1988. Marginalia: Anatomy of the extinction debate. *American Scientist* 76: 59–61.

Turco, R. P., O. B. Toon, T. P. Ackerman, J. B. Pollack & C. Sagan. 23 December 1983. Nuclear winter: Global consequences of multiple nuclear explosions. *Science* 222: 1283–91.

Upchurch, G. R., Jr. 1989. Terrestrial environmental changes and extinction patterns at the Cretaceous-Tertiary boundary, North America. *In* S. K. Donovan, ed., *Mass Extinctions: Processes and Evidence*, pp. 195–216. London: Bellhaven Press.

Urey, H. C. 2 March 1973. Cometary collisions and geological periods. *Nature* 242: 32–33.
Watson, C. 1990. *Big Creatures from the Past: A Pop-Up Book*. New York: G. P. Putnams.
Wilford, J. N. 1985. *The Riddle of the Dinosaur*. New York: Knopf.
Wolbach, W. S., R. S. Lewis & E. Anders. 11 October 1985. Cretaceous extinctions: Evidence for wildfires and search for meteoritic material. *Science* 230: 167–70.
Woolgar, S., & D. Pawluch. 3 February 1985. Ontological gerrymandering: The anatomy of social problems explanations. *Social Problems* 32: 214–27.
Yolen, J. 1990. *Dinosaur Dances*. New York: G. P. Putnam.

4. McLaren, Impacts and Extinctions: Science or Dogma?

Burchfield, J. D. 1975. *Lord Kelvin and the Age of the Earth*. New York: Science History Publications.
Clark, J. W., & T. M. Hughes. 1890. *The Life and Letters of the Reverend Adam Sedgwick*, Vols. I and II. Cambridge, England: Cambridge Univ. Press.
Cook, A. 1990. Sir Harold Jeffreys. *Biological Memoirs of Fellows of the Royal Society* 36: 301–33.
Darwin, C. 1859. *On the Origin of Species*. London: John Murray.
Gillespie, C. C. 1959. *Genesis and Geology*. New York: Harper.
Glen, W. 1982. *The Road to Jaramillo: Critical Years of the Revolution in Earth Science*. Stanford, Calif.: Stanford Univ. Press.
Hallam, A. 1981. The end-Triassic bivalve extinction event. *Palaeogeography, Palaeoclimatology, Palaeoecology* 35: 1–44.
Holland, C. H. 1989. Synchronology, taxonomy and reality. *Phil. Trans. Royal Soc. London* B 325: 263–77.
Hull, D. L. 1973. *Darwin and His Critics*. Cambridge: Harvard Univ. Press.
Hutton, J. 1788. *Theory of the Earth*. Trans. Roy. Soc. Edinburgh 1: 209–304.
Kirwan, R. 1799. *Geological Essays*. Dublin. [Reprint of the 1799 edition printed by T. Bensley for D. Bremner, London.]
McLaren, D. J. 1988. Detection and significance of mass killings. *Historical Biology* 2: 5–15.
McLaren, D. J., & W. D. Goodfellow. 1990. Geological and biological consequences of giant impacts. *Annual Reviews of Earth & Planetary Sci.* 18: 123–71.
Raup, D. M. 1986. *The Nemesis Affair*. New York: W. W. Norton.
Raup, D. M., & J. J. Sepkoski, Jr. 1986. Periodic extinctions of families and genera. *Science* 231: 833–36.
Smith, W. 1815. *A Memoir to the Map and Delineation of the Strata of England and Wales with Part of Scotland*. London: J. Cary.
Thomson, W. 1854. On the mechanical energies of the Solar System. *Philosophical Magazine* Series 4(8): 409–30.
———. 1881. On the sources of energy in nature available to man for the

production of mechanical effect. *British Association Report (York)*: 433–50.
Toit, A. L. du. 1927. *A Geological Comparison of South Africa with South America*. Washington, D.C.: Carnegie Institution, Publ. 381.
———. 1937. *Our Wandering Continents: An Hypothesis of Continental Drift*. Edinburgh & London: Oliver & Boyd.
Wegener, A. 1966 [1929]. *The Origin of Continents and Oceans*. New York: Dover. [English transl. of 4th revised edition of *Die Entstehung der Kontinente und Ozeane*. Friedr. Vieweg & Sohn.]
Woodward, H. B. 1908. *The History of the Geological Society of London*. London & New York: Longmans, Green.

5. Sepkoski, What I Did with My Research Career

Alvarez, L. W., W. Alvarez, F. Asaro & H. V. Michel. 1980. Extraterrestrial cause for the Cretaceous-Tertiary extinction. *Science* 208: 1095–1108.
Alvarez, W., L. W. Alvarez, F. Asaro & H. V. Michel. 1982. Current status of the impact theory for the terminal Cretaceous extinction. *In* L. T. Silver & P. H. Schultz, eds., *Geological Implications of Impacts of Large Asteroids and Comets on Earth, Geol. Soc. Amer. Special Paper* 190: 305–16.
Bretsky, P. W. 1968. Evolution of Paleozoic marine invertebrate communities. *Science* 159: 1231–33.
———. 1969. Evolution of Paleozoic benthic marine invertebrate communities. *Palaeogeography, Palaeoclimatology, Palaeoecology* 6: 45–59.
Carroll, R. 1987. *Vertebrate Paleontology and Evolution*. San Francisco: W. H. Freeman.
Cutbill, J. L., & B. M. Funnell. 1967. Numerical analysis of *The Fossil Record*. *In* W. B. Harland et al., eds., *The Fossil Record*, pp. 791–820. London: The Geological Society of London.
Fischer, A. G., & M. A. Arthur. 1977. Secular variations in the pelagic realm. *In* H. E. Cook & P. Enos, eds., *Deep-Water Carbonate Environments*, pp. 19–50. Society of Economic Paleontologists and Mineralogists Special Publication 25.
Flessa, K. W., & J. Imbrie. 1973. Evolutionary pulsations: Evidence from Phanerozoic diversity patterns. *In* D. H. Tarling & S. K. Runcorn, eds., *Implications of Continental Drift to the Earth Sciences*, pp. 247–85. London: Academic Press.
Gould, S. J., N. L. Gilinsky & R. Z. German. 1987. Asymmetry of lineages and the direction of evolutionary time. *Science* 236: 1437–41.
Gould, S. J., D. M. Raup, J. J. Sepkoski, Jr., T. J. M. Schopf & D. S. Simberloff. 1977. The shape of evolution: A comparison of real and random clades. *Paleobiology* 3: 23–40.
Grieve, R. A. F. 1982. The record of impact on Earth: Implications for a major Cretaceous/Tertiary impact event. *In* L. T. Silver & P. H. Schultz, eds., *Geological Implications of Impacts of Large Asteroids and Comets on Earth, Geol. Soc. Amer. Special Paper* 190: 25–38.

Hallam, A. 1977. Jurassic bivalve biogeography. *Paleobiology* 3: 58–73.
Harland, W. B., C. H. Holland, M. R. House, N. F. Hughes, A. B. Reynolds, M. J. S. Rudwick, G. E. Satterthwaite, L. B. H. Tarlo & E. C. Wiley, eds. 1967. *The Fossil Record*. London: The Geological Society of London.
Holser, W. T. 1994 [in press]. Development of physical and chemical parameters in the Phanerozoic. *In* O. Balliser, ed., *Phanerozoic Bio-events Event-stratigraphy*. Berlin: Springer-Verlag.
Kukalova'-Peck, J. 1973. *A Phylogenetic Tree of the Animal Kingdom (Including Orders and Higher Categories)*. Ottawa: Nat. Museums of Canada Publications in Zoology No. 8.
Moore, R. C., C. Teichert & R. A. Robison, eds. 1953–84. *Treatise on Invertebrate Paleontology*. Lawrence, Kansas: Geol. Soc. America and Univ. of Kansas Press.
Newell, N. D. 1963. Crises in the history of life. *Scientific American* 208: 76–92.
———. 1967. Revolutions in the history of life. *In* C. C. Albritton, Jr., ed., *Uniformity and Simplicity: A Symposium on the Principle of the Uniformity of Nature, Geol. Soc. Amer. Special Paper* 89: 63–91.
Palmer, A. R. 1965a. Biomere, a new kind of biostratigraphic unit. *J. Paleontology* 3: 149–53.
———. 1965b. Trilobites of the Late Cambrian Pterocephalid Biomere in the Great Basin, United States. *U.S. Geol. Survey Professional Paper* 493: 1–105.
Patterson, C., & A. B. Smith. 1987. Is the periodicity of extinctions a taxonomic artifact? *Nature* 330: 248–51.
Raup, D. M. 1986. *The Nemesis Affair*. New York: W. W. Norton.
———. 1991. Periodicity of extinctions: A review. *In* D. W. Muller, J. A. McKenzie & W. Weissert, eds., *Controversies in Modern Geology*, pp. 193–208. New York: Academic Press.
Raup, D. M., S. J. Gould, T. J. M. Schopf & D. S. Simberloff. 1973. Stochastic models of phylogeny and the evolution of diversity. *J. Geology* 81: 525–42.
Raup, D. M., & J. J. Sepkoski, Jr. 1982. Mass extinctions in the marine fossil record. *Science* 215: 1501–3.
———. 1984. Periodicity of extinctions in the geologic past. *Proc. Nat. Acad. Sci.* 81: 801–5.
———. 1986. Periodic extinction of families and genera. *Science* 231: 833–36.
Romer, A. S. 1966. *Vertebrate Paleontology*, 3rd ed. Chicago: Univ. of Chicago Press.
Rozanov, A. Yu. 1967. The Cambrian lower boundary problem. *Geological Magazine* 104: 415–34.
Schopf, J. W., & C. Klein, eds. 1992. *The Proterozoic Biosphere: A Multidisciplinary Study*. Cambridge, England: Cambridge Univ. Press.
Schopf, T. J. M. 1974. Permo-Triassic extinctions: Relation to sea-floor spreading. *J. Geology* 82: 129–43.
Seilacher, A. 1974. Flysch trace fossils: Evolution of behavioral diversity in the deep-sea. *Neues Jahrbuch für Paläontologie Monatshefte* 4: 233–45.
Sepkoski, J. J., Jr. 1976a. A kinetic model of Phanerozoic diversity. *Geol. Soc. America Abstracts with Program* 8: 1098–99.

———. 1976b. Species diversity in the Phanerozoic: Species-area effects. *Paleobiology* 2: 298–303.
———. 1977. The enigma of the Cambrian diversification. *Geol. Soc. America Abstracts with Program* 9: 1168.
———. 1978. A kinetic model of Phanerozoic taxonomic diversity, I. Analysis of marine orders. *Paleobiology* 4: 223–51.
———. 1979. A kinetic model of Phanerozoic taxonomic diversity, II. Early Phanerozoic families and multiple equilibria. *Paleobiology* 5: 222–52.
———. 1981. A factor-analytic description of the Phanerozoic marine fossil record. *Paleobiology* 7: 36–53.
———. 1982a. *A Compendium of Fossil Marine Families. Milwaukee Public Museum Contributions in Biology and Geology* 51: 1–125.
———. 1982b. Mass extinctions in the Phanerozoic oceans: A review. *In* L. T. Silver & P. H. Schultz, eds., *Geological Implications of Impacts of Large Asteroids and Comets on Earth, Geol. Soc. Amer. Special Paper* 190: 283–89.
———. 1986. Global bioevents and the question of periodicity. *In* O. H. Walliser, ed., *Global Bio-Events: A Critical Approach*, pp. 47–61. Berlin: Springer-Verlag.
———. 1989. Periodicity in extinction and the problem of catastrophism in the history of life. *J. Geol. Soc. London* 146: 7–19.
———. 1990. The taxonomic structure of periodic extinction. In V. L. Sharpton & P. D. Ward, eds., *Global Catastrophes in Earth History, Geol. Soc. Amer. Special Paper* 247: 33–44.
———. 1992. *A Compendium of Fossil Marine Animal Families*, 2nd edition. *Milwaukee Public Museum Contributions in Biology and Geology* 83: 1–156.
Sepkoski, J. J., Jr., & D. M. Raup. 1986. Periodicity in marine extinction events. *In* D. Elliott, ed., *Dynamics of Extinction*, pp. 3–36. New York: John Wiley.
Sepkoski, J. J., Jr., & P. M. Sheehan. 1983. Diversification, faunal change, and community replacement during the Ordovician radiations. *In* M. J. S. Tevesz & P. J. McCall, eds., *Interactions in Recent and Fossil Benthic Communities*, pp. 673–717. New York: Plenum Press.
Simberloff, D. S. 1974. Permo-Triassic extinctions: Effects of area on biotic equilibrium. *J. Geology* 82: 267–74.
Stitt, J. H. 1971. Repeating evolutionary pattern in Late Cambrian trilobite biomere. *J. Paleontology* 45: 178–81.
———. 1975. Adaptive radiation, trilobite paleoecology, and extinction, Ptychaspid Biomere, Late Cambrian of Oklahoma. *In* A. Martinsson, ed., *Evolution and Morphology of the Trilobita, Trilobitoidea, and Merostomata: b. Proceedings of a NATO Advanced Study Institute Held in Oslo 1–8 July 1973*, pp. 381–90. Fossils and Strata No. 4. Oslo: Universitetsforlaget.
Valentine, J. W. 1969. Patterns of taxonomic and ecological structure of the shelf benthos during Phanerozoic time. *Palaeontology* 12: 684–709.
———. 1973. *Evolutionary Paleoecology of the Marine Biosphere*. Englewood Cliffs, N.J.: Prentice-Hall.
Van Valen, L. 1973. A new evolutionary law. *Evolutionary Theory* 1: 1–30.

Zhuravleva, I. T. 1970. Marine faunas and Lower Cambrian stratigraphy. *Amer. J. Science* 269: 417–45.

6. Raup, The Debates: A View from the Trenches

Alvarez, L. W., W. Alvarez, F. Asaro & H. V. Michel. 1980. Extraterrestrial cause for the Cretaceous-Tertiary extinction. *Science* 208: 1095–1108.
Brinkmann, R. 1929. Statistisch-biostratigraphische Untersuchungen am mitteljurassischen Ammoniten über Artbegriff und Stammesentwicklung. *Abh., Ges. Wiss. Göttingen, Math-Phys. Kl., N.F.,* 13(3): 1–249.
Darwin, C. 1859. *On the Origin of Species*. London: John Murray.
Lyell, C. 1833. *Principles of Geology*, Vol. III. London: John Murray.
Shoemaker, E. M. 1984. Large-body impacts through geologic time. *In* H. D. Holland & A. F. Trendall, eds., *Patterns of Change in Earth Evolution*, pp. 15–40. Berlin: Springer-Verlag.
Strahler, A. N. 1963. *The Earth Sciences*. New York: Harper & Row.
Wetherill, G. W., & E. M. Shoemaker. 1982. Collision of astronomically observable bodies with the Earth. *Geol. Soc. Amer. Special Paper* 190: 1–13.

7. Clube, Comets: Hazards from Space

See the Bibliographic Note at the conclusion of the chapter.

8. Shaw, The Liturgy of Science

Abraham, F. D. (with R. H. Abraham & C. D. Shaw). 1990. *A Visual Introduction to Dynamical Systems Theory for Psychology*. Santa Cruz, Calif.: Aerial Press. 290 pp.
Ager, D. V. 1980. *The Nature of the Stratigraphical Record*, 2d ed. London: Macmillan/Halstead Press. 117 pp.
———. 1984. The stratigraphic code and what it implies. *In* W. A. Berggren & J. A. Van Couvering, eds., *Catastrophes and Earth History: The New Uniformitarianism*, pp. 91–100. Princeton, N.J.: Princeton Univ. Press.
Alfvén, H. 1980. Cosmology and recent developments in plasma physics: The Dirac Lecture in Theoretical Physics for 1979. *The Australian Physicist*, November: 161–65.
———. 1981. *Cosmic Plasma*. Boston, Mass.: D. Reidel.
Aflvén, H., & G. Arrhenius. 1975. *Structure and Evolutionary History of the Solar System*. Boston, Mass.: D. Reidel.
Allan, R. R. 1967a. Resonance effects due to the longitude dependence of the gravitational field of a rotating primary. *Planetary & Space Sci.* 15: 53–76.
———. 1967b. Satellite resonance with longitude-dependent gravity, II. Effects involving the eccentricity. *Planetary & Space Sci.* 15: 1829–45.
———. 1971. Commensurable eccentric orbits near critical inclination. *Celestial Mech.* 3: 320–30.
Alvarez, L. W. July 1987. Mass extinctions caused by large bolide impacts. *Phys. Today* 40: 24–33.

Alvarez, L. W., W. Alvarez, F. Asaro & H. V. Michel. 1980. Extraterrestrial causes for the Cretaceous-Tertiary extinction. *Science* 208: 1095–1108.

Alvarez, W. 1986. Toward a theory of impact crises. *Eos* 67: 649–58.

Alvarez, W., & F. Asaro. October 1990. An extraterrestrial impact. *Scientific American* 263: 78–84.

Bagby, J. P. 1969. Terrestrial satellites: Some direct and indirect evidence. *Icarus* 10: 1–10.

Bak, P., C. Tang & K. Wiesenfeld. 1988. Self-organized criticality. *Phys. Rev. A* 38: 364–74.

Baker, G. L., & J. P. Gollub. 1990. *Chaotic Dynamics: An Introduction*. New York: Cambridge Univ. Press.

Barinaga, M. 1993. Death gives birth to the nervous system. But how? [Cashing in on cell death, *in* sidebar regarding commercial portent.] *Science* 259: 762–63.

Bate, R. R., D. D. Mueller & J. E. White. 1971. *Fundamentals of Astrodynamics*. New York: Dover.

Beer, A., & P. Beer, eds. 1975. *Kepler—Four Hundred Years: Proceedings of a Conference Held in Honor of Johannes Kepler* (*Vistas in Astronomy* 18). New York: Pergamon Press.

Beer, P. 1979. *Newton and the Enlightenment: Proceedings of a Symposium Held at Cagliari, Italy, 3–5 October, 1977* (*Vistas in Astronomy* 22). New York: Pergamon Press.

Bradley, W. H. 1965. Vertical density currents. *Science* 150: 1423–28.

Burnet, M. 1976. A homeostatic and self-monitoring immune system. *In* F. M. Burnet, ed., *Immunology: Readings from Scientific American*, pp. 158–61. San Francisco: W. H. Freeman.

Chapman, C. R., & D. Morrison. 1989. *Cosmic Catastrophes*. New York: Plenum Press.

Chen, K., & P. Bak. 1989. Is the universe operating at a self-organized critical state? *Phys. Letters A* 140: 299–302.

Chouet, B., & H. R. Shaw. 1991. Fractal properties of tremor and gas piston events observed at Kilauea volcano, Hawaii, *J. Geophys. Res.* 96: 10,177–89.

Clube, S. V. M., & W. M. Napier. 1982. *The Cosmic Serpent: A Catastrophist View of Earth History*. New York: Universe Books.

———. 1990. *The Cosmic Winter*. Oxford: Basil Blackwell.

Cohen, B. 1975. Kepler's century: Prelude to Newton. *In* A. Beer & P. Beer, eds., *Kepler: Four Hundred Years* (*Vistas in Astronomy* 18), pp. 3–36. New York: Pergamon Press.

Courtillot, V. E. October 1990. Debate: What caused the mass extinction? A volcanic eruption. *Scientific American* 263: 85–92.

Crichton, M. 1990 [1991]. *Jurassic Park*. New York: Ballantine Books.

Dauben, J. W. June 1983. Georg Cantor and the origins of transfinite set theory. *Scientific American* 248: 122–31.

Edelman, G. M. 1985. Neural Darwinism: Population thinking and higher

brain function. *In* M. Shafto, ed., *How We Know* [Nobel Conference XX], pp. 1–30. San Francisco: Harper and Row.
———. 1987. *Neural Darwinism: The Theory of Neuronal Group Selection*. New York: Basic Books.
Eldredge, N. 1989. *Macroevolutionary Dynamics*. New York: McGraw-Hill.
Feigenbaum, M. J. 1980. Universal behavior in nonlinear systems. *Los Alamos Science* 1: 4–27.
Freeman, W. J. February 1991. The physiology of perception. *Scientific American* 264: 78–85.
———. 1992. Tutorial on neurobiology: From single neurons to brain chaos. *Internat. Jour. Bifurcation and Chaos* 2: 451–82.
Gamow, G. 1962. *Gravity*. Garden City, N.Y.: Anchor Books/Doubleday.
Gleick, J. 1987. *Chaos: Making A New Science*. New York: Viking Penguin.
Glen, W. 1982. *The Road to Jaramillo: Critical Years of the Revolution in Earth Science*. Stanford, Calif.: Stanford Univ. Press.
———. July–August 1990. What killed the dinosaurs? *American Scientist* 78: 354–70.
———. 12 July 1991. The power of gestalt, standards of appraisal, and the polemic mode. *Abstracts, Biannual Meeting, International Society for the History, Philosophy and Social Studies of Biology*, p. 25.
Gould, S. J. 1982. Darwinism and the expansion of evolutionary theory. *Science* 216: 380–86.
Gruntfest, I. J. 1963. Thermal feedback in liquid flow: Plane shear at constant stress. *Trans. Soc. Rheology* 7: 195–207.
Gruntfest, I. J., & H. R. Shaw. 1974. Scale effects in the study of Earth tides. *Trans. Soc. Rheology* 18: 287–97.
Haase, R. 1975a. Kepler's Harmonies: Between Pansophia and Mathesis Universalis. *In* A. Beer & P. Beer, eds., *Kepler: Four Hundred Years* (*Vistas in Astronomy* 18), pp. 519–33. New York: Pergamon Press.
———. 1975b. Kepler's Harmonies: Past, present and future. *In* A. Beer & P. Beer, eds., *Kepler: Four Hundred Years (Vistas in Astronomy* 18), pp. 535–36. New York: Pergamon Press.
Hall, A. R., & M. B. Hall. 1962. *Unpublished Scientific Papers of Isaac Newton*. London: Cambridge Univ. Press.
Halliday, D., & R. Resnick. 1966. *Physics: Parts I and II*. New York: John Wiley.
Harland, W. B., R. L. Armstrong, A. V. Cox, L. E. Craig, A. G. Smith & D. G. Smith. 1990. *A Geologic Time Scale 1989*. New York: Cambridge Univ. Press.
Helleman, R. H. G. 1980. Self-generated chaotic behavior in nonlinear mechanics. *In* E. G. D. Cohen, ed., *Fundamental Problems in Statistical Mechanics* 5: 165–233. New York: North-Holland.
Hofstadter, D. R. 1979. *Gödel, Escher, Bach: An Eternal Golden Braid*. New York: Vintage Books/Random House.
Jablonski, D. 1986a. Background and mass extinctions: The alternation of macroevolutionary regimes. *Science* 231: 129–33.
———. 1986b. Causes and consequences of mass extinctions: A comparative

approach. *In* D. K. Elliott, ed., *Dynamics of Extinction*, pp. 183–229. New York: John Wiley.

Jensen, R. V. 1987. Classical chaos. *American Scientist* 75: 168–81.

———. 1992. Quantum chaos. *Nature* 355: 311–18.

Judson, H. F. 1979. *The Eighth Day of Creation: Makers of the Revolution in Biology*. New York: Touchstone Book/Simon & Schuster.

Knobloch, E., D. R. Moore, J. Toomre & N. O. Weiss. 1986. Transitions to chaos in two-dimensional double-diffusive convection. *J. Fluid Mechanics* 166: 409–48.

Kramer, E. E. 1970. *The Nature and Growth of Modern Mathematics*. New York: Hawthorn Books.

Laskar, J. 1988. Secular evolution of the Solar System over 10 million years. *Astron. & Astrophys.* 198: 341–62.

———. 1989. A numerical experiment on the chaotic behaviour of the Solar System. *Nature* 338: 237–38.

———. 1990. The chaotic motion of the Solar System: A numerical estimate of the size of the chaotic zones. *Icarus* 88: 266–91.

Laskar, J., T. Quinn & S. Tremaine. 1992. Confirmation of resonant structure in the Solar System. *Icarus* 95: 148–52.

Lerner, E. J. 1991. *The Big Bang Never Happened*. New York: Times Books/Random House.

Lorenz, E. N. 1963. Deterministic nonperiodic flow. *J. Atmospheric Sciences* 20: 130–41.

———. 1964. The problem of deducing the climate from the governing equations. *Tellus* 16: 1–11.

Maran, S. P. 1992. *The Astronomy and Astrophysics Encyclopedia*. New York: Van Nostrand Reinhold.

Marcus, P. 1988. Numerical simulation of Jupiter's Great Red Spot. *Nature* 331: 693–96.

Margulis, L. 1992. Rethinking life on Earth. The parts: Power to the protoctists. *Earthwatch* (September/October): 25–29.

Margulis, L., & D. Sagan. 1986. *Microcosmos: Four Billion Years of Evolution from Our Microbial Ancestors*. New York: Summit Books.

May, R. M. 1976. Simple mathematical models with very complicated dynamics. *Nature* 261: 459–67.

McLaren, D. J. 1970. Presidential address: Time, life, and boundaries. *J. Paleontology* 44: 801–15.

———. 1982. Frasnian-Famennian extinctions. *In* L. T. Silver & P. H. Schultz, eds., *Geological Implications of Impacts of Large Asteroids and Comets on the Earth, Geol. Soc. Amer. Special Paper* 190: 477–84.

———. 1986. Abrupt extinctions. *In* D. K. Elliott, ed., *Dynamics of Extinction*, pp. 37–46. New York: John Wiley.

McLaren, D. J., & W. D. Goodfellow. 1990. Geological and biological consequences of giant impacts. *Annual Reviews of Earth & Planetary Sci.* 18: 123–71.

Milani, A., & A. M. Nobili. 1992. An example of stable chaos in the Solar System. *Nature* 357: 569–71.
Moon, F. C. 1987. *Chaotic Vibrations*. New York: John Wiley.
———. 1992. Coming to grips with chaos. *Nature* 355: 675–76.
Morrison, F. 1991. *The Art of Modeling Dynamic Systems: Forecasting for Chaos, Randomness, and Determinism*. New York: Wiley-Interscience.
Nagel, E., & J. R. Newman. 1958. *Gödel's Proof*. New York: New York Univ. Press.
Officer, C. B., C. L. Drake, J. L. Pindell & A. A. Meyerhoff. April 1992. Cretaceous-Tertiary events and the Caribbean caper. *GSA Today* 2(4): 69–70, 73–76.
Ovenden, M. W. 1975. Bode's Law—truth or consequences? *Vistas in Astronomy* 18: 473–96.
Pierce, J. R. 1980. *An Introduction to Information Theory: Symbols, Signals, and Noise*, 2d ed. New York: Dover.
Poincaré, H. 1952. *Science and Method* [first English translation]. New York: Dover.
Post, E. L. 1965. Absolutely unsolvable problems and relatively undecidable propositions: Account of an anticipation. *In* M. Davis, ed., *The Undecidable: Basic Papers on Undecidable Propositions, Unsolvable Problems, and Computable Functions*, pp. 338–433. Hewlett, N.Y.: Raven Press.
Prigogine, I., G. Nicolis & A. Babloyantz. December 1972. Thermodynamics of evolution [in two parts]. *Physics Today* 25: 23–28, 38–44.
Prigogine, I., & I. Stengers. 1984. *Order out of Chaos*. New York: Bantam Books.
Provine, W. B. 1986. *Sewall Wright and Evolutionary Biology*. Chicago: Univ. of Chicago Press.
Raup, D. M. 1986. Biological extinction in Earth history. *Science* 231: 1528–33.
———. 1990. Impact as a general cause of extinction: A feasibility test. *In* V. L. Sharpton & P. D. Ward, eds., *Global Catastrophes in Earth History: An Interdisciplinary Conference on Impacts, Volcanism, and Mass Mortality, Geol. Soc. Amer. Special Paper* 247: 27–32.
Raup, D. M., & J. J. Sepkoski, Jr. 1984. Periodicity of extinctions in the geologic past. *Proc. Nat. Acad. Sci.* 81: 801–05.
Roy, A. E. 1988. *Orbital Motion*, 3rd ed. New York: Adam Hilger.
Ruelle, D. 1980. Strange attractors. *The Mathematical Intelligencer* 2: 126–37.
———. 1991. *Chance and Chaos*. Princeton, N.J.: Princeton Univ. Press.
Ruelle, D., & F. Takens. 1971. On the nature of turbulence. *Commun. Math. Phys.* 20: 167–92.
Sandfort, J. F. 1962. *Heat Engines: Thermodynamics in Theory and Practice*. New York: Doubleday.
Schrödinger, E. 1992. *What is Life?* with *Mind and Matter*, and *Autobiographical Sketches*. New York: Cambridge Univ. Press [Canto Edition].
Searls, D. B. 1992. The linguistics of DNA. *American Scientist* 80: 579–91.
Sewell, E. 1951. *The Structure of Poetry*. London: Routledge & Kegan Paul. [Reprinted by Richard West, 1977.]

Shannon, C. E., & W. Weaver. 1949. *The Mathematical Theory of Communication*. Urbana, Ill.: Univ. of Illinois Press.

Sharpton, V. L., & P. D. Ward, eds. 1990. *Global Catastrophes in Earth History; An Interdisciplinary Conference on Impacts, Volcanism, and Mass Mortality, Geol. Soc. Amer. Special Paper* 247: 631.

Shaw, H. R. 1965. Comments on viscosity, crystal settling, and convection in granitic magmas. *Amer. J. Sci.* 263: 120–52.

———. 1969. Rheology of basalt in the melting range. *J. Petrology* 10: 510–35.

———. 1970. Earth tides, global heat flow, and tectonics. *Science* 168: 1084–87.

———. 1983a. Magmatic processes and orbital evolution. *Geol. Soc. Amer., Abstracts with Programs* 15: 684.

———. 1983b. Mathematical attractor theory and geological patterns: The distribution of the planets as a resonant mapping of chaos. U.S. Geol. Survey, Unpublished Research Report.

———. 1986. *Ring Vortex Bifurcations, Information Cascades, and Volcanogenic Sources of Prebiotic Evolution*. U.S. Geol. Survey, Unpublished Research Report.

———. 1987a. The periodic structure of the natural record, and nonlinear dynamics. *Eos* 68(50): 1651–64.

———. 1987b. *A Linguistic Model of Earthquake Frequencies Applied to the Seismic History of California. U.S. Geol. Survey, Open-File Report* 87-296. 129 pp. [Available from: Open-File Report Section, U.S.G.S., Box 25425, Federal Center, Denver, Colo. 80225.]

———. 1988a. Terrestrial-cosmological correlations in evolutionary processes. *U.S. Geol. Survey, Open-File Report* 88-43.

———. 1988b. Reply, "On Chaos." *Eos* 69: 669–70.

———. 1991. Magmatic phenomenology as nonlinear dynamics: Anthology of some relevant experiments and portraits. *In* G. V. Middleton, ed., *Nonlinear Dynamics, Chaos and Fractals: With Applications to Geological Systems* (Chap. 8), pp. 97–149. Toronto: Geol. Assoc. Canada, Short Course Notes 9.

———. 1994 [forthcoming]. *Craters, Cosmos, and Chronicles: A New Theory of Earth*. Stanford, Calif.: Stanford Univ. Press.

Shaw, H. R., & B. Chouet. 1988. Application of nonlinear dynamics to the history of seismic tremor at Kilauea volcano, Hawaii, *U.S. Geol. Survey, Open-File Report* 88-539. 78 pp.

———. 1989. Singularity spectrum of intermittent seismic tremor at Kilauea volcano, Hawaii. *Geophys. Res. Letters* 16: 195–98.

———. 1991. Fractal hierarchies of magma transport in Hawaii and critical self-organization of tremor. *J. Geophys. Res.* 96: 10,191–10,207.

Shaw, H. R., R. W. Kistler & J. F. Evernden. 1971. Sierra Nevada plutonic cycle, Part II. Tidal energy and a hypothesis for orogenic-epeirogenic periodicities. *Geol. Soc. Am. Bull.* 82: 869–96.

Shaw, H. R., T. L. Wright, D. L. Peck & R. Okamura. 1968. The viscosity of

basaltic magma: An analysis of field measurements in Makaopuhi lava lake, Hawaii. *Amer. J. Sci.* 266: 225–64.
Silver, L. T., & P. H. Schultz, eds. 1982. *Geological Implications of Impacts of Large Asteroids and Comets on the Earth, Geol. Soc. Amer. Special Paper* 190.
Skarda, C. A., & W. J. Freeman. 1987. How brains make chaos in order to make sense of the world. *Behavioral and Brain Sciences* 10: 161–95.
Sonea, S., & M. Panisset. 1983. *A New Bacteriology*. Portola Valley, Calif.: Jones & Bartlett.
Spencer-Brown, G. 1972. *Laws of Form*. New York: The Julian Press.
Steel, D. 1991. Our asteroid-pelted planet. *Nature* 354: 265–67.
Stern, S. A. 1991. On the number of planets in the outer Solar System: Evidence of a substantial population of 1,000-km bodies. *Icarus* 90: 271–81.
Sussman, G. J., & J. Wisdom. 1992. Chaotic evolution of the Solar System. *Science* 257: 56–62.
Wisdom, J. 1987. Chaotic behavior in the Solar System. *In* M. V. Berry, I. C. Percival & N. O. Weiss, eds., *Dynamical Chaos*, pp. 109–29. Princeton, N.J.: Princeton Univ. Press.
Zirin, H. 1988. *Astrophysics of the Sun*. New York: Cambridge Univ. Press.

9. Van Valen, The Nature of Selection by Extinction

Anstey, R. L. 1978. Taxonomic survivorship and morphologic complexity in Paleozoic bryozoan genera. *Paleobiology* 4: 407–18.
———. 1986. Bryozoan provinces and patterns of generic evolution and extinction in the late Ordovician of North America. *Lethaia* 19: 33–51.
Arnold, A. J., & K. Fristrup. 1982. The theory of evolution by natural selection: A hierarchical expansion. *Paleobiology* 8: 113–29.
Bambach, R. K. 1990. Origination and extinction: Episodic or continuous? *Geol. Soc. America, Abstracts with Programs* 22: A268.
Benton, M. J. 1987. Progress and competition in macroevolution. *Biological Reviews* 62: 305–38.
Boucot, A. J. 1975. *Evolution and Extinction Rate Controls*. Amsterdam: Elsevier.
———. 1983. Does evolution take place in an ecological vacuum? II. *J. Paleontology* 57: 1–30.
Boyajian, G. E. 1986. Phanerozoic trends in background extinction: Consequences of an aging fauna. *Geology* 14: 955–58.
———. 1987. Reply [to Van Valen 1987]. *Geology* 15: 876.
———. 1991. Taxon age and selectivity of extinction. *Paleobiology* 17: 49–57.
Brocchi, G. 1814. *Conchiliologica fossile subapennina, con osservazioni geologiche sugli Apennini e sul suolo adiacenta*. Milan: Stamperia reale. [Not seen; cited from Lyell 1832 and National Union Catalog.]
Charig, A. J. 1984. Competition between therapsids and archosaurs during the Triassic Period: A review and synthesis of current theories. *Symposia of the Zoological Society of London* 52: 597–628.

Cooper, W. S. 1984. Expected time to extinction and the concept of fundamental fitness. *J. Theoretical Biology* 107: 603–29.
Damuth, J., & I. L. Heisler. 1988. Alternative formulations of multilevel selection. *Biology and Philosophy* 3: 407–30.
Darden, L., & K. Bach. 1990. Causal relations of selection events at multiple hierarchical levels. *Fourth International Congress of Systematic and Evolutionary Biology*. [Abstracts volume.]
Darwin, C. 1845. *Journal of Researches into the Natural History and Geology of the Countries Visited During the Voyage of H.M.S. Beagle Round the World*, ed. 2. London: John Murray.
———. 1859. *On the Origin of Species by Means of Natural Selection*. London: John Murray.
Diamond, J. M. 1984. "Normal" extinctions of isolated populations. *In* M. H. Nitecki, ed., *Extinctions*, pp. 191–246. Chicago: Univ. of Chicago Press.
Edinger, E. N., & M. J. Risk. 1990. Corals and coral associates in the Oligocene-Miocene regional mass extinction of Caribbean reef corals. *Fourth International Congress of Systematic and Evolutionary Biology*. [Abstracts volume.]
Erwin, D. H. 1990. The end-Permian mass extinction. *Annual Review of Ecology and Systematics* 21: 69–91.
Fagerstrom, J. A. 1987. *The Evolution of Reef Communities*. New York: John Wiley.
Flessa, K. W., rapporteur. 1986. Causes and consequences of extinction. *In* D. M. Raup & D. Jablonski, eds., *Patterns and Processes in the History of Life*, pp. 235–57. Berlin: Springer-Verlag.
Fortey, R. A. 1989. There are extinctions and extinctions: Examples from the lower Palaeozoic. *Philosophical Trans. Royal Soc. London*, Series B (325): 327–55.
Fowler, C. W., & J. A. MacMahon. 1982. Selective extinction and speciation: Their influence on the structure and functioning of communities and ecosystems. *American Naturalist* 119: 480–98.
Gilinsky, N. L., & R. K. Bambach. 1987. Asymmetrical patterns of origination and extinction in the fossil record. *Paleobiology* 13: 427–45.
Gilinsky, N. L., & I. J. Good. 1991. Problems of origination, persistence, and extinction of families of marine invertebrate life. *Paleobiology* 17: 145–66.
Gilinsky, N. L., & A. E. Hubbard. 1990. Mass extinctions evaluated in families within orders. *Fourth International Congress of Systematic and Evolutionary Biology*. [Abstracts volume.]
Guthrie, R. D. 1982. Mammals of the mammoth steppe as paleoenvironmental indicators. *In* D. M. Hopkins, J. V. Matthews, Jr., C. E. Schweger & S. B. Young, eds., *Paleoecology of Beringia*, pp. 307–26. New York: Academic Press.
Hansen, T. A. 1980. Influence of larval dispersal and geographic distribution on species longevity in neogastropods. *Paleobiology* 6: 193–207.
Heisler, I. L., & J. Damuth. 1987. A method for analyzing selection in hierarchically structured populations. *American Naturalist* 130: 582–602.

Hoffman, A., & J. A. Kitchell. 1984. Evolution in a pelagic planktonic system: A paleobiologic test of models of multispecies evolution. *Paleobiology* 10: 9–33.

Howe, H. F. 1985. Gomphothere fruits: A critique. *American Naturalist* 125: 853–65.

Jablonski. D. 1986. Mass extinctions: New answers, new questions. In L. Kaufman & K. Mallory, eds., *The Last Extinction*, pp. 43–61. Cambridge, Mass.: MIT Press.

———. 1989. The biology of mass extinction: A palaeontological view. *Philosophical Trans. Royal Soc. London*, Series B(325); 357–68.

Janzen, D. H., & P. S. Martin. 1981. Neotropical anachronisms: The fruits the gomphotheres ate. *Science* 215: 19–27.

Johnson, J. G. 1974. Extinction of perched faunas. *Geology* 2: 479–82.

Karr, J. R. 1990. Avian survival rates and the extinction process on Barro Colorado Island, Panama. *Conservation Biology* 4: 391–97.

Kitchell, J. A. 1990. Biological selectivity of extinction. *In* E. G. Kauffman & O. H. Walliser, eds., *Extinction Events in Earth History* [= *Lecture Notes in Earth Sciences* 30], pp. 31–43. New York: Springer-Verlag.

Kitchell, J. A., D. L. Clark & A. M. Gombos, Jr. 1986. Biological selectivity of extinction: A link between background and mass extinctions. *Palaios* 1: 504–511.

Kitchell, J. A., & D. Pena. 1984. Periodicity of extinctions in the geologic past: Deterministic versus stochastic explanations. *Science* 226: 689–92.

Krause, D. W. 1986. Competitive exclusion and taxonomic displacement in the fossil record: The case of rodents and multituberculates in North America. *In* K. M. Flanagan & J. A. Lillegraven, eds., *Vertebrates, Phylogeny, and Philosophy* (= *Contributions to Geology, University of Wyoming, Special Paper* 3), pp. 95–117. Laramie: Univ. of Wyoming.

Levinton, J. S. 1974. Trophic group and evolution in bivalve molluscs. *Palaeontology* 17: 579–85.

Lyell, C. 1832. *Principles of Geology*, Vol. 2. London: John Murray.

Maas, M. C., D. W. Krause & S. G. Strait. 1988. The decline and extinction of Plesiadapiformes (Mammalia: ?Primates) in North America: Displacement or replacement? *Paleobiology* 14: 410–31.

Mac Arthur, R. H., & E. O. Wilson. 1967. *The Theory of Island Biogeography*. Princeton, N.J.: Princeton Univ. Press.

McCauley, D. E. 1991. Genetic consequences of local population extinction and recolonization. *Trends in Ecology and Evolution* 6: 5–8.

McGhee, G. R., Jr. 1989. Catastrophes in the history of life. *In* K. C. Allen & D. E. G. Briggs, eds., *Evolution and the Fossil Record*, pp. 26–50. London: Bellhaven Press.

McKinney, M. L. 1987. Taxonomic selectivity and continuous variation in mass and background extinctions of marine taxa. *Nature* 325: 343–45.

Melzer, A. L., & J. H. Koeslag. 1991. Mutations do not accumulate in asexual isolates capable of growth and extinction: Muller's ratchet re-examined. *Evolution* 45: 649–55.

Morgan, T. H. 1903. *Evolution and Adaptation*. New York: Macmillan.
Newell, N. D. 1971. An outline history of tropical organic reefs. *American Museum Novitates* 2465: 1–37.
Patterson, B. D. 1984. Mammalian extinction and biogeography in the southern Rocky Mountains. *In* M. H. Nitecki, ed., *Extinctions*, pp. 247–93. Chicago: Univ. of Chicago Press.
Patterson, B. D., & W. Atmar. 1986. Nested subsets and the structure of insular mammalian faunas and archipelagos. *Biol. Journal of the Linnaean Society* 28: 65–82.
Phillips, J. 1841. *Figures and Descriptions of the Palaeozoic Fossils of Cornwall, Devon and West Somerset*. London: Longman, Brown, Green and Longman. [Not seen; cited from Newell 1988.]
Pielou, E. C. 1979. *Biogeography*. New York: John Wiley.
Pimm, S. L., H. L. Jones & J. Diamond. 1988. On the risk of extinction. *American Naturalist* 132: 757–85.
Pojeta, J., Jr., & T. J. Palmer. 1976. The origin of rock boring in mytilacean pelecypods. *Alcheringa* 1: 167–79.
Rabinowitz, D. 1981. Seven forms of rarity. *In* H. Synge, ed., *The Biological Aspects of Rare Plant Conservation*, pp. 205–17. Chichester, England: Wiley.
Raup, D. M. 1978. Cohort analysis of generic survivorship. *Paleobiology* 4: 1–15.
———. 1986. Biological extinction in Earth history. *Science* 231: 1528–33.
———. 1990. Scaling background and mass extinction. *Fourth International Congress of Systematic and Evolutionary Biology*. [Abstracts volume.]
Raup, D. M., & G. E. Boyajian. 1988. Patterns of generic extinction in the fossil record. *Paleobiology* 14: 109–25.
Rosen, B. R. 1990. Extinction in the tropical marine biota. *Fourth International Congress of Systematic and Evolutionary Biology*. [Abstracts volume.] Unpaginated.
Schmalhausen [=Shmal'gausen, etc.], I. I. 1946. *Faktory Evolyutsii* (*Teoriya stabiliziruyushchego otbora*). Moscow: Izdatel'stvo Akademii Nauk Soyuza SSR.
———. 1949. *Factors of Evolution*. [Bowdlerized translation of Schmalhausen 1946.] Philadelphia: Blakiston.
Sepkoski, J. J., Jr. 1990. Overview of Phanerozoic diversity patterns. *Fourth International Congress of Systematic and Evolutionary Biology*. [Abstracts volume.]
Sheehan, P. M. 1988. Late Ordovician events and the terminal Ordovician extinction. *New Mexico Bureau of Mines and Mineral Resources, Memoir* 44: 405–15.
———. 1990. Macroevolution and synecologic reorganization following mass extinctions of marine invertebrates. *Fourth International Congress of Systematic and Evolutionary Biology*. [Abstracts volume.]
Sheehan, P. M., & T. A. Hansen. 1986. Detritus feeding as a buffer to extinction at the end of the Cretaceous. *Geology* 14: 868–70.
Signor, P. W., & D. H. Erwin. 1990. Extinction in an extinction-resistant clade:

Evolutionary history of the Gastropoda. *Fourth International Congress of Systematic and Evolutionary Biology*. [Abstracts volume.]

Simpson, G. G. 1953. *The Major Features of Evolution*. New York: Columbia Univ. Press.

Small, J. 1946. Quantitative evolution, VIII. Numerical analysis of tables to illustrate the geological history of species number in diatoms. *Proc. Royal Irish Academy* (B)51: 53–80.

———. 1948. Quantitative evolution: X. Generic sizes in relation to time and type. *Proc. Royal Irish Academy* (B)51: 279–95.

Stanley, S. M. 1990. A definition of species selection. *Fourth International Congress of Systematic and Evolutionary Biology*. [Abstracts volume.]

Stehli, F. G., & S. D. Webb, eds. 1985. *The Great American Biotic Interchange*. New York: Plenum.

Stigler, S. M. 1987. Testing hypotheses or fitting models? Another look at mass extinctions. *In* M. H. Nitecki & A. Hoffman, eds., *Neutral Models in Biology*, pp. 147–59. Oxford: Oxford Univ. Press.

Talent, J. A. 1988. Organic reef-building: Episodes of extinction and symbiosis? *Senckenbergiana Lethaea* 69: 325–68.

Thackeray, J. F. 1990. Rates of extinction in marine invertebrates: Further comparison between background and mass extinctions. *Paleobiology* 15: 22–24.

Thoday, J. M. 1953. Components of fitness. *Symposia of the Society for Experimental Biology* 7: 96–113.

Van Blaricom, G. R., & J. A. Estes, eds. 1988. *The Community Ecology of Sea Otters*. Berlin: Springer-Verlag.

Van Valen, L. M. 1970. Late Pleistocene extinctions. *In* E. L. Yochelson & L. M. Van Valen, eds., *Evolution of Communities* [part E, separately published, of *Proc. North American Paleontological Convention*], pp. 469–85. Lawrence, Kans.: Allen Press.

———. 1971. Group selection and the evolution of dispersal. *Evolution* 25: 591–98.

———. 1973. A new evolutionary law. *Evolutionary Theory* 1: 1–30.

———. 1975. Group selection, sex, and fossils. *Evolution* 29: 87–94.

———. 1984. A resetting of Phanerozoic community evolution. *Nature* 307: 50–52, 660.

———. 1985a. How constant is extinction? *Evolutionary Theory* 7: 93–106.

———. 1985b. A theory of origination and extinction. *Evolutionary Theory* 7: 133–42.

———. 1987. [Age and extinction.] *Geology* 15: 875–76.

———. 1990. Levels of selection in the early Cenozoic radiation of mammals. *Evolutionary Theory* 9: 171–80.

———. 1991. Biotal evolution: A manifesto. *Evolutionary Theory* 10: 1–13.

Van Valen, L. M., & R. E. Sloan. 1966. The extinction of the multituberculates. *Systematic Zoology* 15: 261–78.

———. 1977. Ecology and the extinction of the dinosaurs. *Evolutionary Theory* 2: 37–64.

Vermeij, G. J. 1987. *Evolution and Escalation: An Ecological History of Life*. Princeton, N.J.: Princeton Univ. Press.
———. 1990. Geographical restriction as a guide to the causes of extinction: The case of the cold northern oceans during the Neogene. *Paleobiology* 15 [for 1989]: 335–56.
Vrba, E. S. 1980. Evolution, species, and fossils: How does life evolve? *South African J. of Science* 76: 61–84.
Vrba, E. S., & S. J. Gould. 1986. The hierarchical expansion of sorting and selection: Sorting and selection cannot be equated. *Paleobiology* 12: 217–28.
Wade, M. J. 1978. A critical review of the models of group selection. *Quarterly Review of Biology* 53: 101–11.
Ward, P., & P. Signor. 1983. Evolutionary tempo in Jurassic and Cretaceous ammonites. *Paleobiology* 9: 183–98.
Williamson, M. H. 1989. Natural extinctions on islands. *Philosophical Trans. Royal Soc. London*, Series B(325): 457–68.
Wilson, E. O. 1961. The nature of the taxon cycle in the Melanesian ant fauna. *American Naturalist* 95: 169–93.
Wolfe, J. A. 1990. Long-term biotic effects of major climatic perturbations. *Fourth International Congress of Systematic and Evolutionary Biology*. [Abstracts volume.]
Wright, S. 1945. Tempo and mode in evolution: A critical review. *Ecology* 26: 415–19.
Ziegler, W., & H. R. Lane. 1987. Cycles in conodont evolution from Devonian to mid-Carboniferous. *In* R. J. Aldridge, ed., *Palaeobiology of Conodonts*, pp. 147–63. Chichester, England: Horwood.

10. Hsü, Uniformitarianism vs. Catastrophism

Chamberlin, T. C. 1907. Diastrophism as an ultimate basis of correlation. *J. Geology* 17: 685–93.
Hay, W. W. 1969. Sedimentation rates. *In* R. G. Bader et al., *Initial Reports of the Deep Sea Drilling Project* 4: 668–70.
Hsü, K. J. 1982. Actualistic catastrophism. *Sedimentology* 30: 3–9.
———. 1986. *The Great Dying*. New York: Harcourt Brace Jovanovich.
———. 1989. Time and place in Alpine orogenesis: The Fermor Lecture. *In* M. P. Coward, D. Dietrich & R. G. Park, eds., *Alpine Tectonics*. Geol. Soc. London Special Publication no. 45: 421–43.
Hsü, K. J., & J. E. Andrews. 1969. Lithology and history of the South Atlantic Ocean. *In* A. E. Maxwell et al., *Initial Reports of the Deep Sea Drilling Project* 3: 445–53.
Hsü, K. J., Q. He, J. A. McKenzie, et al. 1982. Mass mortality and its environmental and evolutionary consequences. *Science* 216: 249–56.
Hsü, K. J., & J. A. McKenzie. 1987. A strangelove ocean in the earliest Tertiary. In E. T. Sundqvist & W. T. Broeker, eds., *The Carbon Cycle and Atmospheric* CO_2*: Natural Variations, Archean to Present*. Am. Geophys. Union, Geophysical Monographs 32: 487–92.

———. 1990. Carbon-isotope anomalies at era-boundaries: Global catastrophes and their ultimate cause. *Geol. Soc. Amer. Special Paper* 247: 61–70.

Mandelbrot, B. 1977. *The Fractal Geometry of Nature*. New York: W. H. Freeman.

Marvin, U. B. 1989. Meteorite impact and its consequences for geology. *28th International Geological Congress Abstracts*, v. 2: 381.

McLean, D. 1982. Deccan volcanism and Cretaceous-Tertiary transition scenario. *Syllogeous* 39: 143–44.

Perch-Nielsen, K., McKenzie, J. A. & He Qixiang. 1982. Biostratigraphic and isotope stratigraphy of the "catastrophic" extinction of calcareous nannoplankton at the Cretaceous/Tertiary boundary. *In* L. T. Silver & P. H. Schultz, eds., *Geological Implications of Impacts of Large Asteroids and Comets on Earth, Geol. Soc. America Special Report* 190: 353–71.

Raup, D. M. 1991. *Extinction: Bad Genes or Bad Luck?* New York: W. W. Norton.

Ryan, W. B. F., K. J. Hsü et al. 1973. *Initial Reports of the Deep Sea Drilling Project* 13.

Weisman, E. 1982. Terrestrial impact rates for long- and short-period comets. *In* L. T. Silver & P. H. Schultz, eds., *Geological Implications of Impacts of Large Asteroids and Comets on Earth, Geol. Soc. America Special Report* 190: 15–24.

11. Briggs, Mass Extinctions: Fact or Fallacy?

Allaby, M., & J. Lovelock. 1983. *The Great Extinction*. London: Secker & Warburg.

Alvarez, L. W. 1983. Experimental evidence that an asteroid impact led to the extinction of many species 65 million years ago. *Proc. Nat. Acad. Sci.* 80: 627–42.

Alvarez, W., L. W. Alvarez, F. Asaro & H. V. Michel. 1979. Experimental evidence in support of an extraterrestrial trigger for the Cretaceous-Tertiary extinctions. *Eos* 60: 734.

———. 1980. Extraterrestrial cause for the Cretaceous-Tertiary extinction. *Science* 208: 1095–1108.

Alvarez, W., & F. Asaro. October 1990. An extraterrestrial impact. *Scientific American* 263: 78–84.

Archibald, J. D., & L. J. Bryant. 1990. Differential Cretaceous/Tertiary extinctions of nonmarine vertebrates: Evidence from Northeastern Montana. *In* V. L. Sharpton & P. D. Ward, eds., *Global Catastrophes and Earth History: An Interdisciplinary Conference on Impacts, Volcanism, and Mass Mortality, Geol. Soc. Amer. Special Paper* 247: 549–62.

Archibald, J. D., & W. A. Clemens. 1982. Late Cretaceous extinctions. *American Scientist* 70: 377–85.

Benton, M. J. 1990. End-Triassic. *In* D. E. G. Briggs & P. R. Crowther, eds., *Palaeobiology: A Synthesis*, pp. 194–98. Oxford: Blackwell.

Bramlette, M. N. 1965. Massive extinctions in biota at the end of Mesozoic time. *Science* 148: 1696–99.

Brenchley, P. J. 1989. The Late Ordovician extinction. In S. K. Donovan, ed., *Mass Extinctions: Processes and Evidence*, pp. 104–32. New York: Columbia Univ. Press.

———. 1990. End Ordovician. *In* D. E. G. Briggs & P. R. Crowther, eds., *Palaeobiology: A Synthesis*, pp. 181–84. Oxford: Blackwell.

Briggs, J. C. 1990. Global extinctions, recoveries, and evolutionary consequences. *Evolutionary Monographs*, 13. Chicago: Univ. of Chicago.

———. 1991a. A Cretaceous-Tertiary mass extinction? *Bioscience* 41: 619–24.

———. 1991b. Global species diversity. *J. Natural History* 25: 1403–6.

Browne, J. 1983. *The Secular Ark: Studies in the History of Biogeography*. New Haven, Conn.: Yale Univ. Press.

Burger, W. C. 1981. Why are there so many kinds of flowering plants? *Bioscience* 31: 572, 577–81.

Carroll, R. L. 1987. *Vertebrate Paleontology and Evolution*. New York: W. H. Freeman.

Charig, A. J. 1989. The Cretaceous-Tertiary boundary and the last of the dinosaurs. *Phil. Trans. Royal Soc. London* B325: 387–400.

Clemens, W. A., J. D. Archibald & L. J. Hickey. 1981. Out with a whimper, not a bang. *Paleobiology* 7: 293–98.

Courtillot, V. E. October 1990. What caused the mass extinction? A volcanic eruption. *Scientific American* 263: 85–92.

Darwin, C. 1859. *On the Origin of Species by Means of Natural Selection*. London: John Murray.

Dodson, P. 1991. Maastrichtian dinosaurs. *Geol. Soc. Amer. Abs.*, San Diego meeting, p. 184.

Dunbar, C. O. 1960. *Historical Geology*, 2nd ed. New York: John Wiley.

Hansen, H. J. 1990. Diachronous extinctions at the K/T boundary: A scenario. *In* V. L. Sharpton & P. D. Ward, eds., *Global Catastrophes and Earth History: An Interdisciplinary Conference on Impacts, Volcanism, and Mass Mortality, Geol. Soc. Amer. Special Paper* 247: 417–23.

Herman, Y. 1990. Selective extinction of marine plankton in the Paratethys at the end of the Mesozoic Era; A multiple interaction hypothesis. *In* V. L. Sharpton & P. D. Ward, eds., *Global Catastrophes and Earth History: An Interdisciplinary Conference on Impacts, Volcanism, and Mass Mortality, Geol. Soc. Amer. Special Paper* 247: 531–40.

Hickey, L. J. 1981. Land-plant evidence compatible with gradual, not catastrophic, change at the end of the Cretaceous. *Nature* 292: 529–31.

Hsü, K. J. 1986. *The Great Dying*. San Diego: Harcourt, Brace, Jovanovich.

Kauffman, E. G. 1979. The ecology and biogeography of the Cretaceous-Tertiary event. *In* W. K. Christensen & T. Birkelund, eds., *Cretaceous-Tertiary Boundary Events*, pp. 29–37. Copenhagen: Univ. of Copenhagen.

———. 1986. High-resolution event stratigraphy: Regional and global Cretaceous bio-events. *In* S. Battacharji, G. M. Friedman, H. J. Neugebauer & A. Seilacher, eds., *Global Bio-events*, pp. 279–335. Berlin: Springer-Verlag.

Keller, G. 1988. The K/T boundary mass extinction in the Western Interior Seaway (Brazos River, Texas). *Geol. Soc. Amer., Abstr. Prog.* 20: 370.

May, R. M. 1988. How many species are there on Earth? *Science* 241: 1441–49.

McGhee, G. R., Jr. 1988. The Late Devonian extinction event: Evidence for abrupt ecosystem collapse. *Paleobiology* 14: 250–57.

———. 1989. The Frasnian-Famennian extinction event. *In* S. K. Donovan, ed., *Mass Extinctions: Processes and Evidence*, pp. 133–73. New York: Columbia Univ. Press.

———. 1990. Frasnian-Famennian. In D. E. G. Briggs & P. R. Crowther, eds., *Palaeobiology: A Synthesis*, pp. 184–87. Oxford: Blackwell.

McLaren, D. J., & W. D. Goodfellow. 1990. Geological and biological consequences of giant impacts. *Annual Reviews of Earth & Planetary Sci.* 18: 123–71.

Newell, N. D. 1967. Revolutions in the history of life. *Geol. Soc. Amer. Special Paper* 89: 63–91.

Nichols, D. J., & R. F. Fleming. 1990. Plant microfossil record of the Terminal Cretaceous event in the Western United States and Canada. *In* V. L. Sharpton & P. D. Ward, eds., *Global Catastrophes and Earth History: An Interdisciplinary Conference on Impacts, Volcanism, and Mass Mortality, Geol. Soc. Amer. Special Paper* 247: 445–55.

Niklas, K. J. 1986. Large-scale changes in animal and plant terrestrial communities. *In* D. M. Raup & D. Jablonski, eds. *Patterns and Processes in the History of Life*, pp. 383–405. Berlin: Springer-Verlag.

Officer, C. B., & C. L. Drake. 1985. Terminal Cretaceous environmental events. *Science* 227: 1161–67.

Percival, S. F., & A. G. Fischer. 1977. Changes in calcareous nannoplankton in the Cretaceous-Tertiary boundary crisis at Zumaya, Spain. *Evolutionary Theory* 2: 1–35.

Prothero, D. R. 1989. Stepwise extinctions and climatic decline during the Later Eocene and Oligocene. *In* S. K. Donovan, ed., *Mass Extinctions: Processes and Evidence*, pp. 217–34. New York: Columbia Univ. Press.

Raup, D. M. 1988. Extinction in the geologic past. *In* D. E. Osterbrock & P. H. Raven, eds., *Origins and Extinctions*, pp. 109–19. New Haven, Conn.: Yale Univ. Press.

Raup, D. M., & J. J. Sepkoski, Jr. 1984. Periodicity of extinctions in the geologic past. *Proc. Nat. Acad. Sci.* 81: 801–5.

Raven, P. H. 1990. The politics of preserving biodiversity. *BioScience* 40: 769–74.

Schindewolf, O. 1962. Neokatastrophismus? *Deutsch. Geol. Ges. Zeitschr. Jahrg.* 114: 430–45.

Sheehan, P. M. 1990. Sudden or gradual extinction of the dinosaurs? Patterns of dinosaur diversity at the family level in the Hell Creek Formation. *Geol. Soc. Amer. Abs. Prog.* A, p. 356.

Shoemaker, E. M., & R. F. Wolfe. 1986. Mass extinctions, crater ages and comet showers. *In* R. Smoluchowski, J. N. Bahcall & M. S. Matthews, eds., *The Galaxy and the Solar System*, pp. 338–86. Tucson: Univ. of Arizona Press.

Signor, P. W. 1990. Patterns of diversification. In D. E. G. Briggs & P. R.

Crowther, eds., *Palaeobiology: A Synthesis*, pp. 130–35. Oxford: Blackwell.
Sloan, R. E., K. J. Rigby, L. Van Valen & D. Gabriel. 1986. Gradual dinosaur extinction and simultaneous ungulate radiation in the Hell Creek formation. *Science* 232: 629–33.
Smith, A. B., & C. Patterson. 1988. The influence of taxonomic methods on the perception of patterns of evolution. In M. K. Hecht & B. Wallace, eds., *Evolutionary Biology* 23: 127–216. New York: Plenum Press.
Sweet, A. R., D. R. Braman & J. F. Lerbekmo. 1990. Palynofloral response to K/T boundary events: A transitory interruption within a dynamic system. *In* V. L. Sharpton & P. D. Ward, eds., *Global Catastrophes and Earth History: An Interdisciplinary Conference on Impacts, Volcanism, and Mass Mortality, Geol. Soc. Amer. Special Paper* 247: 457–69.
Thomas, C. D. 1990. Fewer species. *Nature* 347: 237.
Van Valen, L. M. 1988. *Paleocene Dinosaurs or Cretaceous Ungulates in South America*. Univ. of Chicago: Evolutionary Monographs.
Vermeij, G. J. 1987. *Evolution and Escalation*. Princeton: Princeton Univ. Press.
Wallace, A. R. 1855. On the law which has regulated the introduction of new species. *Annual Mag. Nat. Hist.*, Ser. 2(16): 184–96.

12. W. Clemens, On the Debates: An Interview

Alvarez, L. W., W. Alvarez, F. Asaro & H. V. Michel. 1980. Extraterrestrial cause for the Cretaceous-Tertiary extinction. *Science* 28: 1095–1108.
Alvarez, W. 1986. Toward a theory of impact crises. *Eos* 67: 649, 653–55, 658.
Archibald, J. D., & L. J. Bryant. 1990. Differential Cretaceous/Tertiary extinctions of nonmarine vertebrates: Evidence from northeastern Montana. *In* V. L. Sharpton & P. D. Ward, eds., *Global Catastrophes in Earth History: An Interdisciplinary Conference on Impacts, Volcanism, and Mortality, Geol. Soc. Amer. Special Paper* 247: 549–62.
Benton, M. J. 1990. Scientific methodologies in collision: The history of the study of extinction of the dinosaurs. *Evolutionary Biology* 24: 371–400.
De Laubenfels, M. W. January 1956. Dinosaur extinction: One more hypothesis. *J. Paleontology* 30(1): 207–12.
Dingus, L., & P. M. Sadler. 1982. The effects of stratigraphic completeness on estimates of evolutionary rates. *Systematic Zoology* 31: 400–12.
Dott, R. H., Jr. 1983. Episodic sedimentation: How normal is average? How rare is rare? Does it matter? *J. Sedimentary Petrology* 53: 5–23.
Fastovsky, D. E. 1987. Paleoenvironments of vertebrate-bearing strata during the Cretaceous-Paleogene transition, eastern Montana and western North Dakota. *Palaios* 2: 282–95.
Heissig, K. 1986. No effect of the Ries impact event on the local mammals. *Modern Geology* 10: 171–79.
Jastrow, R. 1983. The dinosaur massacre: A double-barreled mystery. *Science Digest* (September): 151–53.
Jepsen, G. 1963. Terrible lizards revisited. *Princeton Alumni Weekly* 64: 6–10, 17–19.

Lowenstam, H. A., & S. Weiner. 1989. On Biomineralization. Oxford: Oxford Univ. Press.

MacLeod, N., & G. Keller. 1991. How complete are Cretaceous/Tertiary boundary sections? A chronostratigraphic estimate based on graphic correlation. *Geol. Soc. America, Bulletin* 103: 1437–57.

Newell, N. D. 1982. Mass extinctions: Illusions or realities? *In* L. T. Silver & P. H. Schultz, eds., *Geological Implications of Impacts of Large Asteroids and Comets on the Earth, Geol. Soc. America Special Paper* 190: 257–64.

Raup, D. M. 1991. *Extinction: Bad Genes or Bad Luck?* New York: W. W. Norton.

Signor, P. W., III, and J. H. Lipps. 1982. Sampling bias, gradual extinction patterns and catastrophes in the fossil record. *In* L. T. Silver & P. H. Schultz, eds., *Geological Implications of Impacts of Large Asteroids and Comets on the Earth, Geol. Soc. Amer. Special Paper* 190: 291–96.

Urey, H. C. 2 March 1973. Cometary collisions and geological periods. *Nature* 242: 32–33.

13. Gould, On the Debates: An Interview

Alvarez, L. W., W. Alvarez, F. Asaro & H. V. Michel. June 1980. Extraterrestrial cause for the Cretaceous-Tertiary extinction. *Science* 208: 1095–1108.

Alvarez, W., L. W. Alvarez, F. Asaro & H. V. Michel. 1979. Experimental evidence in support of an extraterrestrial trigger for the Cretaceous-Tertiary extinctions. *Eos* 60: 734. [Abstract.]

Douglas, M. 1986. *How Institutions Think.* Syracuse, N.Y.: Syracuse Univ. Press.

Durkheim, E. 1893. *De la division du Travail Social: Étude sur l'organisation des sociétés supérieures.* Paris: Alcan. [Transl. by George Simpson as *The Division of Labor in Society*, 1933. Glencoe, Ill.: Free Press.]

Fischer, A. G., & M. A. Arthur. 1977. Secular variations in the pelagic realm. *In* H. E. Cook & P. Enos, eds., *Deep-Water Carbonate Environments, Soc. Econ. Paleontol. and Mineral Special Publ.* 25: 19–50.

Fleck, L. 1935. *Entstehung und Entwicklung einer wissenschaftlichen Tatsache.* [Trans. 1979 by F. Bradley & T. J. Trenn as *The Genesis and Development of a Scientific Fact.* Chicago: Univ. of Chicago Press.]

Gerson E. M. [In press.] *The American System of Research: Evolutionary Biology 1890–1950.* Berkeley, Calif.: Univ. of California Press.

Goodman, N. 1978. *Ways of Worldmaking.* Indianapolis, Ind.: Hackett.

Gould, S. J. March 1965. Is uniformitarianism necessary? *American Journal of Science* 263: 223–28.

———. February 1984. Ediacaran experiment. *Natural History*, 14–23.

———. 1985. Paradox of the first tier: An agenda for paleontology. *Paleobiology* 11(1): 2–12.

———. March 1992. Dinosaurs in the haystack. *Natural History*, 2–14.

Gould, S. J., & N. Eldredge. 1977. Punctuated equilibria: The tempo and mode of evolution reconsidered. *Paleobiology* 3: 115–51.

Hume, D. 1739. In L. A. Selby-Bigge, ed., *A Treatise of Human Nature*, rev. ed., 1958. Oxford: The Clarendon Press.

Jablonksi, D. 1984. Background vs. mass extinctions: The alternation of macroevolutionary regimes. *Geol. Soc. Amer. Abstracts with Programs* 16: 549.

Kuhn, T. S. 1970. *The Structure of Scientific Revolutions*, 2d ed., enl. Chicago: Univ. of Chicago Press.

Latour, B. 1987. *Science in Action*. Cambridge, Mass.: Harvard Univ. Press.

McLaren, D. J. 1970. Time, life and boundaries. *J. Paleontology* 44: 801–15.

Raup, D. M., & J. J. Sepkoski, Jr. 1984. Periodicity of extinctions in the geologic past. *Proc. Nat. Acad. Sciences* 81: 801–5.

Signor, P. W., III, & J. H. Lipps. 1982. Sampling bias, gradual extinction patterns and catastrophes in the fossil record. *In* L. T. Silver & P. H. Schultz, eds., *Geological Implications of Impacts of Large Asteroids and Comets on the Earth, Geol. Soc. Amer. Special Paper* 190: 291–96.

14. Glen et al., A Panel Discussion on the Debates

Ahrens, T. J., & J. D. O'Keefe. 15 February 1983. Impact of an asteroid or comet in the ocean and extinction of terrestrial life. *Proc. Thirteenth Lun. & Plan. Sci. Conf.*, Part 2, *J. Geophys. Res.* 88, Supplement pp. A799–A806.

Alt, D., J. M. Sears & D. W. Hyndman. November 1988. Terrestrial maria: The origins of large basalt plateaus, hotspot tracks and spreading ridges. *J. Geology* 96(6): 647–62.

De Laubenfels, M. W. January 1956. Dinosaur extinction: One more hypothesis. *J. Paleontology* 30(1): 207–12.

Dietz, R. S. 1961. Astroblemes. *Scientific American* Aug., 50–58.

Flessa, K. W. 1979. Extinction. *In* R. W. Fairbridge & D. Jablonski, eds., *The Encyclopedia of Paleontology*, 300–305.

Gallant, R. 1964. *Bombarded Earth*. London: John Baker.

Gould, S. J. 1985. Paradox of the first tier: An agenda for paleontology. *Paleobiology* 11(1): 2–12.

Hull, D. L. 1988. *Science as a Process*. Chicago: Univ. of Chicago Press.

Jablonski, D. 10 January 1986. Background and mass extinctions: The alternation of macroevolutionary regimes. *Science* 231: 129.

Laplace, P. S. de. 1797. *Exposition du Système du Monde*. Paris.

Latour, B. 1987. *Science in Action*. Cambridge, Mass.: Harvard Univ. Press.

Levinton, J. S. 1981. Heisenberg's uncertainty principle: The growth of scientific knowledge and adaptive strategies of natural selection. *Eidema* 1(2). Helsinki, Finland.

Lyell, C. 1830, 1832, 1833. *Principles of Geology* (three volumes). London: John Murray. Pp. 511, 330.

Maupertuis, P. L. M. de. 1742. *Lettre sur la comète*. [Issued anonymously in Paris.]

McLaren, D. J. 1970. Time, life and boundaries. *J. Paleontology* 44: 801–15.

McLaren, D. J., & W. D. Goodfellow. 1990. Geological and biological consequences of giant impacts. *Annual Rev. Earth & Planetary Science* 18: 123–71.

Opik, E. J. 1958. On the catastrophic effects of collisions with celestial bodies. *Irish Astronomical J.* 5: 34–35.

Oro, J. 1963. Studies in experimental organic cosmochemistry. *In* anon., *Life-like Forms in Meteorites and the Problems of Environmental Control on the Morphology of Fossil and Recent Protobionta. New York Acad. Sci. Annals* 108: 464–81.

Palmer, A. R. 1982. Biomere boundaries: A possible test for extraterrestrial perturbation of the biosphere. *In* L. T. Silver & P. H. Schultz, eds., *Geological Implications of Impacts of Large Asteroids and Comets on the Earth, Geol. Soc. Amer. Special Paper* 190: 469–95.

———. 1984. The biomere problem: Evolution of an idea. *J. Paleontology* 58(3): 599–611.

Rampino, M. R. 11 June 1987. Impact cratering and flood basalt volcanism. *Nature* 327: 468.

Russel, D. A., & W. Tucker. 19 February 1971. Supernovae and the extinction of the dinosaurs. *Nature* 229: 553–54.

Schindewolf, O. H. 1954. Über die möglichen Ursachen der grossen erdgeschichtlichen Faunenschnitte. *Neues Jahrb. Geol. Pal.* 10: 457–65.

Urey, H. C. 2 March 1973. Cometary collisions and geological periods. *Nature* 242: 32–33.

Van Valen, L. M. 1973. A new evolutionary law. *Evolutionary Theory* 1: 1–30.

———. 1974. Molecular evolution as predicted by natural selection. *J. Molec. Evol.* 3: 89–101.

Vogt, P. R. 1972. Evidence for global synchronism in mantle plume convection, and possible significance for geology. *Nature* 240: 338–42.

Whewell, W. 1832. Review of Volume 2 of Charles Lyell's *Principles of Geology*. *Quarterly Review* 47: 103–32.

Index

Index

In this index an "f" after a number indicates a separate reference on the next page, and an "ff" indicates separate references on the next two pages. A continuous discussion over two or more pages is indicated by a span of page numbers, e.g., "pp. 57–58." *Passim* is used for a cluster of references in close but not continuous sequence.

Library of Congress Cataloging-in-Publication Data

The Mass-extinction debates : how science works in a crisis / edited by William Glen.
p. cm.
Includes bibliographical references and index.
ISBN 0-8047-2285-4 (acid-free paper)
ISBN 0-8047-2286-2 (pbk. : acid-free paper)
1. Extinction (Biology) 2. Paleontology—Methodology. I. Glen, William.
QE721.2.E97M35 1994
575′.7—dc20 93-26195
CIP

∞This book is printed on acid-free paper.